Mechanik, Werkstoffe und Konstruktion im Bauwesen

Band 66

Reihe herausgegeben von
Ulrich Knaack, Darmstadt, Deutschland
Jens Schneider, Darmstadt, Deutschland
Johann-Dietrich Wörner, Darmstadt, Deutschland
Stefan Kolling, Gießen, Deutschland

Institutsreihe zu Fortschritten bei Mechanik, Werkstoffen, Konstruktionen, Gebäudehüllen und Tragwerken. Das Institut für Statik und Konstruktion der TU Darmstadt sowie das Institut für Mechanik und Materialforschung der TH Mittelhessen in Gießen bündeln die Forschungs- und Lehraktivitäten in den Bereichen Mechanik, Werkstoffe im Bauwesen, Statik und Dynamik, Glasbau und Fassadentechnik, um einheitliche Grundlagen für werkstoffgerechtes Entwerfen und Konstruieren zu erreichen. Die Institute sind national und international sehr gut vernetzt und kooperieren bei grundlegenden theoretischen Arbeiten und angewandten Forschungsprojekten mit Partnern aus Wissenschaft, Industrie und Verwaltung. Die Forschungsaktivitäten finden sich im gesamten Ingenieurbereich wieder. Sie umfassen die Modellierung von Tragstrukturen zur Erfassung des statischen und dynamischen Verhaltens, die mechanische Modellierung und Computersimulation des Deformations-, Schädigungs- und Versagensverhaltens von Werkstoffen, Bauteilen und Tragstrukturen, die Entwicklung neuer Materialien, Produktionsverfahren und Gebäudetechnologien sowie deren Anwendung im Bauwesen unter Berücksichtigung sicherheitstheoretischer Überlegungen und der Energieeffizienz, konstruktive Aspekte des Umweltschutzes sowie numerische Simulationen von komplexen Stoßvorgängen und Kontaktproblemen in Statik und Dynamik.

Miriam Schuster

Characterization of Laminated Safety Glass Interlayers

Thermorheology, Crystallinity and Viscoelasticity

Springer Vieweg

Miriam Schuster
Bau- und Umweltingenieurwesen
TU Darmstadt
Darmstadt, Germany

Vom Fachbereich 13 – Bau- und Umweltingenieurwissenschaften
der Technischen Universität Darmstadt zur Erlangung des
akademischen Grades eines Doktor-Ingenieurs (Dr.-Ing.)
genehmigte Dissertation von Master of Science Miriam Schuster aus Luxemburg.

1. Gutachten: Prof. Dr.-Ing. Jens Schneider
2. Gutachten: Prof. Dr.-Ing. habil. Stefan Kolling

Tag der Einreichung: 27. Oktober 2021
Tag der mündlichen Prüfung: 21. Januar 2022
Darmstadt 2022

ISSN 2512-3238 ISSN 2512-3246 (electronic)
Mechanik, Werkstoffe und Konstruktion im Bauwesen
ISBN 978-3-658-39820-0 ISBN 978-3-658-39821-7 (eBook)
https://doi.org/10.1007/978-3-658-39821-7

Responsible Editor: Ralf Harms
This Springer Vieweg imprint is published by the registered company Springer Fachmedien Wiesbaden GmbH, part of Springer Nature.
The registered company address is: Abraham-Lincoln-Str. 46, 65189 Wiesbaden, Germany

Danksagung

An dieser Stelle möchte ich mich herzlichst bei allen Personen bedanken, die mich während meiner Dissertation unterstützt und begleitet haben.
Prof. Dr.-Ing. Jens Schneider danke ich für die Möglichkeit an seinem Lehrstuhl zu promovieren, die ansteckende Begeisterung für das Forschen im Bereich des Glasbaus und das entgegengebrachte Vertrauen, welches mir ein selbstständiges Arbeiten bei gleichzeitiger fachlicher Unterstützung ermöglichte. Bei Prof.-Dr.-Ing habil. Stefan Kolling bedanke ich mich für die Übernahme des Korreferats, die stets freundliche Unterstützung und motivierenden Diskussionen.

Ein weiterer Dank geht an alle aktuellen Kollegen, die ehemaligen Mitarbeiter des Institutes und meine anderen Forschungskollegen. Besonders hervorheben möchte ich Prof. Dr.-Ing. Johannes Kuntsche und Dr.-Ing. Michael Kraus. Durch das freundschaftliche Verhältnis und die gegenseitige Unterstützung im gesamten Team, habe ich die Zeit als Doktorandin sehr genossen. Besonders an die jährlichen Doktorandenseminare und die gemeinsamen Konferenzbesuche werde ich mich stets gerne zurückerinnern.
Ein großer Dank gilt auch den Studierenden, die mich mit ihren Abschlussarbeiten unterstützt haben, dem Institut für Werkstoffe im Bauwesen und dem Fachbereich Chemie (Makromolekulare und Papier Chemie; Anorganische und Physikalische Chemie) der TU Darmstadt für die Unterstützung bei den DSC und XRD Messungen.

Des Weiteren möchte ich mich bei den Folienherstellern Eastman Chemical Company (besonders Dr. Stevels), Kuraray (besonders Dr. Härth) und Glaast (besonders Dr. Goebel) bedanken, die mich zum einen mit Probekörpermaterialien unterstützt haben, und zum anderen auch stets für einen fachlichen Austausch bereitstanden.

Ein ganz besonderer Dank gilt meiner Familie, meinen Freunden und meinem Partner Lukas, für die Geduld, uneingeschränkte Unterstützung und Motivation während der letzten Jahre.

Darmstadt, Oktober 2021 *Miriam Schuster*

Abstract

Laminated safety glass enables the safe construction of transparent structures. The mechanical behaviour depends on the polymeric interlayer both in the intact and in the post fracture state. In the present work, the mechanical behaviour of ethylene vinyl acetate-based (EVA) and ionoplastic interlayers is investigated for the intact laminated safety glass condition. The mechanical behaviour of the interlayer in the fractured laminated safety glass is carried out on polyvinyl butyral-based (PVB) interlayers.

The interlayer creates a shear coupling between the glass plies in intact laminated safety glass, which varies, among other things, on the temperature- and load duration-dependent shear modulus of the interlayer and is considered by means of Prony series and time-temperature-superposition principles. Since Prony series and time-temperature shifts have so far predominantly been determined for PVB interlayers, the structure of EVA and ionoplastic interlayers is investigated in more detail in this work. Using X-Ray Diffraction, Differential Scanning Calorimetry and Dynamic-Mechanical-Thermal-Analysis, a semicrystalline structure is detected and characterized for different temperature rates and physical ages. The displacement factors required for mastercurve generation are divided into a purely temperature-dependent and a purely crystallinity-dependent part, so that the material behaviour can be predicted for any temperature and crystallinity combination.

In the fractured state, large deformations occur in the interlayer, so that the loading level is also taken into account in the mechanical characterization of the interlayer. First, the temperature and frequency (time) dependent linearity limits of two different PVB interlayers are determined in Dynamic-Mechanical-Thermal-Analyses. The nonlinear viscoelastic material behaviour is investigated with tensile relaxation tests at different temperatures and strain levels. If the initial degree of nonlinearity, the linear viscoelastic material behaviour and the linearity limit function are known, it can be predicted after which time period the linear viscoelastic material behaviour is reached again. In addition, the Schapery model is investigated to represent the nonlinear viscoelastic material behaviour. This model combines the linear viscoelastic Prony series with several strain-dependent nonlinearity factors, which are determined by shifting the test data horizontally and vertically.

Zusammenfassung

Verbundsicherheitsglas ermöglicht das sichere Konstruieren von transparenten Bauwerken. Das mechanische Verhalten ist dabei sowohl im intakten als auch im gebrochenen Zustand von der polymeren Zwischenschicht abhängig. In der vorliegenden Arbeit wird für den intakten Verbundsicherheitsglaszustand das mechanische Verhalten von Ethylen-Vinylacetat-basierten (EVA) und ionoplastischen Zwischenschichten untersucht. Das mechanische Verhalten der Zwischenschicht im gebrochenen Laminat wird an Polyvinylbutyral-basierten (PVB) Folien untersucht.

Durch die Zwischenschicht entsteht im intakten Verbundsicherheitsglas eine Schubkopplung zwischen den Glasscheiben, die unter anderem von dem temperatur- und belastungsdauerabhängigen Schubmodul der Zwischenschicht bestimmt und mittels Prony-Reihe und Zeit-Temperatur-Verschiebungsprinzip berücksichtigt wird. Da Prony-Reihen und Zeit-Temperatur-Verschiebungen bislang überwiegend für PVB Zwischenschichten bestimmt werden konnten, wird die Struktur der EVA und Ionoplastfolie in dieser Arbeit untersucht. Mittels Röntgenbeugung, Differenzkalorimetrie und Dynamisch-Mechanisch-Thermischen-Analysen wird die teilkristalline Struktur nachgewiesen und für unterschiedliche Temperaturraten und physikalische Alter charakterisiert. Die zur Masterkurvenerstellung notwendigen Verschiebungsfaktoren werden unter Berücksichtigung des Kristallisationsgrades in einen rein temperaturabhängigen und einen rein kristallisationsgradabhängigen Anteil geteilt, sodass das Materialverhalten für jede beliebige Kombination aus Temperatur und Kristallisationsgrad prognostiziert werden kann.

Im gebrochenen Zustand entstehen große Verzerrungen in der Folie, sodass die Belastungshöhe bei der mechanischen Charakterisierung der Zwischenschicht berücksichtigt wird. Zunächst werden die temperatur- und frequenz- (beziehungsweise zeit-) abhängigen Linearitätsgrenzen von zwei unterschiedlichen PVB Zwischenschichten in Dynamisch-Mechanisch-Thermischen-Analysen bestimmt. Das nichtlinear viskoelastische Materialverhalten wird mit Zug-Relaxationsversuchen bei unterschiedlichen Temperaturen und Dehnstufen untersucht. Sind der initiale Nichtlinearitätsgrad, das linear viskoelastische Materialverhalten und die Linearitätsgrenzfunktion bekannt, so kann abgeschätzt werden, nach welchem Zeitraum das linear viskoelastische Materialverhalten wieder erreicht wird. Zudem wird das Schapery-Modell zur Abbildung des nichtlinear viskoelastischen Materialverhaltens untersucht. Dieses Modell kombiniert die linear viskoelastische Prony-Reihe mit mehreren dehnungsabhängigen Nichtlinearitätsfaktoren, die durch horizontales und vertikales Verschieben der Versuchsdaten ermittelt werden.

Résumé

Le verre de sécurité feuilleté permet de construire en sécurité des structures transparentes. Le comportement mécanique dépend de la couche intermédiaire polymère, tant à l'état intact qu'à l'état rompu. Dans le présent travail, le comportement mécanique d'intercalaires à base d'éthylène vinyle acétate (EVA) et ionoplastique est étudié pour le verre de sécurité feuilleté intact. Le comportement mécanique de l'intercalaire dans le verre de sécurité feuilleté fracturé est réalisé sur des films à base de polyvinyle butyral (PVB).
Dans le verre de sécurité feuilleté intact, l'intercalaire crée un couplage de cisaillement entre les feuilles de verre, qui dépend notamment du module de cisaillement de l'intercalaire en fonction de la température et de la durée de la charge et peut être pris en compte par la série de Prony et d'un principe d'équivalence entre temps et température. Comme les séries de Prony et les principes d'équivalence n'ont pu être déterminés jusqu'à présent principalement pour les intercalaires de PVB, la structure des films EVA et ionoplastique est étudiée plus en détail dans ce travail. En utilisant la diffraction des rayons X, la calorimétrie à balayage différentiel et l'analyse thermique mécanique dynamique, une structure semi-cristalline est détectée et caractérisée pour différents taux de température et âges physiques. Dans l'état fracturé, de grandes déformations se produisent dans le film, de sorte que le niveau de charge est également pris en compte dans la caractérisation mécanique de la couche intermédiaire. Tout d'abord, les limites de linéarité dépendant de la température et de la fréquence (temps) de deux couches intercalaires de PVB différentes sont déterminées dans des analyses thermiques mécaniques dynamiques. Le comportement viscoélastique non-linéaire du matériau est étudié par des essais de relaxation en traction à différentes températures et à différents niveaux de déformation. Si le degré initial de non-linéarité, le comportement viscoélastique linéaire du matériau et la fonction limite de linéarité sont connus, il est possible de prédire après quelle période de temps le comportement viscoélastique linéaire du matériau est à nouveau atteint. En outre, le modèle de Schapery est étudié pour représenter le comportement viscoélastique non-linéaire du matériau. Ce modèle combine la série viscoélastique linéaire de Prony avec plusieurs facteurs de non-linéarité dépendant de la déformation, qui sont déterminés en décalant les données d'essai horizontalement et verticalement.

Contents

Glossaries

Abbreviations

Ac., ac.	Acoustic
AS	Amplitude sweep
BSP	Boltzmann superposition principle
cEVA	Crosslinked ethylene vinyl acetate
CH_3	Methyl group
DMTA	Dynamic-Mechanical-Thermal-Analysis
DSC	Differential Scanning Calorimetry
EVA	Ethylene vinyl acetate
FEM	Finite element method
FM	Friedman
FTS	Frequency-temperature sweep
FWHM	Full width at half maximum
HDPE	High density polyethylene
JMAK	Johnson-Mehl-Avrami-Kolmogorov
KIS	Kissinger-Akahira-Sunose
LDPE	Low density polyethylene
LPT	Lamberti-Peters-Titomanlio
LSG	Laminated safety glass

MAA	Methacrylic acid
MFI	Melt flow index
OFW	Ozawa-Flynn-Wall
OH	Hydroxil group
PE	Polyethylene
PFLS	Post fracture limit state
PP	1: Polypropylene, 2: Plate-plate configuration
PV	Photovoltaic
PVB	Polyvinyl butyral
RAF	Rigid amorphous phase
RT	Room temperature
SG	SentryGlas®
SLS	Serviceability limit state
SRF	Solid state fixture
Std., std.	Standard
Struct., struct.	Structural
TCSP	Time-crystallinity-superposition principle
TCT	Through cracked tensile
TFS	Temperature-frequency sweep
TPE	Thermoplastic elastomer
TPE-U	Thermoplastic elastomer based on urethane
tPP	Textured plate-plate configuration
TPU	Thermoplastic polyurethane
TS	Temperature sweep
tS	Time sweep
TSSP	Time-stress / strain-superposition principle
TTCSP	Time-temperature-crystallinity-superposition principle
TTSP	Time-temperature-superposition principle
TTSSP	Time-temperature-stress / strain-superposition principle
ULS	Ultimate limit state
UV	Ultraviolet radiation

VAC	Vinyl acetate
WLF	Williams-Landel-Ferry
XRD	X-Ray Diffraction

Symbols

a	Horizontal shift factor
a_1, a_2	Parameters for a_χ if a_χ is expressed in dependency of the relative crystallinity α
a'_1, a'_2	Parameters for a_χ if a_χ is expressed in dependency of the absolute crystallinity χ
a_{cool}	Horizontal shift factor obtained from the cooling data
a_{heat}	Horizontal shift factor obtained from the heating data
a_{T}	Horizontal temperature-dependent shift factor (TTSP)
$a_{\mathrm{T},\varepsilon}$	Horizontal temperature- and strain-dependent shift factor (TTSSP)
$a_{\mathrm{T},\sigma}$	Horizontal temperature- and stress-dependent shift factor (TTSSP)
$a_{\mathrm{T},\chi}$	Horizontal temperature- and crystallinity-dependent shift factor (TTCSP)
a_ε	Schaperys nonlinearity strain-dependent time shift factor
a_χ	Crystallinity dependent shift factor (TCSP)
A	1: Pre-exponential Arrhenius factor, 2: Actual cross section, 3: Area
A_0	Initial cross section
A_{a}	Integrated area of amorphous halo in XRD
A_{c}	Integrated area of Bragg reflection caused by the crystalline part in XRD
b_1, b_2	Parameters for b_χ if b_χ is expressed in dependency of the relative crystallinity α
b_{nm}	Ratio of Avrami coefficient n to Ozawa coefficient m
b_{T}	Vertical temperature-dependent shift factor (TTSP)
$b_{\mathrm{T},\chi}$	Vertical temperature- and crystallinity-dependent shift factor (TTCSP)
b_χ	Vertical crystallinity-dependent shift factor (TCSP)
C	Cooling rate
C_1, C_2	WLF parameters
C_{p}	Isobaric heat capacity

d	Plane distance in XRD
d_1, d_2	Glass thickness of glass pane 1 resp. 2
d_Q	Interlayer thickness
d_s	Distance of the centers of gravity of the glass panes
d_w	Equivalent thickness for deformations calculations
d_σ	Equivalent thickness for stress calculations
D	Nucleation density
e_i	With E_0 normalized linear elastic spring stiffness of the ith Maxwell element of the generalized Maxwell model
E	Relaxation function in tensile mode, for large strains E is calculated with the true values σ_t and ε_t
E_0	Tensile modulus at t_0, initial stiffness
E_∞	Linear elastic equilibrium spring stiffness of the generalized Maxwell model, tensile modulus at $t \rightarrow \infty$, equilibrium stiffness
E'_{amorph}	Value of storage modulus in amorphous state
E'_{T_g}	Value of storage modulus at the beginning of the glass transition
E'_{T_m}	Value of storage modulus at the melting temperature
E_a	Activation energy
$E_{a,c}$	Activation energy of the crystallization process
$E_{a,m}$	Activation energy of the melting process
ΔE	Transient tensile stiffness E-E_∞
E_{glass}	Young's modulus of glass
E_V	Extended volume or expectation value
$\boldsymbol{E}_{\mathbf{LVE}}$	Matrix, dimension n x 2, containing the time-invariant and transient part of the linear viscoelastic tensile modulus
$\boldsymbol{E}_{\mathbf{LVE}}^{\mathbf{T}}$	Transpose of $\boldsymbol{E}_{\mathbf{LVE}}$
$\boldsymbol{E}_{\mathbf{NL}}$	Vector of length n, containing the values of the nonlinear relaxation test
f	Frequency
f_g	Fractional free volume at the glass transition
f_k	Characteristic glass strength
$f_{ref,i}$	Frequency of data point i on the reference curve
f_v	Fractional free volume
Δf_ε	Strain induced change in fractional free volume
Δf_T	Temperature induced change in fractional free volume
F	1: Nucleation rate, 2: Force

g_i	With G_0 normalized linear elastic spring stiffness of the ith Maxwell element of the generalized Maxwell model
G	1: Free enthalpy, also called Gibbs energy, 2: Shear modulus
G_0	Shear modulus at t_0, initial stiffness
G_∞	Linear elastic equilibrium spring stiffness of the generalized Maxwell model, shear modulus at $t \to \infty$, equilibrium stiffness
G'	Storage modulus
G'_{amorph}	Value of shear storage modulus in the cooling curve, just before the crystallization process begins
G'_{LVE}	Storage modulus in the linear viscoelastic region
G'_{m}	Storage modulus in the molten amorphous state
G'_{s}	Storage modulus at the end of crystallization
G'_{T}	Value of shear storage modulus at temperature of interest T
G'_{T_g}	Value of shear storage modulus at the glass transition temperature
G'_{T_m}	Value of shear storage modulus at the melting temperature
$G'_{\gamma_{\text{lin}}}$	Storage modulus at the linearity limit γ_{lin}
G''	Loss modulus
G_{c}	Crystallization rate
G^*	Complex modulus
$\|G^*\|$	Amount of complex modulus
ΔG	Transient shear stiffness
G_{edge}	Edge shear modulus in FEM
G_{glass}	Shear modulus of glass
G_{i}	Linear elastic spring stiffness of the ith Maxwell element of the generalized Maxwell model
G_{r}	Reduced shear modulus
h_1	Schaperys first nonlinearity factor for the time-dependent stiffness
$h_{1,2}$	Schaperys combined nonlinearity factor for the time-dependent stiffness
$h^*_{1,2}$	Schaperys combined nonlinearity factor for the time-dependent stiffness, when the relaxation function is expressed with the initial stiffness
h_2	Schaperys second nonlinearity factor for the time-dependent stiffness
h_0	Schaperys nonlinearity factor for the time invariant initial stiffness
h_∞	Schaperys nonlinearity factor for the time invariant equilibrium stiffness
H	1: Enthalpy, 2: Height of Gaussian or Lorentzian peak function

ΔH_c	Crystallization enthalpy / fusion enthalpy
ΔH_m	Melting enthalpy
$\Delta H_{m,100\%}$	Melting enthalpy of 100 % crystalline material
$\Delta H_{m,100\%PE}$	Melting enthalpy of a 100% PE crystallite
$\boldsymbol{H}_{\mathbf{NL}}$	Vector containing the nonlinearity factors h_0 and $h^*_{1,2}$ resp. h_∞ and $h_{1,2}$
I	Scattering intensity
I_s	Area moment of inertia of the composite part of a sandwich
k	1: Crystallization rate constant, 2: Ratio of deformation due to shear force and due to bending in Wölfel's theory, which can be transformed in a shear transfer coefficient $\Gamma = 1/(k+1)$
k_c	1: Reduced crystallization rate constant, 2: Construction factor
k_m	Pre-exponential factor of Arrhenius equation for the melting process
$k_{mod,snow}$	Load duration factor for snow
$k_{mod,wind}$	Load duration factor for wind
k_s	Pre-exponential factor of Arrhenius equation for the cooling process
K	1. Temperature dependent rate constant of Ozawa model, 2: Different expression of the crystallization rate constant ($k = K^n$)
K_x	Calibration constant that corrects the signal intensity of the amorphous phase to the crystalline phase
l_0	Initial length
Δl	Change in length
m	Ozawa coefficient
m_{amorph}	Mass of the amorphous part
m_{cryst}	Mass of the crystalline part
m_{total}	Total mass
M	Molar mass
$\overline{M_n}$	Number average molar mass
$\overline{M_W}$	Weight average molar mass
n	Avrami coefficient
N	1: Degree of polynomial function, 2: Number of Maxwell elements in the generalized Maxwell model

p	Pressure
P_i	Polynomial term
Q	Heat
r	Spherulite radius
R	Universal gas constant
s_d	Design snow load
S	Entropy
t	Time
t_0	1: Time at which the crystallization or melting process starts, 2: Time at which the relaxation test starts
t_{end}	Time at which a process (crystallization, melting, etc.) ends
t_{ref}	Reference time
T	Temperature
T_0	Temperature at which the crystallization or melting starts
$T_{begin,hysteresis}$	Temperature at the beginning of the hysteresis
T_c	Crystallization temperature
ΔT	1: Supercooling temperature, Temperature step height
T_{end}	Temperature at which the crystallization process ends
$T_{end,hysteresis}$	Temperature at the end of the hysteresis
T_f	Flow temperature
T_g	Glass transition temperature
$T_{g,cool}$	Glass transition temperature determined with cooling data
$T_{g,heat}$	Glass transition temperature determined with heating data
$T_{g,o}$	Upper glass transition temperature
$T_{g,u}$	Lower glass transition temperature
T_i	Temperature step i
T_m	Melting temperature
T_m^0	Equilibrium melting temperature
T_p	Peak temperature
$T_{p,c}$	Crystallization peak temperature
$T_{p,m}$	Melting peak temperature
$T_{p,m1}$	Low temperature melting peak
T_{ref}	Reference temperature

T_z	Decomposition temperature
T_β	Transition temperature of short polymer chain sections, local or secondary relaxation
U	Internal energy
v	Specific volume
v_{amorph}	Specific volume of amorphous part
v_{cryst}	Specific volume of crystalline part
v_f	Free volume
v_o	Volume occupied by molecules or polymer segments
v_{total}	Total specific volume of semicrystalline material
V	Volume
V_{amorph}	Volume of amorphous part
V_{cryst}	Volume of crystalline part
V_{total}	Total volume of semicrystalline material
w	1: Displacement, 2: Width of Gaussian or Lorentzian peak function
w_d	Design wind load
w_{max}	Maximum deflection
w_{Rd}	Limit deflection for design purposes
w_{ref}	Reference displacement
x_c	Center of Gaussian or Lorentzian peak function
y_0	Baseline offset of Gaussian or Lorentzian peak function
α	1: Relative degree of crystallization, relative crystallinity 2: Factor respecting structural system in Wölfel's sandwich theory
α_M	Peak position of the y-function
α_P^∞	Peak position of the z-function
α_v	Coefficient of thermal volume expansion
γ	Shear strain
γ_0	Shear strain at time t_0
$\gamma_1,\gamma_2,\gamma_3$	Shear strain at mesh point sample 1, 2, 3 in FEM
γ_{lin}	Linearity limit in shear mode
γ_M	Partial safety factor
Γ	1: Rheological function, 2: Shear transfer coefficient

δ	Loss angle
ε	Uniaxial strain
ε_0	Engineering strain
ε_{lin}	Linearity limit in tensile mode
ε_{t}	True strain
η^*	Complex viscosity
$\lvert\eta^*\rvert$	Amount of complex viscosity
η_i	Viscosity of the ith Maxwell element of the generalized Maxwell model
θ	Bragg angle
κ	Isobaric compressibility
λ	1: Stretch, 2: Wavelength
μ	Poisson's ratio
μ_{glass}	Poisson's ratio of glass
ρ	Density
$\rho_{\text{amorphous}}$	Density of amorphous part
ρ_{cryst}	Density of crystalline part
ρ_{ref}	Reference density
ρ_{total}	Density of semicrystalline material
σ	Tensile stress
σ_0	Engineering stress
σ_{glass}	Maximum glass stress
σ_{l}	Yield stress dependent nonlinearity parameter
$\sigma_{\text{Rd,s}}$	Limit stress for design purposes, snow load
$\sigma_{\text{Rd,w}}$	Limit stress for design purposes, wind load
σ_{ref}	Reference stress
σ_{t}	Truc strcss
σ_{y}	Yield stress
τ	Shear stress
τ_i	Relaxation time of the ith Maxwell element of the generalized Maxwell model
Φ	Heat flow or heat flow normalized per sample mass
χ	Absolute degree of crystallization, absolute crystallinity expressed as mass fraction
χ_{c}	Absolute crystallinity calculated with the total crystallization peak area
χ_{cool}	Absolute crystallinity calculated with cooling data
χ_{f}	Absolute crystallinity at the end of crystallization
χ_{heat}	Absolute crystallinity calculated with heating data

$\chi_{initial}$	Initial absolute crystallinity
χ_m	Absolute crystallinity calculated with the total melting peak area
χ_{max}	Maximum difference in absolute crystallinity of two different temperature programs
χ_{ref}	Reference absolute crystallinity
$\chi_{t,T}$	Absolute crystallinity at a certain time or temperature
χ_V	Absolute degree of crystallization, absolute crystallinity expressed as volume fraction
ψ	Reduced time
ω	Angular frequency

1 Introduction

1.1 Motivation

Due to its high transparency, glass has become increasingly popular as a structural element in the building industry in recent years. Figure 1.1 shows three impressive glass constructions. The Oriental Pearl Tower from Shanghai in Figure 1.1 a) offers visitors an observation deck with a glass floor at a height of 259 m, providing a view of the streets of Shanghai and of the Huangpu Jiang river below. Figure 1.1 b) and c) show the cylindrical entrance to the Apple Store and the staircase leading from the entrance to the actual store in Pudong, Shanghai. The high level of transparency is in accordance with the minimalist style and offers visitors an unobstructed view of the surrounding skyscrapers.

The examples also illustrate that for some applications, the brittle failure of glass must be counteracted in order to fulfil safety-relevant requirements with regard to fracture and post fracture behaviour, which leads to the usage of laminated safety glass (LSG). According to EN ISO 12543-2 (2011), LSG consists of at least two glass panes, connected by a polymeric interlayers (laminated glass according to EN ISO 12543-3 (2011)), which additionally is able to bind glass fragments after glass fracture and hence enables a residual load-bearing capacity, also known as post fracture behaviour. Applications that require LSG are for example, when the glass is used for fall protection, can be walked on, is used as overhead glazing or as a structural component (for example as a glass beam).

(a) (b) (c)

Figure 1.1 Constructions with laminated safety glass in Shanghai (a) Oriental Pearl Tower, (b) Apple Store, (c) Stairs in Apple Store.

M. Schuster, *Characterization of Laminated Safety Glass Interlayers*,
Mechanik, Werkstoffe und Konstruktion im Bauwesen 66,
https://doi.org/10.1007/978-3-658-39821-7_1

Different interlayer materials are available for different applications. The most common interlayers are based on polyvinyl butyral (PVB). PVB-based interlayers can be further classified into standard PVB, structural PVB, which has enhanced mechanical properties, and acoustic PVB, which absorbs ambient noise particularly well. In recent years, however, other interlayer materials have also been increasingly used. These include for example, ethylene vinyl acetate-based interlayers (EVA), which are often used as PV module encapsulant material, and the ionoplast SentryGlas® (SG), which was developed for high structural performance.

1.2 State of the art and state of research

1.2.1 Load bearing behaviour of intact laminated safety glass

If adhesion between glass and polymer interlayer is ensured, the interlayer provides shear coupling of the individual glass panes of the intact laminated safety glass under bending stress. The amount of shear coupling is limited by the two limit states *full shear transfer* and *no shear transfer*. In the limit case of *full shear transfer* (Figure 1.2 (a)), the load bearing behaviour is similar to the one of a monolithic glass pane with a thickness that corresponds to the summation of the individual glass pane thicknesses ($d_1 + d_2 + ...$). The interlayer thickness d_Q is neglected (DIN 18008-2). Contrary, in the limit state *no shear transfer* (Figure 1.2 (c)), the load is distributed on the individual glass panes according to their bending stiffness and the dimensioning is done separately for each glass pane. The amount of shear coupling (Figure 1.2 (b)) depends on the geometry and cross-section of the laminate (height, width, glass thicknesses, interlayer thicknesses), the structural system (e.g. single span beam, plate supported linearly on all sides), the type of load (e.g. concentrated load, uniformly distributed load) and the shear modulus of the interlayer.
Laminated safety glass interlayers are polymer-based. Hence, their shear modulus is dependent on the prevailing temperature T and load duration t. In intact laminated safety glass, the interlayer usually is subjected only to small strains, so that the relaxation function $G(T,t)$ is independent on the load magnitude. Often it is assumed that the shear modulus must be within the range of glass (G_{glass} = 28500 MPa) in order to reach the limit case of *full shear transfer*. However, for most cases with glass dimension, glass thicknesses, boundary conditions and load situations, interlayer shear modulus values of $G = 1$ - 10 MPa are sufficient (Kuntsche et al., 2019; Galuppi and Royer-Carfagni, 2013). The effect of considering the shear coupling during glass design is illustrated in a short but detailed commented example in the appendix A.1, where a vertical 1.0 m x 1.0 m large, two-side linearly supported laminated safety glass, loaded with wind is considered. The interlayer thickness is $d_Q = 0.76$ mm and the glass thickness is $d_1 = d_2 = 6$ mm. The shear modulus value of the interlayer is assumed to be $G = 2$ MPa. Further examples showing that the consideration of the shear coupling effect can have a positive effect on the glass

design are shown in (Kuntsche et al., 2019).

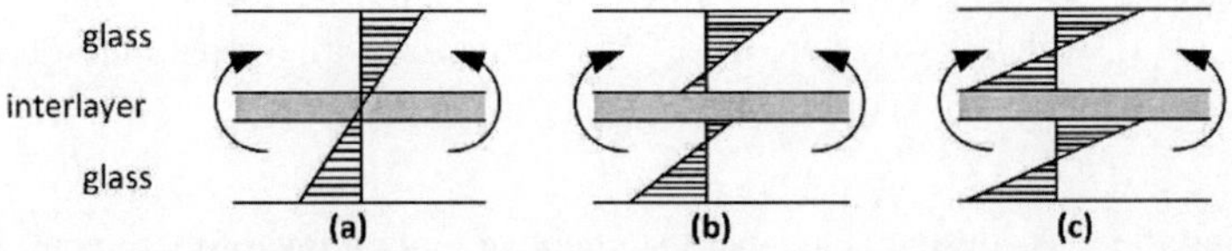

Figure 1.2 Load bearing behaviour of intact LSG subjected to bending. (a) limit state *full shear transfer*, (b) *partial shear bond*, (c) limit state *no shear transfer*.

Many national but also European standards allow the consideration of partial shear transfer during glass design: the Dutch NEN 2608, Belgium's NBN S 23-002-2, the Norwegian NS 3510, the Australian AS 1288, the Austrian ÖNORM B 3716-1, the Italian guideline CNR-DT 210 and the American ASTM E1300 as well as European standards, e.g. EN 16612 and the CEN/TS 19100-2. This is usually done with simplified approaches of effective thickness methods, which can be derived from commonly known sandwich theories, e.g. the sandwich theory of Wölfel (1987). The exact calculation method to determine the effective thickness depends on the respective standard. However, they all have in common, that constant, but load case specific shear modulus values (example: load case snow, T = 0°C and t = 30 d) are used in linear elastic calculations. A few standards also allow to take the temperature- and time-dependent material behaviour of the interlayer into account by using appropriate material models with Finite Element Method (FEM). Mostly, the generalized Maxwell model is used, which is mathematically described by a Prony series.

The current design standard for glass in the German building industry DIN 18008-1 is rather conservative. It specifies that a shear bond which is favourable for the design of the glass structure must not be taken into account. Hence, for load-induced stresses, the limit state of *no shear transfer* has to be considered. If the shear bond appears unfavourable, which is the case with deformation-induced stresses (for example climatic loads in insulating glass units), the limit state *full shear transfer* has to be taken into account. Hence, only the two limit cases are considered. To overcome this, some manufacturers applied for a general building authority approval (abZ in Germany) for their products, which permit to take a favourable shear modulus for the glass dimensioning into account for certain load cases. The use of temperature-dependent linear viscoelastic material models is also possible for some materials via Prony series and time-temperature-superposition principles (TTSP).

A detailed comparison of the design standards can be found in (Kuntsche et al., 2019). In addition, the different methods were evaluated. In summary, it can be said that the simplified method is a good approximation for many load cases if the equivalent thickness is calculated appropriately (taking into account all relevant parameters and complying with the application conditions of the underlying sandwich theory). For complex structures, however, this approximation calculations may overestimate the maximum deflections, which is not necessarily on the safe side, e.g. in the case of impulsive loads (Galuppi and Royer-Carfagni, 2014), which is why here the FEM calculation with temperature-dependent linear viscoelastic models is recommended.

Considering the glass design standards listed above, only a few of them contain information on the shear relaxation function $G(T,t)$ of the interlayer or on load specific shear modulus values: The Dutch NEN 2608 has included temperature dependent linear viscoelastic material parameters for two different materials (standard PVB and ionoplast) and the Norwegian NS 3510 has included load case dependent shear modulus values for three different materials (standard PVB, structural PVB and ionoplast). This is probably due to the large variety of available interlayer materials. The Italian guideline therefore refers to manufacturer specifications or the possibility of experimentally determining the temperature dependent linear viscoelastic material behaviour. Currently, EN 16613 is the only standard that deals specifically with the temperature dependent and linear viscoelastic characterization of laminated safety glass interlayers. Different test methods are proposed for different interlayer types: For isotropic interlayers, Dynamic-Mechanical-Thermal-Analyses (DMTA) are proposed on the pure interlayer material, whereas for anisotropic materials (e.g. multilayers) and for interlayers from which no small specimens can be made, flexural creep tests in the four-point bending mode on laminates (dimensions: 1100 mm x 360 mm) are recommended.
According to EN 16613, a constant load is applied in the flexural bending tests, the time-dependent deflection is measured and a shear transfer coefficient is retro-calculated. However, it would be more advisable to retro-calculate shear modulus values, since contrary to the shear transfer coefficient, shear modulus values are real material parameters and hence independent on the structural system, geometry and load conditions. A method to derive shear modulus values from four-point bending creep tests consists of the usage of metamodels, that have been generated with the help of FEM. In (Kraus, 2018) and (Botz, 2020), also torsional-relaxation tests on laminates are described as alternative test method. The conventional creep and relaxation tests on the laminates have some disadvantages. First of all, the measured interlayer shear modulus values are limited by the limit state *full shear transfer*. This means that shear modulus values that produce a full shear coupling for the given system and the given load cannot be specified unambiguously. Second, in order to take the temperature dependence into account, EN 16613 suggests to carry out measurements between -20°C and +60°C in 10°C steps. For this, large climatic chambers are necessary. Finally, those tests are very time consuming.

In DMTA, the load duration (or frequency) and temperature-dependent material behaviour can be determined in a single measurement, the so-called temperature-frequency sweep.

In addition, the pure interlayer is usually tested, whereby the specimen geometries are in the cm range. The specimen is loaded with different frequencies at a certain constant temperature. This is then repeated for many temperatures in the temperature interval to be investigated, so that the material behaviour is known for many temperatures at different load durations (or frequencies). EN 16613 suggests to investigate at least 15 different temperatures in between -20°C and +60°C. The time or frequency intervals investigated usually extend over only a few decades (e.g. [0.1:100] Hz according to EN16613). In order to make a statement about the long-term behaviour, the time-temperature-superposition principle (TTSP) is applied, in which the individual shear modulus curves are shifted horizontally along the logarithmic frequency (or time) axis until a continuous curve at a reference temperature is obtained (according to EN 16613 T_{ref} = 20°C). This curve is called the mastercurve. The subsequent evaluation of the test data consists of determining shear transfer coefficients for specific load cases. As already explained above, this procedure is not ideal. Instead, deviating from EN 16613, it is more accurate to use the measured data to approximate them with a linear viscoelastic material model, e.g. the generalized Maxwell model (Prony series). This was for example done in some of the general building approvals in Germany (e.g. Z-70.3-230; Z-70.3-236; Z-70.3-256) and in many research papers and dissertations, e.g. (Kuntsche, 2015; Kraus, 2018).
The temperature-dependent shift factors required to generate the mastercurve can be approximated by certain functions, the time-temperature-superposition models or time-temperature-superposition principles TTSP. Those can then be used to shift the mastercurve to other temperatures. EN 16613 only suggest the TTSP according to William Landel and Ferry (WLF). However, according to the original paper (Williams et al., 1955), WLF is only applicable in specific temperature ranges. Other well known shift functions are for example the Arrhenius equation (Schwarzl, 2013) or polynomial functions (used for example in Z-70.3-236 or Z-70.3-256). With the Prony series and the TTSP, the material behaviour of the interlayer is completely described in case of small strains.
DMTA can be conducted in different modes, e.g. tensile mode, shear mode, bending mode or torsional mode. EN 16613 only refers the tensile mode according to ISO 6721-4. However, this seems unusual, since the main stress of the interlayer is shear, resulting from the bending stress of the laminate. An overview of different DMTA modes is given in EN ISO 6721-1. Different loading modes were also compared and evaluated in (Kuntsche, 2015) on the basis of tests carried out on different interlayer materials.
Since the DMTA tests are carried out on relatively small samples of the pure interlayer, it is reasonable to validate the Prony series and the TTSP on the laminate (Kraus, 2018; Kuntsche, 2015). For this purpose, e.g. the previously described four-point-bending creep tests or torsion-relaxation tests at different temperatures can be used.

The disadvantage with DMTA is that the described method can so far only be successfully applied to thermorheologically simple materials. A material behaves thermorheologically simply if the shape of the relaxation curve is always the same regardless of temperature and a change in temperature only leads to a horizontal shift along the time or frequency axis. A wide variety of interlayer materials are represented on the market. The most common one is polyvinyl butyral-based (PVB) and behaves thermorheologically simple in the temperature range of approx. [-40:+80]°C and time range relevant to building practice (up to approx. 50 years), so that the creation of Prony series and TTSP functions is state of the art for these materials, see for example (Kuntsche, 2015; Kraus, 2018; Sackmann, 2008; Duser et al., 1999; Barredo et al., 2011; Hooper et al., 2012b; Z-70.3-230; Eastman Chemical Company, 2017; Z-70.3-236; NEN 2608; NS 3510). The Prony series specified in the general building authority approval (abZ in Germany) and standards are shown in Figure 1.3. This also shows that interlayers of similar material (e.g. PVB Saflex DG 41 and PVB Trosifol ES, both developed for structural applications) can differ significantly in material behaviour and therefore the classification into product families proposed by EN 16612 and EN 16613 is difficult.

In recent years, however, other interlayer materials have also been increasingly used. These include, for example, ethylene vinyl acetate-based interlayers (EVA), which, unlike PVB, still produce comparatively high shear couplings even at high temperatures. Another interlayer is the ionoplast SentryGlas® (SG). This behaves much more stiffly than PVB up to a temperature of $\approx$ 30°C - 50°C. The different material behaviour of all three materials is based on different molecular structures. Simply put, all three materials are different types of polymers. Contrary to PVB, EVA and SG behave in a thermorheologically complex manner, which complicates the application and evaluation of DMTA. Hence, the general building authority approvals Z-70.3-197 and Z-70.3-170 only contain load specific shear modulus values, as shown exemplary in Figure 1.3 for the load case snow on an overhead glazing with a heated interior.
In literature, only isolated studies on the temperature dependent linear viscoelastic material behaviour of EVA-based laminated glass interlayers can be found. Eitner et al. (2010) and Pander et al. (2011) investigated EVA interlayers used in solar modules in relaxation and creep tests on the pure interlayer at different temperatures. Mastercurves were generated, but no Prony series were published. Hána et al. (2020), Kuntsche (2015) and Kraus (2018) investigated EVA using DMTA. Due to strong scattering of the mastercurves and displacement factors depending on different loading modes (torsional DMTA, double shear on the pure material, double shear on the laminate) and the selected temperature steps, Kuntsche (2015) could not derive Prony parameters and TTSP. Hána et al. (2020) recently published a Prony series and WLF parameters derived from DMTA experiments on a single slap small glass laminate. However, the investigations do not cover the entire temperature range relevant to construction practice. Kraus (2018) published a Prony series and a TTSP for a wide temperature range. The DMTA tests were performed in tensile mode on the plain interlayer and subsequently compared with flexural creep data on the laminate. Additional vertical and horizontal displacements were required to bring the DMTA data into

agreement with the flexural creep data.
For SG interlayers, even less information on the temperature dependent linear viscoelastic material behaviour can be found in literature. Besides the Prony series in NEN 2608, Kuntsche (2015) shows results from DMTA. Here, however, no Prony series and no TTSP could be derived due to strong scattering of the mastercurves and displacement factors as a function of the stress mode and the temperature step size.
Overall, it can be stated that little literature exists with data on the temperature dependent linear viscoelastic material behaviour of EVA-based interlayers and SG interlayers, and thus the creation of Prony series and TTSP is still an open research question. Some of the mentioned authors have pointed out that both EVA and SG materials have semicrystalline regions. However, the influence on the temperature dependent linear viscoelastic behaviour has not been investigated in detail.

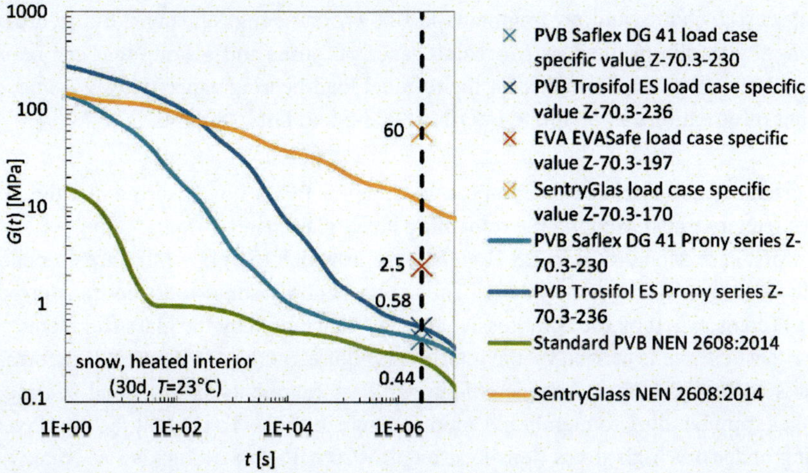

Figure 1.3 Relaxation function $G(T,t)$, resp. load specific shear modulus values for different interlayers according to different standards and technical approvals. The load case *snow load on an overhead glazing with heated interior* is considered. T = 23 ℃, t = 30 d according to the general building authority approval (abZ) in Germany.

1.2.2 Load bearing behaviour of fractured laminated safety glass

Laminated safety glass must provide a residual load-bearing capacity in the case of fracture of one or more glass panes. From a mechanical point of view, two post fracture states can be distinguished, which are labeled *Post Fracture Limit State I* (PFLS I) and *Post Fracture Limit State II* (PFLS II) in the following. In PFLS I, Figure 1.4 (a), at least one glass pane remains intact and in PFLS II, Figure 1.4 (c), all glass panes are fractured. According to DIN 18008-1, the residual load-bearing capacity is the "ability of a glazing structure to remain stable/load-bearing for a sufficient period of time in the event of a specified state of destruction under defined external influences (load, temperature, etc.)". Sufficient residual load-bearing capacity can be proved by complying with design specifications (restricted to PVB interlayers, defined glass types, defined support conditions, etc.), by experimental verification or by numerical calculation.

According to DIN 18008-1, the experimental verification has to be carried out on components with the dimensions and support conditions as-installed. In general, the external glass panes of horizontal and vertical laminated safety glass structures have to be destroyed in a defined manner and then be loaded at approximately 20°C with a defined surface-load. The test is considered successful if the laminate does not fall out of the support structure for at least 24 hours and no fragments fall down that could endanger traffic areas. For glazing that can be walked on for maintenance purposes and glazing that can be walked on regularly, other requirements for the residual load-bearing capacity tests are necessary. Further information can be found in DIN 18008-5 and DIN 18008-6.

The proof via numerical calculation can be done for PFLS I by ensuring that the load can be completely transferred via the remaining intact glass pane.
The worst case scenario in PFLS II is the case in which the glass plies are co-centrically fractured (i.e. one above the other). The acting bending stress must be transferred via a pair of forces, whereby the compressive force is transmitted by the glass fragments and the entire tensile force is carried by the interlayer (Figure 1.4 (b)). With greater deformations, there is no further contact between the upper glass fragments and the load-bearing effect changes from bending to membrane action (Figure 1.4 (c)). To prevent the interlayer from tearing under this high stress (tensile strength > 20 MPa and elongation at break > 250% according to the German MVV TB), a certain local delamination of the glass is necessary, as shown in Figure 1.4 (b) and (c). Delamination can occur when the adhesive strength of the interlayer to the glass is less than the tensile strength of the interlayer (Franz, 2015). The interlayer hence locally undergoes large deformations in the post fracture state, so that in addition to the influence of temperature and time, the influence of the load level (stress or strain) on the material behaviour of the interlayers must be taken into account. Furthermore, the interlayer is exposed to the environment due to the glass fracture, so that moisture and other environmental influences can also have an effect. Studies on the influence of moisture content in the interlayer on its mechanical behaviour have e.g. recently been published by Botz (2020).

The post fracture behaviour is thus dependent on the structural system including geometries, cross sections, support conditions; the glass, especially the glass type and fracture

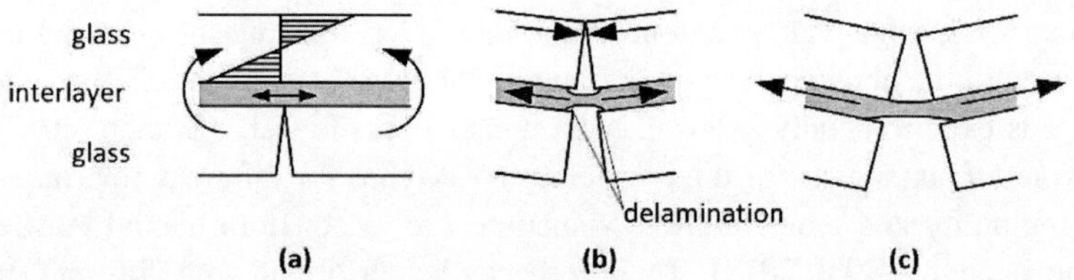

Figure 1.4 Post fracture load bearing behaviour of LSG subjected to bending. (a) the lower glass pane is fractured, (b) both glass panes are fractured on top of each other: The external bending is transferred via a pair of forces and (c) both glass panes are fractured on top of each other: the upper glass fragments have no contact, the load is transferred via membrane forces in the interlayer.

structure, see further e.g. (Kraus, 2018) and (Pour-Moghaddam, 2019); the mechanical behaviour of the interlayer and the adhesion between interlayer and glass. The high number of influencing parameters illustrates why, in addition to compliance with design specifications, experimental verification of the residual load-bearing capacity has been the primary method used up to date.

A few approaches to the numerical representation of the post fracture behaviour of laminated safety glass exist in literature, although these are predominantly concerned with short blast loads. An approximate calculation of the residual load-bearing capacity is also available in the current version of the glass design software MEPLA SJ (version 5.0). Here, the calculation of the fracture state is carried out via an investigation of the main tensile stresses of the glass panes. If they exceed a certain limit value, a fracture state is activated and a crack surface is generated, which then only permits to transmit compressive forces. Delamination and the influence of large deformations in the material behaviour of the interlayer are not taken into account. In (Alter, 2019), the glass breakage of windshields was also simulated by erosion (deleting glass elements). For the interlayer, a hyper-viscoelastic material model determined by Rühl (2017) was applied. Delamination between interlayer and glass was not taken into account. Kuntsche (2015) has numerically modeled laminated safety glass under blast loading. Glass breakage was simulated by erosion and the material behaviour of the interlayer was modeled with a strain rate dependent hyperelastic model. In addition, delamination was simulated using cohesive zone model. Pelfrene (2016) also depicted laminated safety glass under blast loading. To do this, he modeled glass breakage by deleting elements, chose a combination of hyperelastic and viscoelastic material behaviour for the interlayer, and accounted for delamination with a cohesive zone model (from Through Cracked Tensile (TCT) tests and peel tests). Franz (2015) used both hyperelastic and linear viscoelastic material models to numerically replicate the Through Cracked Tensile (TCT) test. In addition, he mapped delamination behaviour using cohesive zone models.

Regarding the mechanical behaviour of the interlayer, hyperelastic material models are mainly used in the above presented literature. The hyperelastic material behaviour of the interlayer is experimentally determined in uniaxial tensile test. (Ohmen, 2018) contains results from uniaxial tests for three different PVB types for different environmental conditions (humidity and temperature). Sometimes, test data from biaxial tensile tests are taken into account (Rühl, 2017). These tests can be performed with different strain rates. The higher the strain rate, the stiffer the material response. In order to depict the time-dependent material behaviour for certain constant strain or stress levels that exceed the linear range, hyper-viscoelastic models, that combine hyperelastic and linear viscoelastic material behaviour are commonly used in literature, see e.g. (Rühl, 2017; Pelfrene, 2016). The linear basic spring of the generalized Maxwell model (linear viscoelastic model, Prony series) is replaced by a nonlinear spring (with hyperelastic material behaviour). This model assumes that the stress can be represented by the product of a strain dependent function (hyperelastic model) and a time dependent function (linear viscoelastic generalized Maxwell model, Prony series). Hence it is assumed that the time-dependent part of the material behaviour is independent of the applied strain level (Hooper et al., 2012a). However, this is not necessarily fulfilled.
Hooper et al. (2012a) performed uniaxial tensile tests with different strain rates (0.2 - 400 s^{-1}) on standard PVB. By plotting stress-time curves for different engineering strain levels (0.11 - 1.72) using the data of the strain rate tests, he showed that the material response is not separable for the investigated engineering strain range. However, the experimental data were not further analyzed. Contrary, in (Botz, 2020), it was shown in creep tests that the large strain level relaxation curves of a standard PVB were purely vertically shifted in the log-log-scale up to engineering strains of 20%.

Due to the limited literature on the time-dependent large strain behaviour of PVB, it is advisable to perform relaxation or creep tests over a wide strain or stress range to systematically investigate the material behaviour and to check whether the simplified hyper-viscoelastic material model can be applied, or more complex nonlinear viscoelastic material models must be used.

1.3 Objectives

The overall aim of the present work is to mechanically characterize a wide variety of laminated safety glass interlayers and thus to provide important knowledge for the design of laminated safety glass in the intact as well as in the post fracture state. The work is concerned with three different interlayers: Polyvinyl butyral-based interlayers (PVB), ethylene vinyl acetate-based interlayers (EVA) and ionoplasts (SentryGlas®, SG).

The state of the art of PVB-based interlayers in intact laminated safety glass is significantly more advanced than that of EVA-based interlayers and ionoplasts.
In this work, therefore, the molecular structure of EVA and SG is studied in more detail, and the influence of the molecular structure on the time- and temperature-dependent material behaviour is investigated. The goal is to design a DMTA test procedure and evaluation method that leads to reproducible results. In addition, shear modulus values for the usual load cases and/or Prony parameters as well as time-superposition principles are to be determined.

The investigations on the PVB-based interlayers focus on the mechanical behaviour in the nonlinear viscoelastic range, which especially plays a role when the interlayer is undergoing large deformations in the post fracture behaviour. One objective in this work is therefore to determine the linearity limits that describe the transition from the linear to the nonlinear viscoelastic range. Subsequently, the nonlinear viscoelastic material behaviour is to be described with a mechanical model. For this purpose, the Schapery model is to be used, which extends the linear viscoelastic constitutive law by means of nonlinearity factors.

1.4 Outline

All the fundamentals relevant to this thesis are summarized in chapter 2. First, the polymers in general are discussed with regard to their differences in intra- and intermolecular structure and their thermo-mechanical behaviour. Subsequently, the interlayers PVB, EVA and ionoplast investigated in the present work are assigned to a polymer class.
Since EVA-based and ionoplastic interlayers have a semicrystalline structure, basic information on the crystallization process from the melt, the degree of crystallization and crystal morphologies is given. Finally, models are presented that can be used to mathematically describe the crystallization process.
All three materials investigated exhibit time-dependent material behaviour typical of polymers. The fundamentals of viscoelasticity include the introduction of a linear viscoelastic and a nonlinear viscoelastic material model. In addition, information is given on the linearity limits which distinguish the linear from the nonlinear viscoelastic range.
The last part of the fundamentals deals with time-superposition principles. These include the basics of the time-temperature-superposition principle, as well as a time-crystallinity-superposition and a time-strain-superposition principle.

Chapter 3 describes all experimental methods used in this work. These include Differential Scanning Calorimetry (DSC), Dynamic-Mechanical-Thermal-Analysis (DMTA), X-Ray Diffraction (XRD) and uniaxial tensile relaxation tests. In each case, the general test principle, the specific test setup and the evaluation methods used are presented. Finally, a test overview provides clarity about which tests were carried out with which interlayers.

Chapter 4 and 5 are dedicated to the investigations that were performed on the EVA-based and SG interlayers. Chapter 4 deals in detail with the semicrystalline structure. The degrees of crystallization, characteristic temperatures, enthalpies of melting and fusion are determined experimentally with XRD, DSC and DMTA. In addition, the crystallization kinetics are described and the influence of physical aging is shown. Based on this, chapter 5 deals more specifically with the influence of the semicrystalline structure on the linear viscoelastic material behaviour. Prony series and time-superposition principles are determined based on DMTA and DSC, which take into account both the influence of temperature and the influence of the degree of crystallization.

Chapters 6 and 7 deal with the investigations of PVB based interlayers. In chapter 6, the Prony parameters are determined and compared with the results of four-point-bending creep tests on laminates. The linearity limits are identified in dependency of temperature and frequency (or load duration) with DMTA. Then, in chapter 7, the nonlinear viscoelastic material behaviour is experimentally investigated in uniaxial relaxation tests at high strains and approximated by the Schapery model.

In chapter 8, the knowledge gained during this work is prepared for engineering application. This includes a guide to the mechanical characterization of interlayers in intact laminated safety glass. Improvements and additions to the current EN 16613 are proposed. The special focus is on semicrystalline polymers.
In addition, a guide to determine the linearity limits and to characterize the nonlinear viscoelastic material behaviour is given. This chapter is completed by practical design examples.

Chapter 9 summarizes the present work and gives an outlook for further research topics.

2 Theoretical background

In the following, all the fundamentals required to achieve the objectives of the thesis are presented. First, the intra- and intermolecular structure of polymers in general and their thermo-mechanical behaviour are presented. Based on this, the investigated interlayer materials PVB, EVA and ionoplast are assigned to polymer classes.
As both, EVA and ionoplastics exhibit semicrystalline regions, the crystallization process is discussed in more detail subsequently.
The polymer-typical time-dependent material behaviour is introduced next. A distinction between linear and nonlinear viscoelastic material behaviour is made.
Finally, different time-superposition principles are presented. These include the basics of the time-temperature shift principle, as well as a time-crystallization shift principle and a time-strain-superposition principle.

2.1 Polymers

Most polymeric materials used in construction are synthetic organic and are developed to meet specific requirements. They mainly consist of macromolecules. Additives are used to specifically change the processing or application properties (Brinson et al., 2008). Macromolecules include more than 1000 atoms, which can be grouped into smaller repeating parts (building blocks). Hence, they are also called polymer-molecules or simply polymer (*poly* = a lot, *mer/meros* = parts). They are created in polymerization processes (e.g. chain, condensation and addition polymerization) starting from monomers (*mono* = one; *mer* = part) (Kaiser, 2011). As the term *organic* in the definition of polymeric materials already implies, the most common atom of the macromolecule is carbon (symbol C), the second is hydrogen (symbol H) (Kuntsche, 2015).
In order to understand the mechanical behaviour, the polymer structure is discussed first. Furthermore, the incorporation of foreign atoms as well as the influence of the molar mass and the molar mass distribution on the mechanical properties are explained.

2.1.1 Molecular and intermolecular structures

A distinction is made between the molecular structure within one macromolecule and the connection between different macromolecules.

M. Schuster, *Characterization of Laminated Safety Glass Interlayers*,
Mechanik, Werkstoffe und Konstruktion im Bauwesen 66,
https://doi.org/10.1007/978-3-658-39821-7_2

Consideration of the single macromolecular chain

The chemical structure of a macromolecule is characterized by its constitution, configuration and conformation.
The *constitution* describes the atomic basic structure of the macromolecule. This includes the type, number and combination of the basic building blocks including end groups and side chains, type and length of branches, molar mass and mass distribution. The repeating unit of polyethylene (PE) macromolecule consists exclusively of carbon and hydrogen atoms. The carbon atoms from the main chain are connected via simple chemical bonds. Each of these carbon atoms has two hydrogen atoms attached to it. The main chain of polypropylene (PP) also consists of simply chemically bonded carbon atoms. However, methyl group (-CH_3) side chains are attached on every second carbon atom, see Figure 2.1 (a).

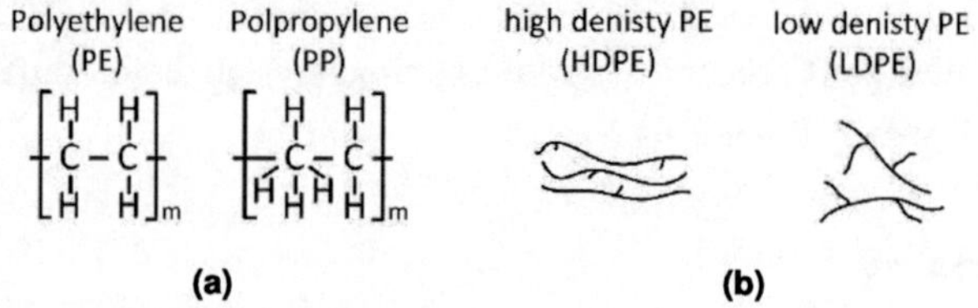

Figure 2.1 (a) building blocks of PE and PP, (b) linear HDPE and branched LDPE.

The simplest form of combining the basic blocks into a macromolecule is the creation of a long, linear main chain. Besides the simple form of a long linear chain, there also exist branched macromolecules. Branch points are defined in a way that three or more chains originate from them. They can be produced by specifically changing the process parameters during polymerization. For example, a distinction is made between the linear HDPE (high density polyethylene) and the branched LDPE (low density polyethylene), both types consisting of the same basic building block, see Figure 2.1 (b). Another possibility is the formation of ring-shaped connections.
For polymers with different side atoms or side chains, an additional distinction must be made between a head-head or a head-tail arrangement. The head-tail sequence corresponds to a consecutive arrangement of the basic mer-unit, whereas in the head-head configuration the second mer-unit is mirrored (adjacent (R) groups), see Figure 2.2 (a)-(c).
The *configuration* describes the spatial arrangement of the side groups (e.g. methyl group of PP) along the main chains. Molecules with different configurations but identical constitutions are called isomers. Three tacticities are differentiated. In the isotactic configuration the side groups are arranged regularly and equilaterally. In the syndiotactic configuration, however, the side groups are regular but reciprocal, and in the atactic configuration the

side groups are irregularly arranged. Figure 2.2 (d)- (f) shows all three configurations for head-tail sequences.

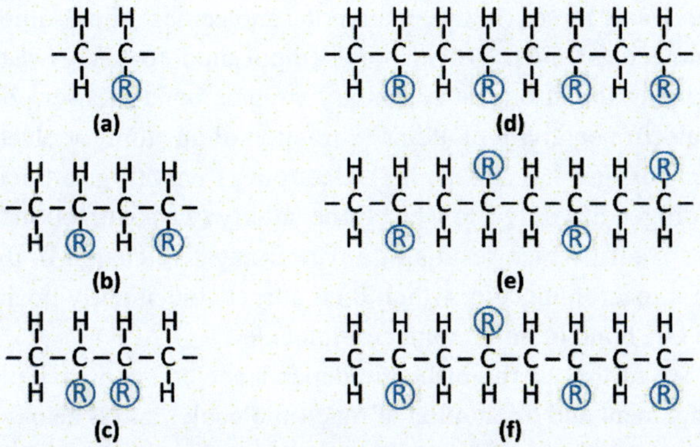

Figure 2.2 Configurations (a) basic mer unit (b) head to tail (c) head to head (d) isotactic (e) syndiotactic (f) atactic.

Due to rotations around the binding axes (= microbrownian molecular movements), the macromolecule can take on different shapes in space, also called *conformations*, with the same configuration. The conformation can change considerably depending on the state (solid, liquid, gas). Basically, a distinction is first made between the tangled ball (entropically favourable state) and the straight chain (entropically unfavourable state). The former are preferably achieved in melts and solutions. In solids, zig-zag or helicoidal chain structures are also possible, see Figure 2.3. The prerequisite for this is a regular chemical constitution and configuration along the chain. Additionally, a low degree of branching and a chain length that is not too long are necessary.

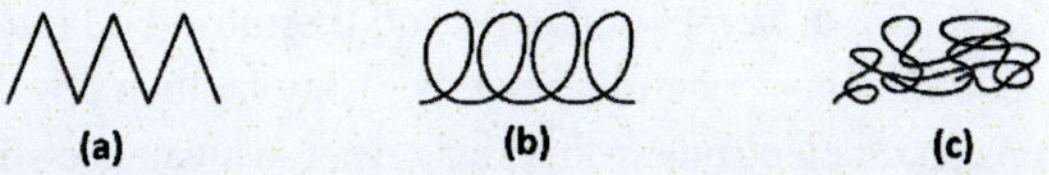

Figure 2.3 Conformations (a) zig-zag, (b) helicoidal, (c) tangled ball.

Consideration of multiple macromolecular chains

There are different ways of combining several macromolecules. The main difference is the type of bonding between the chains (Brinson et al., 2008).

Thermoplastics have only secondary bonds and/or mechanical bonds between chains. As the name reflects, thermoplastics deform plastically under the influence of temperature (they can be melted or molded, hence the bonds are reversible). When they are cooled

down, they can become again solid and resilient.
Secondary bonds, also known as physical bonds or intermolecular forces, are much weaker than primary bonds. They are based on electrostatic interactions. A distinction is made between van-der-Waals forces (between nonpolar molecules), dipol-dipol forces (between partially polar molecules) and hydrogen bonds (dipol-dipol forces between hydroxil groups OH), the latter being the strongest secondary valence bond (Kaiser, 2011). This is explained via the electronegativity of atoms (attraction of an atom for electrons in the outermost shell). While oxygen has a really high electronegativity (high attraction for additional electectron), hydrogen has only a low electronegativity (weak attraction for its single electron on the outer shell), which results in a partial negative charge in the oxygen side of the OH bond in a macromolecule, which then attracts the slightly positive charge of the hydrogen of the OH bond of another macromolecule.
Thermoplastics are further distinguished in dependence on the degree of order, which describes the arrangement and orientation of macromolecules and is also referred to as physical structure, or morphology. If the macromolecules are present in a disordered state, one speaks of an *amorphous* structure. If the chains are structured in a regular way (constitution and configuration), partially ordered areas can be created (see zig-zag or helicoidal conformations in 2.3), especially in linear macromolecules. Those ordered areas are called crystallites. Since polymeric materials, that include crystallites also always include disordered, amorphous areas, they are called *semicrystalline.* Several crystallites can form superstructures, such as spherulites. The degree of crystallization describes the proportion of crystalline regions within the polymer material (Kaiser, 2011). It depends on various parameters, which are further discussed in chapter 2.2.

Thermosets have primary bonds between chains. While thermoplastics deform plastically under the influence of temperature, this is hindered in the case of thermosets by the chemical bonds between the molecule chains (permanent bond). Accordingly, thermoplastics are also referred to as *linear* or *non-crosslinked* and thermosets as *crosslinked* polymers.
Primary bonds are also knows as chemical bonds. They include covalent bonds (electron pair bonds, two atoms share electrons from their respective valence shell).
Depending on the crosslinking density, thermosets are further divided into *elastomers* (weakly crosslinked) and *duromers* (high crosslinking density).

Figure 2.4 shows the molecular structures of the different groups.

Modification through incorporation of foreign atoms

Since it is almost impossible to develop new plastics nowadays (Domininghaus, 2013), the modification of properties through the incorporation of foreign atoms or molecules plays

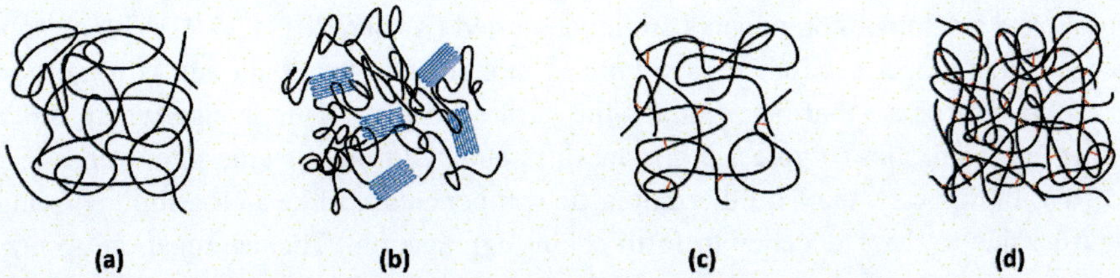

Figure 2.4 Polymer groups (a) linear amorphous thermoplastic, (b) linear semicrystalline thermoplastic, (c) chemically crosslinked elastomer, (d) chemically crosslinked duromer.

a major role. If the macromolecules of a plastic are made of one type of monomer, it is called a *homopolymer* (Domininghaus, 2013).
Copolymers are created, when different monomer types are chemically incorporated in the macromolecules. If the different monomer types are distributed purely randomly in the macromolecule, this is called a statistical copolymer (random copolymer). However, if the monomer types in the macromolecule occur alternately, this is called an alternating copolymer. In block copolymers, longer chain pieces of one monomer alternate with those of the other monomer. Finally, a distinction is made between graft copolymers, in which side chains grow on an already existing macromolecule chain by polymerization of an additional monomer (Kaiser, 2011). The presented copolymer types are shown in Figure 2.5 (a). Polymer mixtures are also known as polymer blends (Figure 2.5 (b)). There are physical bonds between the different macromolecule chains (Kaiser, 2011). It should be noted that in block copolymers, graft copolymers and polymer blends, the individual polymer characteristics (such as the glass transition temperatures of both polymers) are retained. Only in the case of random copolymers, smeared polymer characteristics occur (e.g. only one, but shifted glass transition temperature) (Domininghaus, 2013).

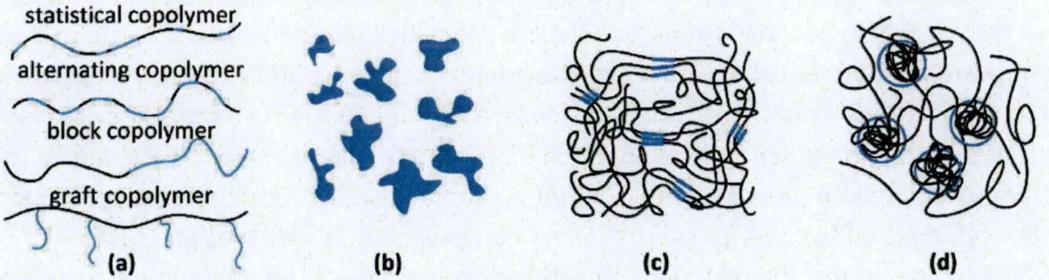

Figure 2.5 (a) different copolymer types, (b) polymer blend, (c) block copolymer with crystalline hard segments, (d) block copolymer with amorphous hard segments.

Block copolymers and polymer blends of a hard thermoplastic and a soft elastomeric phase (and hence a multiphase system) are called *thermoplastic elastomer* (TPE). A TPE polymer blend consists of a hard thermoplastic polymer matrix in which non-crosslinked or crosslinked elastomer components are incorporated (Kaiser, 2011). A TPE block copolymer contains longer sections of one type of monomer in the chain and is usually made up of the soft phase that incorporates the hard segments. This is also where the crystalline thermoplastics (Figure 2.5 (c)) are differentiated from the amorphous ones (Figure 2.5 (d)). In this way, the hard thermoplastic components act like a crosslinking, reinforcing filler due to their tendency to form chains aggregation. This aggregation occurs due to glassy solidification, crystallization, hydrogen bonds or ion association (Kaiser, 2011). Copolymers with ion associations are also called *ionoplasts*.

2.1.2 Classification based on thermo-mechanical behaviour

The previous considerations lead to different ways of classifying polymers into groups based on the chemical and physical structures of the macromolecules. These included:

- Mer units: homopolymer or copolymer (heteropolymer);
- Polymerisation process: chain, condensation or addition polymerization;
- Bonding: linear thermoplastic or crosslinked thermoset;
- Morphology: amorphous or semicrystalline.

The intra- and intermolecular structures have an influence on the thermo-mechanical behaviour of the polymeric material. This is due to the fact that temperature influences the mobility of the molecule chains or individual segments of the molecule chains. In case of high mobility intermolecular forces possibly can be overcome. Depending on temperature (and hence mobility), the polymeric material is in a hard and stiff (glassy), a soft flexible (rubbery) or even a viscous liquid state.
At low temperatures, the material is in a glassy state with high stiffness. As a result of a load, i.e. energy input, the atoms in adjacent molecules are deflected from their equilibrium position. Mainly the weak, intermolecular forces are stretched. A rearrangement of molecule sections, a rotation of the C-C bonds, a slipping of loops as well as shape changes are not possible in the range of small strains. When the load is removed, the atoms return to their energetically more favorable initial position, which is why the material behaviour in this region is called *energy-elastic* (Domininghaus, 2013; Rösler et al., 2012).
At high temperatures, the material behaves rubbery and has compared to the glassy state a relatively low stiffness. The mechanical behaviour is called *entropy-elastic* in this range. In case of a load, the entangled molecule chains are stretched. This stretching is hindered by entanglements and/or crosslinking points, so that it only takes place between the anchoring points. Stretching the molecular chain requires changes in position and rotation of entire molecular chains, which must be thermally activated and therefore takes a certain amount of time. When released, the polymer strives for the state of greatest entropy,

which corresponds to the shape of the entangled ball. The anchoring points ensure that the polymer returns to its original shape (Rösler et al., 2012).
If the temperature is increased even further to the *flow* temperature, thermoplastics reach another phase change point, the melting point. The plastic starts to flow. This leads to irreversible loosening of the loops and thus to molecular segregation.

Considering the stiffness-temperature curves, the following thermo-mechanical classification can be made (see also Figure 2.6):

- Amorphous thermoplastic;
- Semicrystalline thermoplastic;
- Chemically crosslinked elastomer;
- Chemically crosslinked duromer;
- Thermoelastics;
- Physically crosslinked elastomer (thermoplastic elastomer).

Amorphous thermoplastics

Completely disordered and entangled polymer chains are in a glassy, energy-elastic state at temperatures below the glass transition T_g. The microbrownian movement is largely frozen. At very low temperatures there is no change in the shape of the molecules (no change in conformation). At slightly higher temperatures, but still in the glassy state, secondary dispersion areas can occur. Here side groups or short parts of the main chains become mobile and can carry out conformational changes. The higher the temperature, the greater the mobility. At temperatures above the glass transition temperature T_g, the polymer is in the rubbery, entropy-elastic state. Here, the microbrownian movements are fully developed and position changes and rotations of whole molecule sections are possible (when the condition of sufficient high free volume is fulfilled), without loosening the tangles. The transition between the energy-elastic and entropy elastic state is called glass transition or softening/hardening transition, which is characterized by the glass transition temperature T_g. This range can be up to 60 K wide in amorphous polymers. If the temperature is raised even further to the flow temperature T_f, thermoplastics start to flow and behave like a viscous liquid. This leads to irreversible loosening of the loops and thus to molecule segregation.

Semicrystalline thermoplastics

While for amorphous thermoplastics both, the glass transition T_g temperature and the flow temperature T_f, are of great importance, for semicrystalline thermoplastics the melting temperature of the crystallites T_m is also decisive. This is always above the glass transition temperature. Considering a two-phase model, a glass transition temperature exists for the amorphous phase at which a transition from glassy to rubbery occurs. Below this glass transition temperature, both the amorphous and the crystalline phase are in a solid state and thus in the energy-elastic range. In the range between T_g and T_m, the semicrystalline polymer can be seen as a dispersion (heterogeneous mixture) of hard crystalline particles in a very soft rubber (rubber-like amorphous phase). The crystallites consist of ordered macromolecules (e.g. folded). In these areas the intermolecular forces are particularly high, which is associated with a high stiffness compared to amorphous thermoplastics, resulting in an overall leather-like aggregate. When the melting temperature T_m is reached, the crystalline parts melt and the thermoplastic behaves like a purely amorphous thermoplastic.
According to Wunderlich (2006), semicrystalline polymers can have a third phase in addition to the crystalline and amorphous phases. It represents the interface between the crystalline and amorphous regions within a macromolecule. Due to the closeness of the crystallites, the mobility of this interface phase (also known as mesophase) is limited, which is why it is called the *rigid amorphous phase*. It widens the glass transition region or can even cause a second glass transition of its own.
More detailed information about semicrystalline polymers can be found in chapter 2.2. However, it should be noted already here, that the melting and formation of crystallites are two different processes, characterized by the two different temperatures: the melting temperature T_m and the crystallization temperature T_c.

The operating temperature range of thermoplastics is usually at temperatures below T_g and T_m respectively, which corresponds to the high stiffness range.

Elastomers

Elastomers have a low crosslinking density and hence a wide-meshed network. After the energy-elastic and the glass transition range follows an extended entropy-elastic (rubber-elastic) range until thermal decomposition of the polymer occurs. The chemical crosslinking points ensure the original arrangement of the molecules in the entropy-elastic range after relief, so that no plastic deformations occur. In the entropy-elastic range, the shear modulus is mainly dependent on the amount of crosslinking points and increases slightly with increasing temperature due to higher entropy at higher temperatures (Schwarzl, 2013).

The operating temperature range of elastomers is usually at temperatures above T_g, which corresponds to the entropy-elastic range.

Duromers

Duromers have a high crosslinking degree and hence a close-meshed network. This prevents molecular movements due to external loads. Duromers hence show only a very low degree of dispersion (softening) in the glass transition range (if a glass transition occurs at all). They are dimensionally stable until the decomposition temperature T_z is reached.

Thermoelastics

Thermoelastics are a special class as they include very long non-chemically crosslinked macromolecules (high polymerization and high molar mass) and only very slightly crosslinked macromolecules. Like thermoplastics, they soften at temperatures above the glass transition temperature T_g. However, they are not meltable (Kaiser, 2011; Burr and Harsch, 2005).

Thermoplastic elastomers

Thermoplastic elastomers include block copolymers, polymer blends and ionoplasts. The macromolecules are crosslinked physically and soften when the transformation temperature is reached (T_g for amorphous segments, T_m for crystalline segments) (Kaiser, 2011). Polymer blends and block copolymers often have two glass transitions due to phase separation. In contrast, only one glass transition can be detected in very short phases.

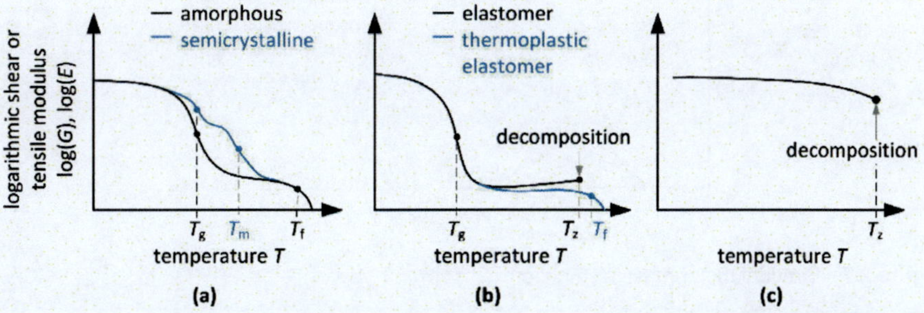

Figure 2.6 Shear or tensile modulus in dependency of temperature (a) amorphous and semicrystalline thermoplastics, (b) chemically and physically crosslinked elastomers, (c) duromers.

Influence of molecular weight and molecular weight distribution

The molecular weight and distribution of the polymer or copolymer influences its mechanical properties. The degree of crystallization decreases with increasing chain length. This is due to the fact that the transition to the ordered semicrystalline state, starting from the amorphous state in the melt is more difficult with increasing chain length. Long chains increase the impact strength and the dimensional stability at elevated temperatures. At the same time, the flow behaviour in the melt deteriorates with increasing chain length (higher viscosity), which leads to more difficult processability. The flowability is indicated by the melt flow index (MFI). This is a measure of how much of the molten thermoplastic moulding compound can be forced through a certain nozzle at a certain pressure at a certain temperature within 10 minutes (Kaiser, 2011). The molar mass M is the sum of the masses of all atoms in the molecule. Substances consisting of different molecules are described by an average molar mass (usually between 10^4-10^6 g/mol) and a corresponding molar mass distribution (narrow or wide). A distinction is also made between the number average $\overline{M_\mathrm{n}} = \sum n_\mathrm{i} \cdot M_\mathrm{i} / \sum n_\mathrm{i}$ and the weight average $\overline{M_\mathrm{W}} = \sum m_\mathrm{i} \cdot M_\mathrm{i} / \sum m_\mathrm{i}$. The molar mass distribution width (polydispersity) can be derived from these two values: $\overline{M_\mathrm{W}}/\overline{M_\mathrm{n}}$. Narrow molar mass distributions (synonymous with many identical molecules) lead to narrow thermal softening ranges, while wide molar mass distributions also have wider softening ranges. Molar mass distributions with two maxima are called bimodal (Figure 2.7). Such a distribution can be used, for example, to combine good processability (rather with smaller molar masses) with high mechanical properties (rather with greater molar masses) (Kaiser, 2011). The low-molecular fractions have a softening effect at higher temperatures (Henderson, 1993).

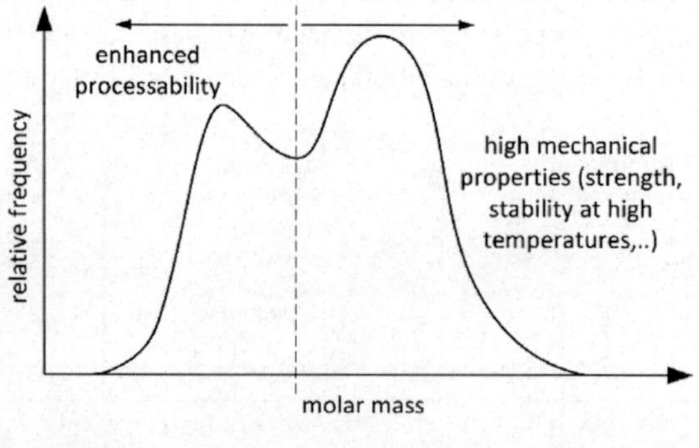

Figure 2.7 Bimodal molar mass distribution.

In this section, the thermo-mechanical behaviour of different polymer groups was explained. In the following sections of this chapter, the transition regions are discussed in more detail. First, a short introduction to thermodynamics and kinetics is given.

2.1.3 Thermodynamics and kinetics

Thermodynamics is the investigation of the behaviour of physical systems under temperature changes, especially during the supply and removal of heat. It deals with the relationships between energy, work and temperature.
Kinetics deals with the time sequence of chemical reactions (reaction kinetics) or physical-chemical processes (e.g. diffusion, deposition of substances on surfaces).

In order to better understand the behaviour of the polymers under investigation, the basic thermodynamic quantities *free enthalpy G*, *enthalpy H*, *entropy S* and the term *thermodynamical equilibrium* are introduced. The terms *thermal transformation* and *relaxation* are also described. The explanations are mainly based on (Stephan et al., 2010).

Enthalpy

Enthalpy H (unit: Joule [J]) is a measure for the energy of a thermodynamic system. The enthalpy is composed of the internal energy U and the volume work pV. The internal energy U consists of the thermal energy (based on the undirected movement of the molecules; kinetic energy, rotational energy, vibrational energy), the chemical binding energy and the potential energy of the atomic nuclei. It increases approximately in proportion to the temperature of the system. Only the difference of the enthalpies ΔH between two states can be measured. ΔH is the heat quantity emitted or absorbed by a reaction. A positive ΔH describes an endothermic (supply heat) reaction, while a negative ΔH is obtained for exothermic (heat is released) reactions.

Entropy

Entropy S (unit: Joule per Kelvin [J/K]) is a thermodynamic quantity that describes the equivalent value of transformation of heat in work. The entropy change ΔS is calculated as the quotient of the amount of heat which is exchanged with the environment during a reversible change of state and the absolute temperature prevailing during this process (transported enthalpy). A heat supply corresponds to an increase in entropy and a heat loss to a decrease in entropy. However, the entropy not only allows the determination of the maximum usable work of a state change, but also the degree of irreversibility or the driving force of a state change. If an isolated (adiabatic) system and a reversible change of state is considered, no new entropy is produced and the entropy change is 0. If an isolated system and an irreversible change of state (dissipation of work) is considered, entropy always increases (entropy is produced). The latter can also be explained with a statistical approach, in which the number of microstates (realization possibilities for arrangement) that cause the same macrostate are considered. The more microstates there are, the higher is the entropy and the more probable is the macrostate. Thus the state of highest entropy is always strived for. Entropy is also often interpreted as the degree of disorder or randomness in the system.

Free enthalpy

Free enthalpy G (unit: Joule [J]), also known as Gibbs-Helmholtz enthalpy or Gibbs-energy, is a measure of the driving force of a process (thermodynamical potential). Reactions with a negative ΔG (G after reaction < G before reaction) release energy, which means that they can take place without energy supply (spontaneously, exergonic reactions). In contrast, reactions with a positive ΔG need an energy supply to run (not spontaneously, endergonic reaction).
The change of free enthalpy ΔG, enthalpy ΔH and entropy ΔS are connected via the Gibbs-Helmholtz equation, shown in eq. (2.1). T corresponds to the temperature in [K].

$$\Delta G = \Delta H - \Delta S \cdot T \tag{2.1}$$

Thermodynamic equilibrium

A system is in thermodynamic equilibrium when it is in a stationary state in which all macroscopic flows of matter and energy within the system disappear. These include:

- Thermal equilibrium, which means that there is no macroscopic heat flow.
- Mechanical equilibrium, which means that a macroscopic subsystem does not carry out work on another subsystem.
- Chemical equilibrium, which means that the composition of the system from different phases remains the same.

Real processes always take place in the direction of equilibrium. A state of equilibrium is called *stable* if only such states exist in its neighbourhood that the system returns to the initial state by itself after all temporary forced displacements (Figure 2.8 (a)). It is called *unstable* if there are only such states in its environment that the system, after a temporary forced displacement, moves further and further away from the initial state and does not return there by itself (Figure 2.8 (b)). A state of equilibrium is called *neutral* or *reversible* if the system remains in its new position after a temporary forced shift (Figure 2.8 (c)). *Metastable* equilibrium prevails if the system is stable with respect to immediately adjacent states but unstable with respect to finitely distant states (Figure 2.8 (d)). Metastable states are stable in terms of the thermodynamic stability theory. A finite perturbation is required to transform them into a stable equilibrium. Metastable phases can form in technical or natural processes because a certain *potential threshold* or *activation energy* must be exceeded during phase transitions. This is a kinetic process, which cannot be fully described by means of equilibrium thermodynamics. This potential threshold is however arbitrarily small if so called germination points are present in the system.

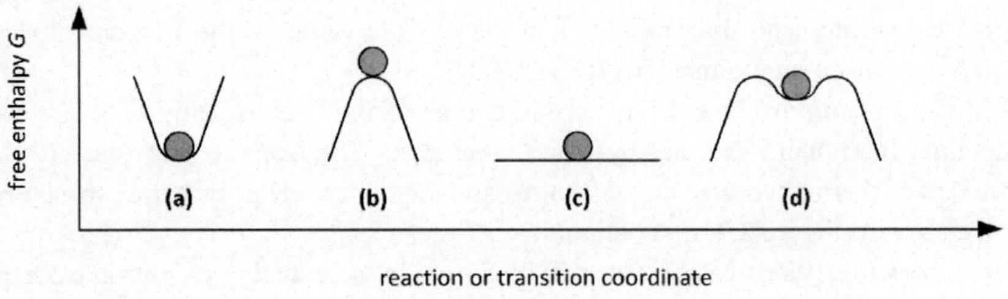

Figure 2.8 States of Equilibrium (a) stable, (b) unstable, (c) neutral, (d) metastable.

The following two general and equivalent equilibrium criterion apply under the conditions of constant entropy, volume and molar quantity:

- The internal energy U of a closed system exhibits a minimum for given values of entropy S and volume V.
- The entropy S of a closed system exhibits a maximum for given values of the internal energy U and the volume V.

For constant T or / and p, the equilibrium condition can also be defined as:

- The free enthalpy G of a closed system exhibits a minimum for given values of temperature T and pressure p or pressures of the individual phases (subsystems).
- The enthalpy H of a closed system exhibits a minimum for given values of entropy S, pressure p or pressures of the individual phases (subsystems).

It is convenient to express the thermodynamic energy levels of the various possible states in terms of a potential energy landscape, where the x-axis represents the reaction or transition coordinate and the y-axis the free enthalpy change ΔG.

Thermal transformations

Thermal transformations describe a change of state of a substance, in which the substance is in thermodynamic equilibrium both above and below the transformation temperature. A thermal transformation hence takes place at a specific temperature (and pressure) and is independent on time effects.

The free enthalpy G and the quantities derived from it (enthalpy H, entropy S, volume V, isobaric heat capacity C_p, coefficient of thermal volume expansion α_V and the isobaric

compressibility κ, see eq. (2.2) and eq. (2.3) below) show a very characteristic behaviour at thermal transitions (Lechner et al., 2014). Depending in which derivation from G to T or p a discontinuity occurs, different orders of thermal transitions are distinguished, e.g.: thermal transitions with discontinuities in the Nth derivative of the free enthalpy G are called N^{th} order thermal transitions (Elias, 2009).
First order transitions show a kink in the course of the free enthalpy G and show discontinuities in enthalpy H, entropy S and volume V. The isobaric heat capacity C_{p}, the coefficient of thermal volume expansion α_{V} and the compressibility κ thus strive towards infinity, see also Figure 2.9, first column.
Second order transitions can be detected by a kink in the enthalpy H, entropie S and volume V and an abrupt change in the isobaric heat capacity C_{p}, thermal volume expansion coefficients α_{V} and isobaric compressibility κ, see also Figure 2.9.

First derivation of free enthalpy G:

$$H = G - T\left(\frac{\mathrm{d}G}{\mathrm{d}T}\right)_p \quad S = -\left(\frac{\mathrm{d}G}{\mathrm{d}T}\right)_p \quad V = -\left(\frac{\mathrm{d}G}{\mathrm{d}p}\right)_T \tag{2.2}$$

Second derivation of free enthalpy G:

$$C_{\mathrm{p}} = -T\left(\frac{\mathrm{d}^2G}{\mathrm{d}T^2}\right)_p \quad \alpha_{\mathrm{V}} = \frac{1}{V}\left(\frac{\mathrm{d}^2G}{\mathrm{d}T\mathrm{d}p}\right)_p \quad \kappa = -\frac{1}{V}\left(\frac{\mathrm{d}^2G}{\mathrm{d}p^2}\right)_T \tag{2.3}$$

Thermal relaxations

Thermal relaxations are kinetic effects. They occur when a thermal activation leads to movements in or between molecular chains (Elias, 2009; Rösler et al., 2012). Examples in polymers are rearrangements and conformational changes of polymer chains (Rösler et al., 2012). The thermal activation represents an energy barrier that must be overcome, which takes time. Thermal relaxations therefore respond to the frequency of the method and thus also to the time scale. Unlike thermal transformations, the system is not in thermodynamic equilibrium below and possibly also above the relaxation temperature (Elias, 2009).
The *theory of free volume* is often used to describe molecular mobility. For this purpose, the specific volume v is considered first. This is the reciprocal value of density ρ. On a molecular level, the specific volume is composed of the volume occupied by molecules or polymer segments v_{o} and the free volume v_{f}, see eq. (2.4). The free volume can be considered as empty volume, which occurs for example due to chain ends, entanglements, etc. and hinder a tight packing of the chains (Naumann, 2012). The free volume is a function of temperature; v_{f} is large for a liquid (melt) and small for a solid. This can be explained

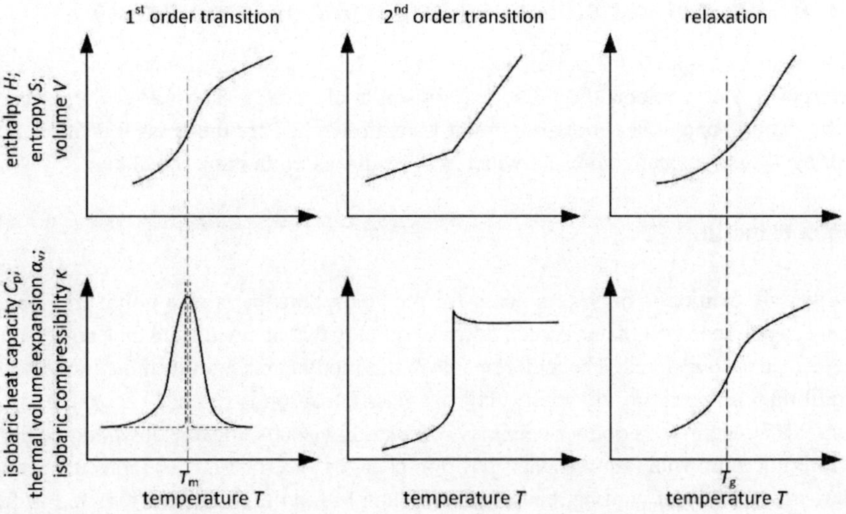

Figure 2.9 Thermal transformations and relaxations, based on (Elias, 2009).

by the fact that, due to thermal movement, neighbouring molecules occasionally swing so far apart that new, stable, empty spaces, which themselves have molecular dimensions, are created (Schwarzl, 2013).

$$v = v_0 + v_f \tag{2.4}$$

The mobility of the molecules (microbrownian movement) is dependent on the free volume and the thermal energy. Due to the large free volume and thermal energy at high temperatures, the polymer molecules can quickly return to their thermodynamic state of equilibrium when a thermodynamic variable changes. As the temperature drops, the free volume fraction is reduced by diffusing excess empty spaces from the material. The mobility of the molecules and the speed with which the free volume can change also decrease (Schwarzl, 2013; Naumann, 2012). If the temperature decreases further, microbrownian movement movements and diffusion of superfluous empty spaces are no longer possible. The free volume is frozen. The material is then in a thermodynamic imbalance, as there is more free volume than corresponds to the thermodynamic equilibrium state (Schwarzl, 2013). The specific volume of the polymer continues to try to reach its state of equilibrium which leads to very slow reorganisation effects and thus to further relaxation (Schwarzl, 2013).

2.1.4 Thermal transitions and relaxations in polymers

As shown in chapter 2.1.2 polymers can be in a hard and stiff (glassy), a soft and flexible (rubbery) or a viscous liquid state (Brinson et al., 2008). Those states are separated by transition zones. The most important transition zones are the glass transition T_g, the melting T_m and crystallization T_c, which will be discussed in more detail here.

Glass transition

The glass transition is only significant for the amorphous parts of a polymeric material. Hence, a polymer melt that is cooled down so quickly that no nuclei and thus no crystallites are formed is considered. The melt remains in thermodynamic equilibrium (and retains the equilibrium properties of the melt) until the glass transition is reached (Zografi and Newman, 2017). Here, the polymer becomes sufficiently viscous, to slow down molecular diffusion (microbrownian movements; position changes and rotations of molecule sections) relative to the rate of cooling, causing the system to fall out of equilibrium and to freeze into a glassy state. However, the structure does not change; both before and after the glass transition the substance is in an amorphous state. The glass transition temperature is not a single temperature, but a range that is usually specified by an average glass transition temperature T_g. This can be explained by the fact that the glass transition temperature is directly proportional to the molar mass. Polymers, however, consist of macromolecules of different lengths or different molar masses (Kaiser, 2011). The value of the glass transition temperature depends on the cooling rate. If the cooling is infinitely slow, there is no glass transition temperature at all. Accordingly, depending on the manner in which the glassy state was reached (e.g. cooling rate), different glassy structures with different thermodynamic properties (molar volume, density, enthalpy, and entropy) are feasible. The assignment of T_g to a phase transformation level is difficult. Although there is some evidence of a 2nd degree transformation (jump from C_p, α_V and κ, see also Figure 2.9 third column), it is not a true thermodynamic transformation because there is no equilibrium on either side of T_g. In addition, T_g is dependent on the cooling rate or heating rate and is therefore influenced by the kinetics (Lechner et al., 2014). An increase of the cooling rate by a factor of 10 increases the glass temperature by about 3 K (Schwarzl, 2013; Elias, 2009). Furthermore, due to different kinetic effects, the softening temperature from the glassy state does not correspond to the solidification temperature from the melt. In practice, however, the difference is only slight, which is why no distinction is made here between the two temperatures and only the glass transition temperature is generally referred to (Elias, 2009).

Gibbs and DiMarzio (1958) denote the glass transition as a kinetic manifestation of a thermodynamic transition that is sought but not achieved. Strictly speaking, then, this is not a thermal transformation, but a *relaxation process* (Elias, 2009; Brinson et al., 2008).

A common way to determine T_g is to examine the specific volume v (reciprocal of density ρ). T_g is the temperature at which the discontinuous change in the slope of the specific volume (Figure 2.10) occurs. According to the free volume theory (eq. (2.4)), T_g is the temperature at which the free volume of the material freezes and leaves its equilibrium state.

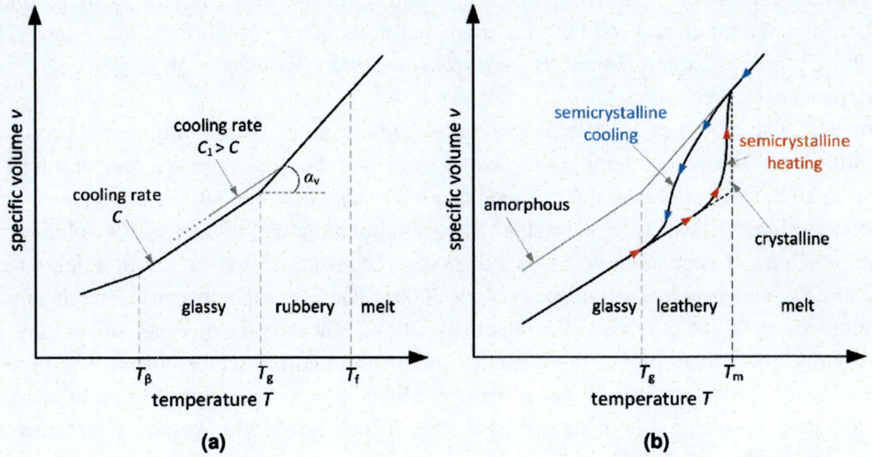

Figure 2.10 Specific volume (a) amorphous polymer, (b) semicrystalline polymer.

In some semicrystalline polymers, two glass transitions occur, a lower glass transition at $T_{g,u}$ and an upper one at $T_{g,o}$. The temperature $T_{g,u}$ covers the purely amorphous regions and $T_{g,o}$ covers the amorphous regions in whose vicinity crystallites are located (rigid amorphous phase). In most cases $T_{g,o} \approx 1.2 \cdot T_{g,u}$, $T_{g,u} \approx 0.575 \cdot T_m$ and $T_{g,o} \approx 0.7 \cdot T_m$ (Lechner et al., 2014).

In addition to the glass transition, local relaxations often occur in many polymers. It involves the relaxation of a very short section of a polymer chain and is called β relaxation. $T_\beta < T_g$ and $T_\beta \approx 0.75 \cdot T_g$ (Lechner et al., 2014).

Melting

Melting is the thermal transformation of a solid crystalline or partially crystalline state of matter into the less ordered, amorphous liquid / melt state of matter. The amount of heat required for this is called the heat of fusion. The solid state components (ions, atoms or molecules) are excited to oscillate by the heat supplied. These vibrations are so strong that the crystal lattice suddenly breaks down (Lechner et al., 2014). With perfect, infinitely large crystals without impurities, this thermal transformation occurs at a special temperature, the thermodynamic equilibrium melting temperature T_m^0. Below this temperature, the material is in a solid, crystalline state, above this temperature, it is in a molten, amorphous state. In terms of free enthalpy: At temperatures $T < T_m^0$ the free enthalpy of the crystal is lower than that of the amorphous phase, which is why the crystal state is preferred. Analogously, for temperatures $T > T_m^0$, the free enthalpy of the amorphous phase is lower than that of the crystalline phase so that melting occurs (Lechner et al., 2014). At the moment of melting, the liquid and solid phases are in thermodynamic equilibrium. ΔG is zero. The melting process is a *first degree thermal transformation* (Brinson et al., 2008) and hence independent on frequency.

However, due to imperfect crystal structures (only semicrystalline polymers exists), the melting process covers a temperature range characterized by an average temperature T_m. The smaller, less perfect crystallites melt at a lower temperature than the larger, more perfect crystallites (Elias, 2009). The steps (discontinuous change) of the specific volume v or the enthalpies H degenerate to S-shaped curves. The sharp signals of the first derivatives α_V and C_p, now become bell-shaped (Elias, 2009). The heat capacity runs through a maximum for crystalline polymers (Lechner et al., 2014). The average melting temperature T_m is usually lower than the thermodynamic equilibrium melting temperature T_m^0 of perfect crystals. But it can also be higher if the crystallites are overheated during rapid heating. In this case the material is in a metastable state (Elias, 2009). One speaks of superheated crystals (Wunderlich, 2006).

Considering the phase diagram, the metastable, imperfect crystals have higher free enthalpy than perfect equilibrium crystals, whereby the enthalpy temperature behaviour is approximately parallel, see Figure 2.11. The more imperfect the crystal (smaller crystal) the higher its free enthaply. At the point of intersection of the free enthalpy of the metastable crystal with the free enthalpy of the (supercooled) melt, both phases have the same degree of metastability. The globally metastable system hence contains local areas which have a melting and crystallization equilibrium. Hence, partially melted polymer chains, may crystallize and melt reversibly. This is only valid as long as the molecular nucleus is hindered to melt itself (Wunderlich, 2006).

The empirical Beaman-Bayer rule gives the following relation between the glass transition temperature and the melting temperature of a semicrystalline homopolymer: $T_g = 2/3 \cdot T_m$ in [K]. For copolymers however, T_g and T_m are independently variable (Lechner et al., 2014).

Crystallizaton

The reversal of the melting is the crystallization process. The crystallization process consists of nucleation and crystal growth, which only takes place when a critical nuclei size has been reached. This represents an energy barrier in the energy landscape, which has to be

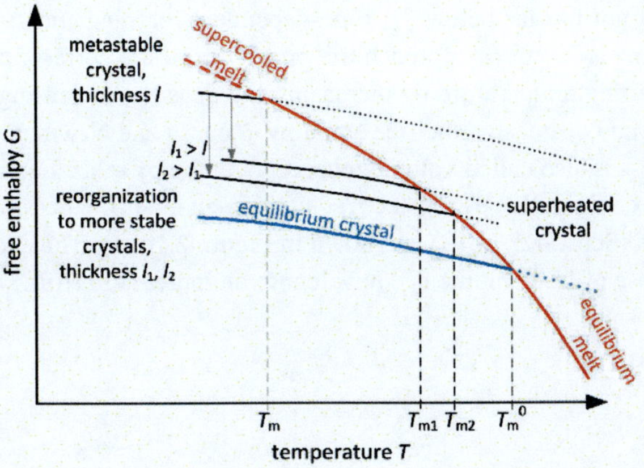

Figure 2.11 Free enthalpy as a function of temperature, based on (Wunderlich, 2006).

overcome. Crystallization is a *kinetically controlled* process and depends mainly on nucleation. Therefore the crystallization temperature T_c is always below the thermodynamically controlled melting temperature T_m. Looking at the specific volume of a semicrystalline polymer, the volume-temperature curve is approximately composed of that of the purely amorphous and the purely crystalline state. At the melting point of the crystallites, the specific volume in the semicrystalline polymer increases in a relatively narrow temperature range (s-curve). However, the formation of crystallites during the cooling process extends over a much wider temperature range, limited by the glass transition temperature T_g and the crystallite melting temperature T_m. A hysteresis is created in the specific volume, see Figure 2.10 (b) (Schwarzl, 2013). More detailed information about crystallization processes are explained in chapter 2.2.
However, there are also other important effects that are not immediately apparent from Figure 2.6. These include enthalpy relaxation, crystal perfection and secondary crystallization processes, which are the result of thermodynamical imbalance after cooling from the melt. Since those processes are mainly very slow kinetic phenomena, they are collectively referred to as physical aging phenomena in the sense of DIN 50035 (Kaiser, 2011). The physical aging processes are accelerated by the effect of temperature.

Enthalpy relaxation

As the glass transition temperature T_g is reached, the free volume freezes and specific volume leaves its thermodynamic equilibrium state and behaves thermodynamically unstable relative to the supercooled liquid, see Figure 2.12 (a). In terms of energy landscape, the polymer in the glassy state has a number of relatively shallow local energy minima. When held not too far below T_g, time-dependent rearrangements of the molecules or molecule segments occur to approach the properties of the corresponding supercooled state at the prevailing temperature by increasing the density (decreasing free volume and specific volume) and decreasing the free enthalpy (Zografi and Newman, 2017). This rearrangement process is also called volume relaxation, enthalpy relaxation or physical aging. The stronger the cooling, the greater the free volume and thus the possibility of the physical aging process (Schwarzl, 2013), as shown in Figure 2.10 (a). The decrease in mobility due to the decrease of free volume is equivalent to an increase in stiffness.

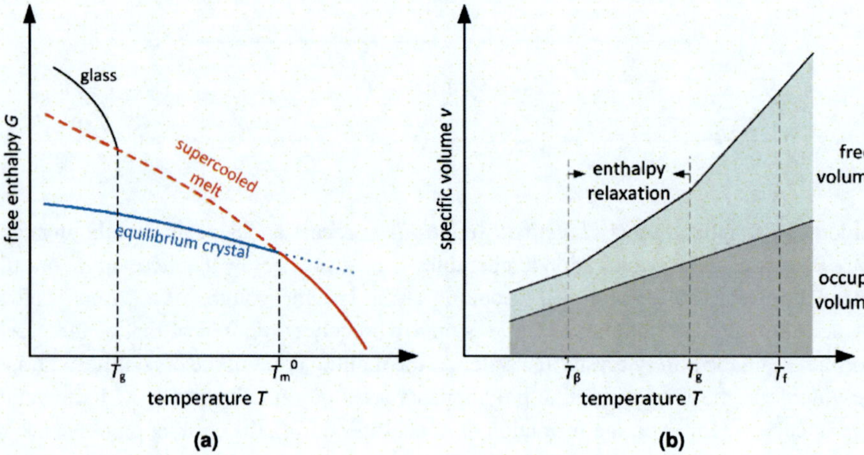

Figure 2.12 Enthalpy relaxation (a) free enthalpy in dependency of temperature, (b) free volume of an amorphous polymer.

The physical age comprises the time span from the cooling process to the present (Schwarzl, 2013) and a temperature range from the glass transition temperature T_g to the β-relaxation (local relaxation of polymer chain segments) temperature, see Figure 2.12 (b). By reheating the polymer to a temperature above the glass transition temperature, the physical aging process is reversed (Sackmann, 2008).

Secondary crystallization and crystal perfection

Secondary crystallization and crystal perfection modify the existing structures. The degree of crystallization and / or the thickness of the lamellae may increase with time. In Figure 2.11 this corresponds to a vertical step. Crystal perfection and secondary crystallization can only occur at temperatures above T_g and below T_m. More detailed information about the crystallization process can be found in chapter 2.2.

2.1.5 Laminated glass interlayers

Polyvinyl butyral PVB

The most commonly used interlayer in construction is polyvinyl butyral (PVB)-based. Manufacturers are for example: Kuraray, Eastman, Everlam and Sekisui. PVB-based interlayers mainly consist of PVB-resin ($\approx$ 75 wt%), additives ($\approx$ 1 wt%) and plasticizers / softeners ($\approx$ 25 wt% for standard PVB-based interlayers) (Eastman Chemical Company, 2003; Kuntsche, 2015).

The homopolymer polyvinyl butyral (PVB) is obtained through acetalization of polyvinyl alcohol with butyraldehyde (butanol) in an acidic environment. Polyvinyl alcohol, in turn, results from the hydrolysis (saponification) of polyvinyl acetate with methanol (Kuraray Trosifol World of Interlayers, 2012). The three stages of the manufacturing process are shown in Figure 2.13 .

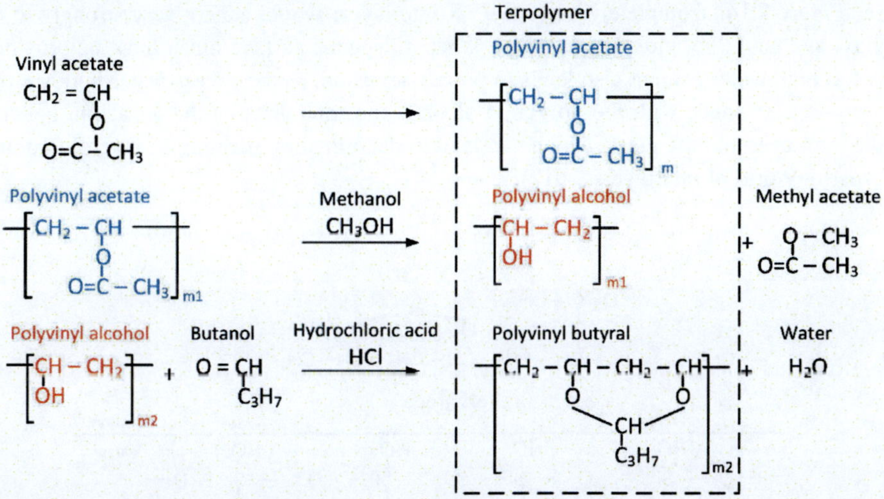

Figure 2.13 Production of PVB-resin.

Depending on the degree of hydrolysis of polyvinyl acetate to polyvinyl alcohol and the amount of butyraldehyde used for acetalization, different resins can be obtained. Thus, the PVB resin does not only consist of the pure homopolymer PVB, but is a terpolymer of vinyl acetate ($m < 2\%$), vinyl alcohol ($m_1 \approx 20\%$) and vinyl butyral ($m_2 \approx 80\%$) (Kuraray Trosifol World of Interlayers, 2012; Kuntsche, 2015; Dhaliwal and Hay, 2002).

PVB resin is a white powder with high stiffness, so that plasticizers are necessary to increase elasticity for further processing to the laminated safety glass interlayer. A well-tolerated plasticizer is for example triethyleneglycol di-2-ethylhexanoate (= ester of poly-

ethylene glycol with aliphatic carboxylic acids) (Eastman Chemical Company, 2003; Kuraray Trosifol World of Interlayers, 2012).
Plastizisers reduce the glass transition temperature of PVB. Hence, by varying the plasticizer content, the PVB-based interlayers can be adapted to the requirements (Kuraray Trosifol World of Interlayers, 2012). PVB-based interlayers for structural applications have a lower plasticizer content (higher glass transition temperature) and PVB-based interlayer for acoustic applications have a higher plasticizer content (lower glass transition temperature) than the PVB-based interlayers for standard applications (Härth et al., 2019). For acoustic applications, not only monolayered but also multilayered PVB-based interlayers are available on the market (Everlam; Eastman Chemical Company, 2012; Kuraray Trosifol World of Interlayers, 2021). They consist of a very soft middle layer with high damping properties and stiffer top layers that facilitate processability (D'Haene and Stevels, 2019). Additionally, additives are needed to control UV transmission, color and adhesion (Schneider et al., 2016). Glass adhesion is created by hydrogen bonds, i.e. solely by adhesion, between the clean glass surface and the polymer, more precisely the polyvinyl alcohol, see Figure 2.14. Completely smooth PVB interlayers would adhere very strongly to the glass and could not move on it, which is why a special surface finish must be provided. A too high water content also leads to poorer adhesion, as the excess water builds up intermolecular bonds with the polyvinyl alcohol and thus disturbs the attraction between the interlayer and the glass, which can lead to delamination (peeling of layers) (Kuraray Trosifol World of Interlayers, 2012).

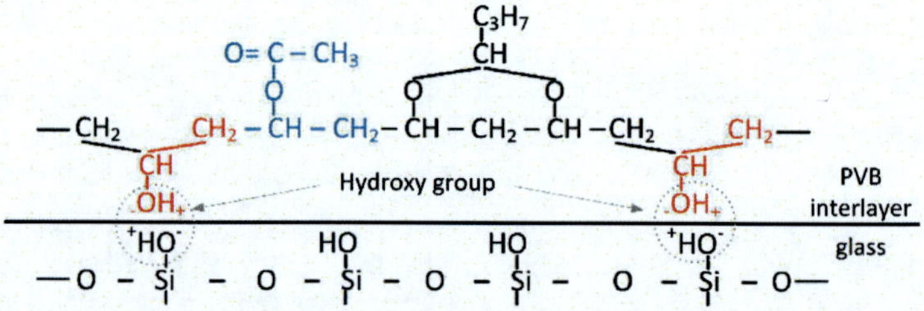

Figure 2.14 Adhesion between PVB and glass, based on (Kuraray Trosifol World of Interlayers, 2012).

PVB-based interlayers can be classified as follows:

- Random terpolymer of vinyl acetate, vinyl alcohol and polyvinyl butyral;
- Gained from chain polymerization of vinyl acetate, hydrolysis to polyvinyl alcohol, and acetalization with butyraldehyde;
- No crosslinks, thermoplastic;

- Amorphous morphology;
- Amorphous thermoplastic thermo-mechanical behaviour.

Ethylene vinyl acetate EVA

Ethylene vinyl acetate (EVA)-based interlayers are mainly used in the solar industry, but also in glass construction. Manufacturers are for example: Folienwerk Wolfen, Glaast and Pujol. In addition to the EVA chains, the EVA-based interlayers also contain additives.
EVA is gained from chain copolymerization of ethylene and vinyl acetate (VAC), see Figure 2.15. It is also denoted as modified PE. Since the reactive ratios of both monomers are similar (dependent on temperature T and pressure p, but always close to 1), the monomers do not have a clear preference of the reaction partner (react similarly probable with a monomer of the same type than with a monomer of the other type), so statistical copolymers are expected (Arsac et al., 1999). Nevertheless, VAC-poor and VAC-rich areas can develop.

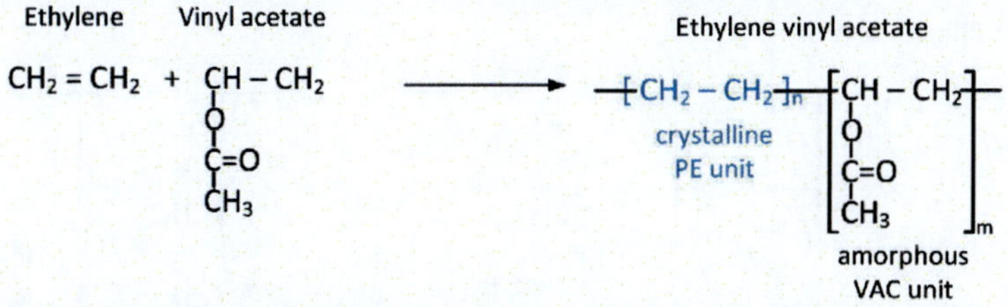

Figure 2.15 Production of EVA-resin, based on (Domininghaus, 2013)

Different amounts of VAC result in a versitility of materials with different mechanical properties. Low and medium density polyethylene (0 wt% VAC) typically have crystallinities of 46-65%. VAC progressively disturbs the crystalline regions formed by polyethylene components with increasing quantity. From a VAC content of approximately 40-50 wt%, no crystalline areas are visible and the copolymer is classified as amorphous (Domininghaus, 2013; Henderson, 1993).
Since increasing VAC amounts decrease crystallinity, transparency is increased. Additionally, density, elongation at break, adhesion / glass bond, weathering stability, crosslinkability and gas permeability are increased. Stiffness and tensile yield strength decrease. The maximum tensile strength is achieved at 25 wt% VAC (Domininghaus, 2013; Henderson, 1993). Typical VAC contents in the building construction industry are about 25% (Schneider et al., 2016) and in the solar industry between 28-33% (wt%) (Hirschl et al., 2016).

The additives include e.g. crosslinker (peroxide), bonding agents (silan) and UV absorbers (Czanderna and Pern, 1996; Schneider et al., 2016). Regarding the mechanical behaviour, the crosslinkers play a particularly important role. During the lamination process ($T \approx$ 120°C, $t \approx$ 50 min in vacuum lamination according to evguard (2014)), the EVA chains are interlinked. This crosslinking reaction is triggered by peroxides, which are split into two oxy radicals when heated. The oxy radicals split a hydrogen atom from the EVA chain (usually from the methyl CH_3 group of the side chain) to form EVA radicals, which then create chemical bonds between the chains via radical recombination, as shown in see Figure 2.16. Only after this crosslinking reaction, EVA becomes a mechanically resistant material, even at the higher temperature (up to 80°C) (Hirschl et al., 2016; Hammer, 2010). The crosslinking degree is dependent on the amount of crosslinkers and on the lamination temperature and time (Hirschl et al., 2016).

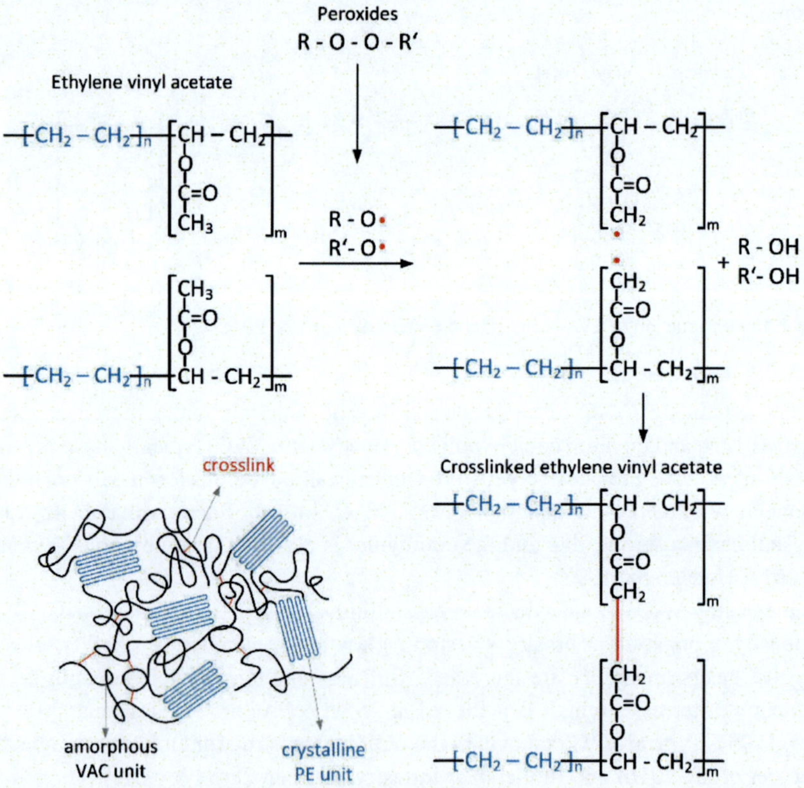

Figure 2.16 Crosslinking reaction, based on (Sharma et al., 2020; Hirschl et al., 2016).

The degree of crosslinking can be measured in the so called Soxhlet test via the gel content. The samples are first dried and weighed. They are then placed in a solvent. At low net density, the sample dissolves, whereas at higher net density a gel is formed. The gel is dried and the mass is put into relation with the original mass (Hirschl et al., 2013, 2016; EN ISO 6427; ASTM D2765). Typically, gel contents of crosslinked EVA-interlayers exceed 80% (Kuntsche, 2015; Hirschl et al., 2013).

Laminated safety glass interlayers based on EVA can be classified as follows:

- Random copolymer of ethylene and VAC;
- Gained from chain polymerization;
- Chemically widely crosslinked (hence often classified as elastomer);
- Semicrystalline morphology in PE regions;
- Elastomeric or thermoelastic thermo-mechanical behaviour.

A schematic representation of a typical EVA-based laminated glass interlayers is shown in Figure 2.16. The adhesion to the glass is made possible by the additive bonding agents, e.g. silanes, which enable a covalent bond to the glass. This suggests a stronger adhesion compared to PVB (Kuntsche, 2015).

Ionoplasts

Ionoplastic interlayers (sometimes also referred to as ionomers) are used for structures with particularly high structural requirements. SG, for example, was specially developed for hurricane glazing (Kuraray Trosifol World of Interlayers, 2019).
Ionoplasts are copolymers that possess a relatively small number of ionized groups, which are distributed along the polymer backbone (Wakabayashi and Register, 2006). In general, they are produced from the copolymerization of a non-polar with a polar monomer and a subsequent (partial) neutralization in an alkaline medium (MacKnight and Earnest Jr, 1981). In the case of SG, the monomers are ethylene and organic methacrylic acid (MAA) (Trösch, 2015), which results in the formation of carboxy groups on the macromolecules, see Figure 2.17. The subsequent neutralization or ionomerization is performed by adding metal cations (Wakabayashi and Register, 2006). According to Kuraray (2014), the metal cations may be sodium, zinc or potassium. The degree of neutralization and the exact neutralization process of SG are unknown. One possible neutralization process would be be the addition metal hydroxides to the ethylene-methacrylic acid copolymer. As a result, the hydrogen is split off from the carboxy group to form water and the carboxyl groups become ionized. The negatively charged oxygen atom of the carboxy group and the remaining positively charged metal atoms of the metal hydroxide then create an ionic pair.

Due to the energy difference between the nonpolar carbon chains and the polar functional groups, clustering of the ion rich zones appears. This results in a nano-separation of the two regions, where the ion-rich domains act as thermoreversible crosslinks (Gómez et al., 2019; Elias, 2009). In these regions the movement of the amorphous areas is restricted.
Similar to EVA-based interlayers, SG is also referred to as modified polyethylene PE, which can develop semicrystalline structures. The polyethylene crystallites are located between the amorphous regions and ion clusters (Wakabayashi and Register, 2006). However, since the ion bonds are thermoreversible, a semicrystalline thermoplastic material or thermoplastic elastomeric behaviour is expected.
With higher amount of MAA properties, the yield point, tensile strength, stiffness, tear strength or hardness increase. However, elongation, freeze point or haze decrease. Similarly, with higher degrees of neutralization, higher tensile strength, stiffness and hardness, etc. are achieved. At the same time other properties (adhesion, haze, freeze point and tear strength) are reduced (Zadrapa, 2011).

Only UV absorbers are used as additives. No softeners or bonding agents are used. Adhesion to the glass is made possible on the one hand by means of a hydrogen bond via the polar OH groups, similar to PVB. In addition, ion bonds are formed with the tin ions on the tin bath side of the float glasses (Kuntsche, 2015).

SG interlayers can be classified as follows:

- Random copolymer of ethylene and methacrylic acid;
- Gained from chain polymerization;
- Widely crosslinked via thermoreversible ion bonds;
- Semicrystalline morphology in PE regions;
- Semicrystalline thermoplastic or thermoplastic elastomeric thermo-mechanical behaviour.

A schematic representation of SG is shown in Figure 2.17.

Other

Besides the mentioned interlayers PVB, EVA and ionoplast, other materials are also used in glass laminates. These include, for example, *thermoplastic polyurethane* (TPU) or *casting resins*. The latter are not laminated as films, but poured between the glass panes. There the crosslinking of the reaction partners takes place, making them particularly suitable for lamination-bent laminated glass. Casting resins also find their use in sound insulation or to

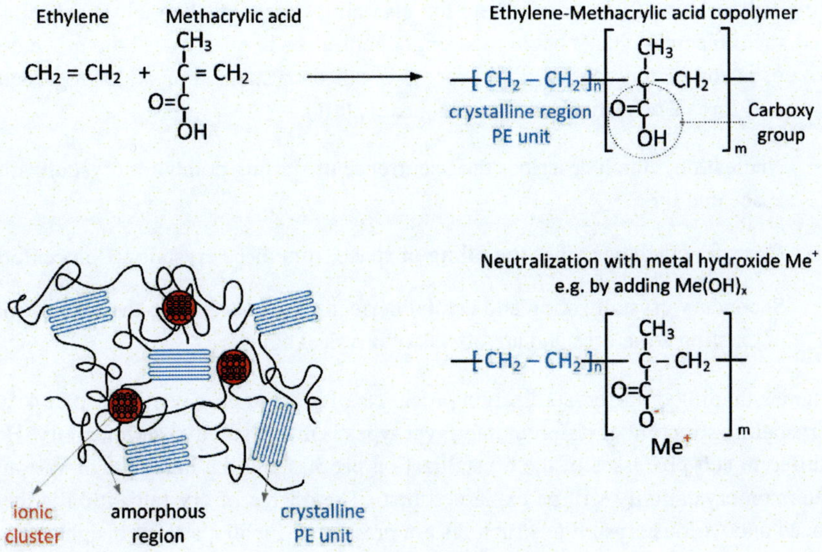

Figure 2.17 Ionoplast production, based on (Trösch, 2015; Zadrapa, 2011).

compensate for high tolerances (Schneider et al., 2016). One-component resins crosslink by UV irradiation. The components of multi-component casting resins are mixed shortly before filling the space between the glass. The crosslinking reaction then proceeds independently. Casting resins consist of acrylate or methacrylate resins, which are assigned to the thermosets (Schneider et al., 2016).

TPU is mainly used in protective glazing such as anti-burglary and bullet-proof glass and polycarbonate glazing. In these protective glazing systems, glass and polycarbonate sheets are laminated together with a TPU interlayer. The high flexibility of the interlayer serves to compensate the different thermal expansion coefficients of glass and polycarbonate. TPU is assigned to the thermoplastic elastomers and sometimes referred to as TPE-U. The block copolymer (Domininghaus, 2013) consists of hard segments which behave energy-elastically and soft segments, which behave entropy-elastically in the service temperature range. The hard segments of different macromolecules form crystalline areas via hydrogen bonds and thus physical crosslinking points (Kuntsche, 2015).

2.2 Crystallization

In the previous section it was shown that the crystallites have a strong impact on the stiffness of semicrystalline polymers. A strong decrease in stiffness can be observed when the melting temperature T_m is reached and the crystallites melt. If the melt is subsequently cooled, the molecular chains begin to arrange themselves regularly until the movement of the chains is completely frozen. This process is called crystallization and is regarded as a sequence of three stochastic processes (Moneke, 2001):

- Nucleation, which describes the occurrence of starting points for crystallization in space and time.
- Growth, which describes the filling of space from these crystallization centers.
- Secondary crystallization and crystal perfection, which lead to the transformation of existing structures and are effects of physical aging.

The understanding of the crystallization process is of great importance in chapter 4, where the crystalline structure of different interlayer types is investigated experimentally. Hence, the different sub-processes of the crystallization process and the influence of various parameters on crystallinity will be explained first. The degree of crystallization is then introduced and various crystallite structures are presented. Finally, a short insight into crystallization kinetics is given, which makes it possible to describe the crystallization process mathematically.

2.2.1 Crystallization from the melt

The crystallization process consists of nucleation, crystal growth, crystal perfection and secondary crystallization.

Nucleation

Nucleation is caused by fluctuations in the melt or solution (Lechner et al., 2014). As a result of the molecular movement, individual chains assemble into short-lived, very small crystal-like structures, the embryos. Above the melting temperature the embryos are unstable. They decay again. Below the melting temperature there is a critical embryo size. Embryos, which are larger than the critical embryo size have a free enthalpy, which is smaller than that of the melt. They continue to grow and become germs. The other embryos dissolve again (Lechner et al., 2014).

There are different types of nucleation. In *homogeneous* nucleation, several polymer chains are randomly arranged in a cluster. No other substances are involved. For this purpose, enough perfect chain segments have to find each other to form a sufficiently large germ. This is a very slow process. Homogeneous nucleation is also referred to as spontaneous nucleation, e.g. in (Piorkowska and Galeski, 2013; Elias, 2009), sporadic nucleation, e.g. in (Elias, 2009), progressive nucleation or temperature induced nucleation,

e.g. in (Elias, 2009).
Heterogeneous nucleation is much more common. Here, the polymer chains attach themselves either to foreign substances such as dust particles or other low-molecular impurities, e.g. catalyst residues, nucleating agents that can extremely promote the formation of germs or to unmelted crystals (Lechner et al., 2014; Bonnet, 2016; Elias, 2009). Since this nucleation process is much faster and occurs simultaneously in the entire sample at the beginning of crystallization, it is also called instantaneous nucleation, e.g. in (Piorkowska and Galeski, 2013), or simultaneous nucleation, e.g. in (Elias, 2009). Another synonym is athermal nucleation (Elias, 2009).
In the following, the terms spontaneous and instantaneous nucleation are used.
If all other factors are kept constant, the number of nuclei formed depends on the crystallization temperature T_c. If T_c is only slightly below the melting temperature T_m, only a few germs are formed. Hence, only a few, but large crystals can then be formed. If, on the other hand, T_c is much smaller than T_m, many nuclei are formed and hence only relatively small crystallites can be formed (Lechner et al., 2014). The temperature difference ($\Delta T = T_m - T_c$) is called supercooling temperature.

Growth

If germs have formed that have exceeded the minimum critical size, germ growth occurs. To achieve this, the molecules must be transported to the interface via diffusion processes and incorporated there (Bonnet, 2016).
Crystal growth is achieved by folded addition of further polymer chain sections. This occurs in a temperature range below the melting temperature T_m and above the glass transition temperature T_g. If the temperature is too high, the attached chains would be detached again by thermal movements; the free enthalpy of the melt is lower than the free enthalpy of the crystallites. Below the glass transition temperature, the mobility of the chains is too low and the movement of the molecular chains is frozen (Bonnet, 2016).

Crystallization rate

Since nucleation and crystal growth come together for crystallization the crystallization rate (G_c) is the product of the nucleation rate and transport (assuming interface-controlled growth). The lower temperature boundary for crystallization is the glass transition temperature T_g, since below T_g no molecular movement and therefore no diffusion takes place. The upper temperature boundary for crystallization, is the melting temperature T_m, because there, no stable germs arise (Bonnet, 2016).
While the diffusion rate and thus the transport process increases with rising temperature, the nucleation rate increases with increasing supercooling, i.e. with decreasing temperature. The crystallization rate thus acquires the bell-shaped course. Since the diffusion rate

decreases strongly with increasing molecular mass M, the curve for the same plastic with higher molecular mass correspondingly flatter. An identical crystallization rate $G_{c,1}$ can therefore be achieved at two different temperatures ($T_{1,1}$ and $T_{1,2}$, see Figure 2.18). However, whereas $T_{1,1}$ is a structure consisting of many small spherulites (fine structures high transparency, improved mechanical properties), a microstructure of a few large approximately radially symmetrical spherulites is formed at $T_{1,2}$. The maximum crystallization rate is in good approximation centered between glass transition and melting temperature: $T_{c,max} = 1/2 \cdot (T_m + T_g)$ (Bonnet, 2016; Lechner et al., 2014).
If the cooling rate is very high, crystallization can be completely suppressed. However, crystallization can be induced by subsequent annealing at a temperature below the melting temperature (Lechner et al., 2014).

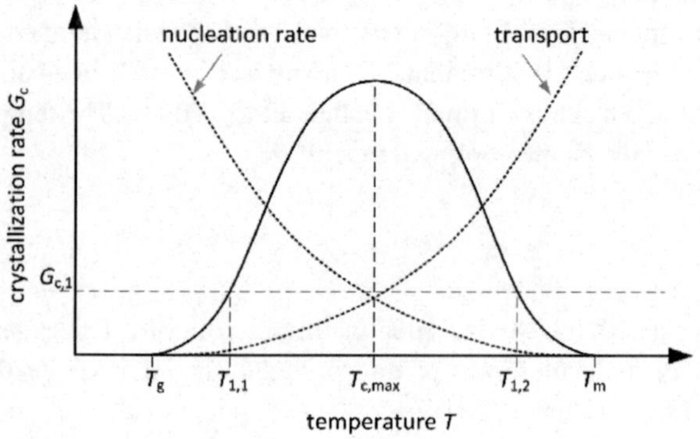

Figure 2.18 Temperature dependent crystallization rate.

Secondary crystallization and crystal perfection

Secondary crystallization is also known as post crystallization or cold crystallization and refers to the subsequent crystallization of a partially crystalline material when the temperature is increased (Frick and Stern, 2013; Hammer, 2010). If a melt is cooled very quickly, only a few molecules crystallise. The polymer is in a thermodynamically unstable state. Elevated temperatures above T_g and below T_m increase the mobility of the molecular chains, so that the initially insufficiently crystalline areas increase their crystallinity due to slow reorganisation effects (Frick and Stern, 2013). This often increases the degree of crystallization by about 10-20% (Lechner et al., 2014). New ordered structures (crystallites) grow at the transition zone between the existing crystalline structures and the amorphous areas. These newly formed crystals can differ from the existing ones by a lower melting temperature (Bonnet, 2016).
By annealing just below the melting temperature T_m, crystal perfection occurs. Small crys-

tallites melt while larger ones grow and increase their lamellar thickness, which is why this process is also called recrystallization (Hammer, 2010). Due to the higher amount of large crystallites, the melting point shifts towards higher temperatures, see also section 2.1.4.

2.2.2 Degree of crystallization

The terms crystallinity, degree of crystallinity and degree of crystallization are used as synonyms in literature. Most evaluations of the degree of crystallization for semicrystalline thermoplastics are based on a two-phase model in which there are perfect crystals (characterized by the melting temperature T_m resp. crystallization temperature T_c) and distinct amorphous areas (characterized by the glass transition temperature T_g and the flow temperature T_f). Deviations due to defects, transition areas between amorphous and crystalline are neglected. Typically, degrees of crystallization of 10 to 80 % are technically realized, (Ehrenstein, 2011). In addition to the already mentioned higher stiffness, a higher degree of crystallization also implies a higher strength, a higher heat distortion temperature, a higher density and a higher processing temperature. The light transmission and gas permeability decreases (Kaiser, 2011).

There exist different definitions for the degree of crystallization, which may lead to different results. A general distinction is made between the mass fraction χ and the volume fraction χ_v (Mo and Zhang, 1995; Lechner et al., 2014; Wunderlich, 2005). In macromolecular chemistry, the mass fraction is usually used as the degree of crystallization (Lechner et al., 2014).
With the total mass being composed of the masses of the crystalline and amorphous part, $m_{total} = m_{cryst} + m_{amorph}$, the mass fraction can be calculated according:

$$\chi = \frac{m_{cryst}}{m_{total}}. \tag{2.5}$$

Using the densities $\rho_i = m_i/V_i$ with i = [total, cryst, amorph] and the total volume $V_{total} = V_{cryst} + V_{amorph}$, the mass fraction χ can be determined from the densities:

$$\chi = \frac{\rho_{cryst}}{\rho_{total}} \cdot \frac{\rho_{total} - \rho_{amorph}}{\rho_{cryst} - \rho_{amorph}}, \tag{2.6}$$

or by the specific volumes ($v = \rho^{-1}$):

$$\chi = \frac{v_{amorph} - v_{total}}{v_{amorph} - v_{cryst}}. \tag{2.7}$$

The mass fraction can be converted into volume fraction with:

$$\chi_{\mathrm{v}} = \frac{V_{\mathrm{cryst}}}{V_{\mathrm{total}}} = \chi \cdot \frac{\rho_{\mathrm{total}}}{\rho_{\mathrm{cryst}}}. \tag{2.8}$$

Other methods to determine the degree of crystallization are Differential Scanning Calorimetry (DSC), which is based on the measurement of the melt or crystallization enthalpies, or X-Ray Diffraction (XRD). Those methods will be explained in chapter 4.1.1 and 4.1.2.

The degree of crystallization depends on various factors. These are:

- Chemical composition: molar mass of the polymer (diffusion rate decreases with increasing molecular weight), tacticity (symmetrical, isotactical chains favour a regular packing of the chains), degree of chain branching (long side chains difficulty the crystallization process), additives like nucleating agents (Lechner et al., 2014), polar side chains with strong intermolecular attraction (see chapter 2.1.1).
- Cooling rate: quenching leads to a low degree of crystallization with a lot of small crystallites, whereas slow cooling results in high crystallinity with only a few but large crystallites (Schwarzl, 2013).
- Thermal prehistory: secondary crystallization may occur during reheating of a fast cooled polymer or through annealing it at a temperature just below its melting point. The rearrangement of the polymer chains increases its crystallinity (Schwarzl, 2013).

2.2.3 Morphology

The first model of crystallites in polymers was the *fringe micelle model* (Figure 2.19 (a)). A fringe crystallite consists of several polymer chains, which are arranged parallel to each other. The ends of the chains hang out of the crystallite like fringes and form an amorphous phase. Each individual polymer chain passes through several crystallites and several amorphous zones (Lechner et al., 2014).
However, when single crystals could be produced for the first time in 1957, it became apparent that the fringe model could not be maintained (Bonnet, 2016). Since the thickness of polymer single crystal is smaller than the conventional contours of the chains, the chains must be folded back in the crystal (Elias, 2009). 1957, A. Keller postulated the structure of crystallites in the form of the folded molecular chains (*folded micelles*, Figure 2.19 (b).The surface of folded micelles is usually blurred. In addition to sharp folds, it also contains smaller and larger loops at the folding arches and chain ends that are hanging out. It is also common for a molecular chain to leave a crystallite and re-enter another. Each crystallite therefore consists of ordered (crystalline) and disordered (amorphous) sub regions. Folded

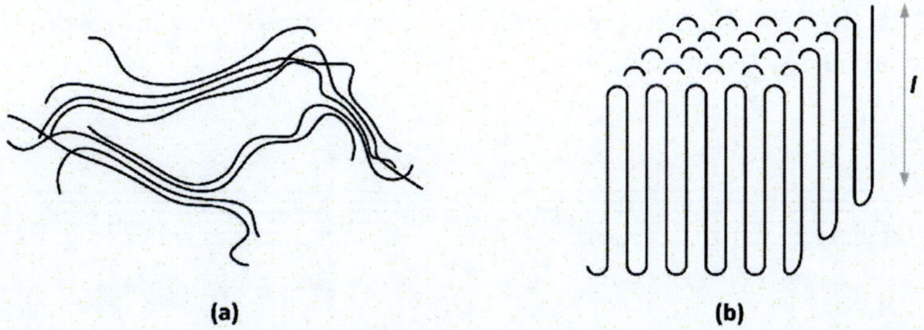

Figure 2.19 (a) fringing micelles, (b) folded micelles, based on (Lechner et al., 2014).

molecular chains are the basic structures of the larger lamellar structures.

The entirety of the orientations of the crystallites present in a material is called texture or morphology. Starting from a crystal nucleus in a relaxed melt, lamellar crystals grow radially in all spatial directions, resulting in globular (spherical) *spherulites*, see Figure 2.20, (Bonnet, 2016). The spherulites are in the initial stage of crystallization (before they touch) spherical and then grow to a polygonal structure with plane or slightly curved interfaces (Lechner et al., 2014). This leads to a two-stage crystallization process. During primary crystallization, the actual spherulite is formed, which is growing radially starting from the nucleus. During secondary crystallization, further lamellas form in the free areas of the spherulites, (Bonnet, 2016).

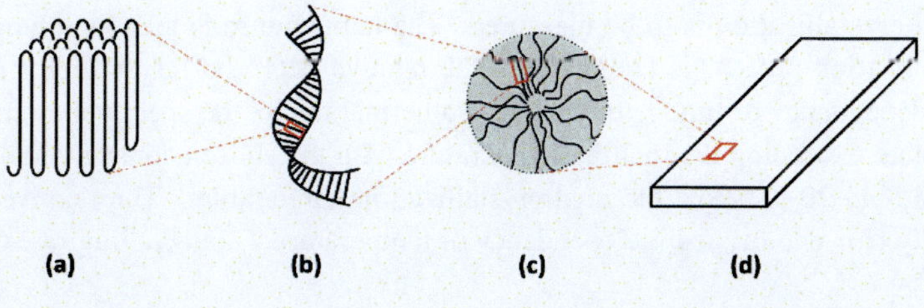

Figure 2.20 Crystallization growth (a) folded micelles, (b) lamellae, (c) spherulite, (d) component

If the melt is not relaxed, crystal morphologies other than spherulites may develop. *Fibre* structures are formed due to a temperature gradient, which may for example occur when the melt solidifies on cooled molded parts in injection moulding. The nucleus are formed preferably in the colder regions but they growth towards the warmer regions, since here the diffusion is faster. Other crystal morphologies are the *needle/shish* structure (Figure 2.21 (a)), which consists of crystals with stretched chains, and the *shish-kebab* morphology

(Figure 2.21 (b)), in which relaxed molecules grow onto the needle (shish) crystals in lamellar morphology (Bonnet, 2016).

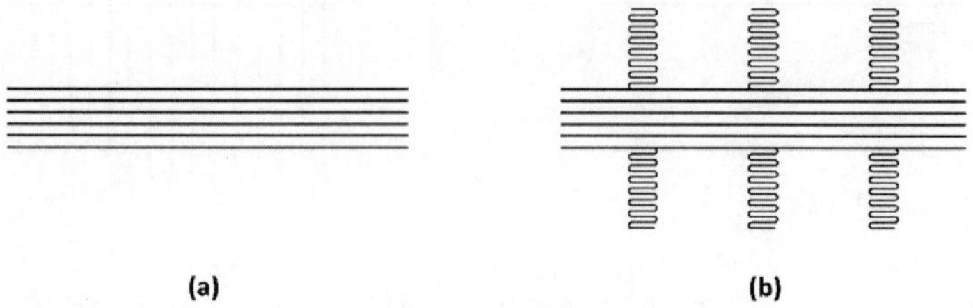

Figure 2.21 (a) shish /needle structure, (b) shish-kebab structure.

2.2.4 Crystallization kinetics

Crystallization kinetics describes the crystallization rate of a material taking into account the effect of variables such as time, temperature, pressure, stress and molecular structure (EN ISO 11357-7). The relative crystallinity $\alpha = \chi_{t,T}/\chi_{\mathrm{f}}$ is the ratio between the crystallinity at a certain time or at a certain temperature $\chi_{t,T}$ and the crystallinity at the end of the crystallization χ_{f}.

The investigation of the crystallization kinetics of polymers is mostly done with Differential Scanning Calorimetry (DSC, see further chapter 3) and can be performed in an isothermal or non-isothermal mode. In the isothermal method, the sample is cooled as quickly as possible at the end of the melting phase to the selected temperature at which the isothermal crystallization is to be measured. The temperature is then kept constant until the crystallization process is completed. The relative degree of crystallinity α is determined in dependency of time t. In the non-isothermal mode, the specimen is cooled from a temperature exceeding the melting temperature with the chosen constant cooling rate to at least 10 K to 20 K below the final crystallization temperature. The relative degree of crystallinity α is determined in dependency of temperature T and cooling rate C (EN ISO 11357-7).

2.2.4.1 Fundamentals about isothermal kinetic models

The two best known approaches to describe the overall crystallization in infinite samples at isothermal conditions are those of Avrami and Evans; both based on the iso-volume assumption.

Johnson, Mehl, Avrami and Komogorov elaborated similar contributions to the description of isothermal crystallization kinetics. Their model is known in literature as the *JMAK* model. However, the equation was popularized by Melvin Avrami in the series of papers published in the Journal of Chemical Physics (Avrami, 1939, 1940, 1941), which is why it is often only referred to as Avrami equation. Avrami utilized the concept of the so called extended volume. As shown in Figure 2.22, the extended volume E_V at the time t is equal to the sum of volumes of all domains as if they were growing without impingement from all nucleation attempts occurred up to the time t (Piorkowska and Galeski, 2013). Hence, an increment of the extended volume fraction dE_V occurring in the unconverted region $1-\alpha$ results in an increase of the real converted volume fraction $d\alpha$, which leads to

$$\frac{d\alpha}{dE_V} = 1 - \alpha. \tag{2.9}$$

Eq. (2.9) can be transformed into

$$1 - \alpha = \exp(-E_V). \tag{2.10}$$

In order to evaluate expressions for the extended volume E_V, the crystallization process is reconsidered. The conversion of melt into spherulites is dominated by the nucleation and growth rate of spherulites. The models of formation of spherulite structure utilize concepts of instantaneous and spontaneous nucleation. As already explained, instantaneous nucleation occurs simultaneously in the entire sample (heterogeneous nucleation) at the beginning of crystallization and spontaneous nucleation is prolonged in time (homogeneous nucleation). The models also differentiate between three-dimensional (3D) and two-dimensional (2D) structures, the first occurring in bulk and the second being found in thin films (Piorkowska and Galeski, 2013).

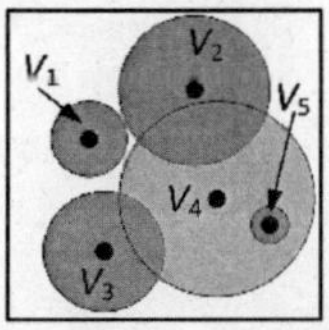

Figure 2.22 Extended volume $E_V = V_1 + V_2 + V_3 + V_4 + V_5$, based on (Piorkowska and Galeski, 2013). The black dots represent the nucleis.

Instantaneous nucleation is characterized by the nucleation density D, which is the number of nuclei per unit volume (or surface area) of a polymer. Spontaneous nucleation is described by the nucleation rate F, which is equal to the number of nuclei appearing in a unit volume (or surface) of an not converted phase per unit time. In general, the nucleation in a given polymer depends on the crystallization temperature, time, and thermal history. For isothermal conditions however, the nucleation rate can be described as a function of time only. The spherulite growth rate G_c, depends primarily on the crystallization temperature

(Piorkowska and Galeski, 2013). Figure 2.23 shows exemplary the crystallization process with spontaneous nucleation.

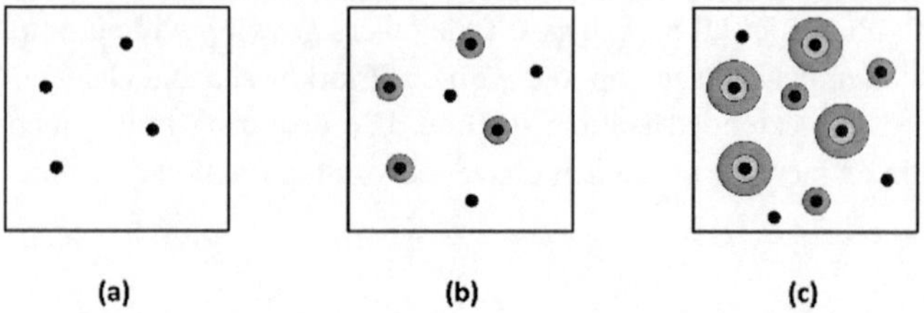

Figure 2.23 Crystallization process with spontaneous nucleation (nucleation rate F) and spherulite growth (growth rate G_c). The black dots represent the nucleis. The growth during Δt is marked with the darker grey color. (a) t_1: first nucleis have formed, (b) $t_2 = t_1 + \Delta t$, (c) $t_3 = t_2 + \Delta t$.

For instantaneous nucleation, the extended volume E_V equals:

$$E_V(t) = D \cdot V(0,t). \tag{2.11}$$

For spontaneous nucleation eq. (2.12) holds.

$$E_V(t) = \int_0^t F(s) \cdot V(s,t)\,\mathrm{d}s \tag{2.12}$$

In case of a 3D structure, the volume V at the time t of a single unimpinged spherulite nucleated at time s, equals the volume of a sphere having the radius $r(s,t)$:

$$V(s,t) = \frac{4\pi}{3} \cdot [r(s,t)]^3. \tag{2.13}$$

In case of a 2D structure, the volume V (corresponds to surface area) at the time t of a single unimpinged disc nucleated at time s, equals the area of a circle having the radius $r(s,t)$:

$$V(s,t) = \pi \cdot [r(s,t)]^2 \tag{2.14}$$

with:

$$r(s,t) = \int_s^t G_c(\tilde{s})\mathrm{d}\tilde{s}. \tag{2.15}$$

Since the growth rate G_c is constant for isothermal conditions, eq. (2.15) can be simplified to:

$$r(s,t) = (t-s) \cdot G_c. \tag{2.16}$$

If eq. (2.16) and eq. (2.13) for the formation of a 3D structure resp. eq. (2.14) for the formation of a 2D structure are substituted in eq. (2.11) (instantaneous nucleation),

$$E_V(t) = D \cdot \frac{4\pi \cdot G_c^3 \cdot t^3}{3} \quad \text{and} \tag{2.17}$$

$$E_V(t) = D \cdot \pi \cdot G_c^2 \cdot t^2 \tag{2.18}$$

are obtained. For spontaneous nucleation (see eq.(2.12)),

$$E_V(t) = \frac{4\pi \cdot G_c^3}{3} \cdot \int_0^t F(s) \cdot (t-s)^3 \, \mathrm{d}s \quad \text{and} \tag{2.19}$$

$$E_V(t) = \pi \cdot G_c^2 \cdot \int_0^t F(s) \cdot (t-s)^2 \mathrm{d}s \tag{2.20}$$

result.
While Avrami used the concept of the extended volume, Evans (1945) took advantage of the probability theory, which is based on the probability of a number of nucleation events occurring in a finite volume. A detailed documentation of Evans theory can be found in (Piorkowska and Galeski, 2013). Here it was also shown, that Evan came to the same results as Avrami.

Both theories can be reduced to the well-known form in eq. (2.21) in the case of instantaneous nucleation with density D or spontaneous nucleation with a constant rate $F(t) = F$.

$$E_V(t) = kt^n \tag{2.21}$$

n is a mechanism constant called Avrami exponent and k is the composite Avrami rate constant (EN ISO 11357-7). k and n depend on the dimension of crystallization and on the nucleation mode (Piorkowska and Galeski, 2013; Elias, 2009; Lechner et al., 2014).

- For eq. (2.17), instantaneous nucleation and 3D structure: $k = 4\pi D G_c^3/3$ and $n = 3$.
- For eq. (2.18), instantaneous nucleation and 2D structure: $k = \pi D G_c^2$ and $n = 2$.
- For eq. (2.19), spontaneous nucleation and 3D structure: $k = \pi F G_c^3/3$ and $n = 4$.

- For eq. (2.20), spontaneous nucleation and 2D structure: $k = \pi F G_c^2/3$ and $n = 3$.

The most common representation of the isothermal crystallization kinetics is:

$$1 - \alpha = \exp(-kt^n). \tag{2.22}$$

This results in a sigmoidal shape for α, as shown in Figure 2.24 (a).

In double logarithmic scale, eq. (2.23) is obtained.

$$\ln[-\ln(1-\alpha)] = \ln(k) + n\ln(t) \tag{2.23}$$

n and k are determined from the slope and intercept of the plot of $\ln[-\ln(1-\alpha)]$ against $\ln(t)$, which is called Avrami plot. The model validity can be checked by a linear relation in the Avrami plot as exemplary shown in Figure 2.24 (b).

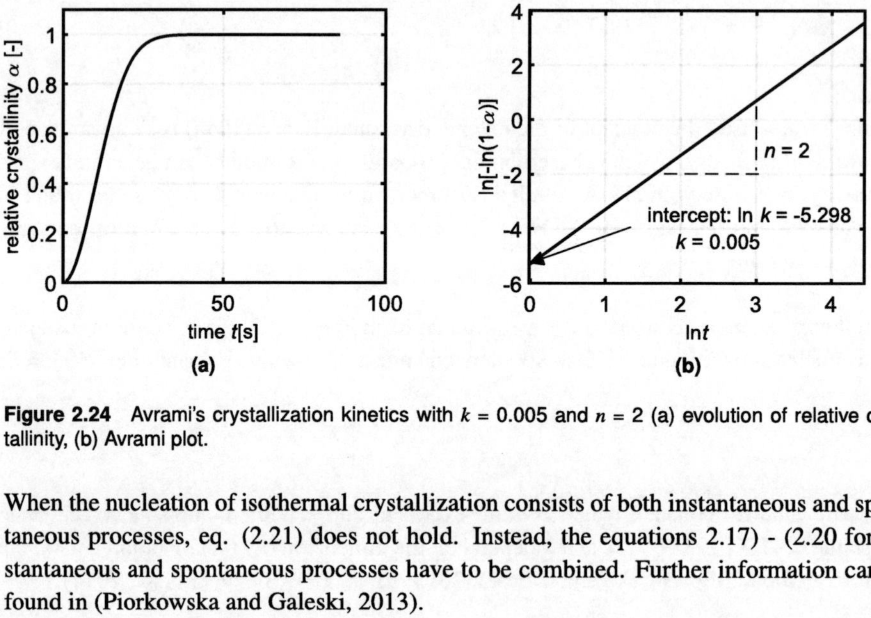

Figure 2.24 Avrami's crystallization kinetics with k = 0.005 and n = 2 (a) evolution of relative crystallinity, (b) Avrami plot.

When the nucleation of isothermal crystallization consists of both instantaneous and spontaneous processes, eq. (2.21) does not hold. Instead, the equations 2.17) - (2.20 for instantaneous and spontaneous processes have to be combined. Further information can be found in (Piorkowska and Galeski, 2013).

2.2.4.2 Fundamentals about non-isothermal kinetic models

Ozawa model

Avrami and Evans studies were carried out for isothermal conditions. The most popular analysis of nonisothermal crystallization is Ozawa's equation for a constant cooling rate C (or heating rate). This model is also part of EN ISO 11357-7. Ozawa (1971) derived his model, which in fact is a modification of Avrami's model, based on Evans probability theory. According to Ozawa, a non-isothermal crystallization can be divided into infinitesimal small isothermal steps (quasi-isothermal crystallization) (Piorkowska and Galeski, 2013).

After taking the temperature dependent growth rate $G_c(T)$ and the temperature dependent nucleation rate $F(T)$ into account and assuming that the temperature T is a linear function of time t, Ozawa obtained:

$$1-\alpha(T)=\exp\left(-\frac{K(T)}{C^m}\right), \tag{2.24}$$

wherein α is the relative crystallinity, C is the cooling rate, $K(T)$ is the Ozawa crystallization rate constant and m is the Ozawa exponent, which depends on the dimension of crystal growth and is generally interpreted in the same way than the Avrami exponent (Keridou et al., 2019).

A detailed derivation of eq. (2.24) starting from $G_c(T)$ and $F(T)$, as well as the specification of $K(T)$ can be found in (Ozawa, 1971; Piorkowska and Galeski, 2013). Eq. 2.24 delivers relative crystallinities at specific temperatures in dependence of the cooling rate as depicted in Figure 2.25 (a).

In double logarithmic scale, eq. (2.24) results in:

$$\ln[-\ln(1-\alpha)]=\ln(K(T))-m\ln(C). \tag{2.25}$$

m and $K(T)$ are determined from the slope and intercept of the plot of $\ln[-\ln(1-\alpha)]$ against $\ln(C)$ for relative crystallinities determined at a given temperature and different cooling rates. The procedure is exemplary shown in Figure 2.25 (b). Alternatively to eq. (2.25), eq. (2.26) is also found in literature (Málek, 2000).

$$\log[-\ln(1-\alpha)]=\log(K(T))-m\log(C) \tag{2.26}$$

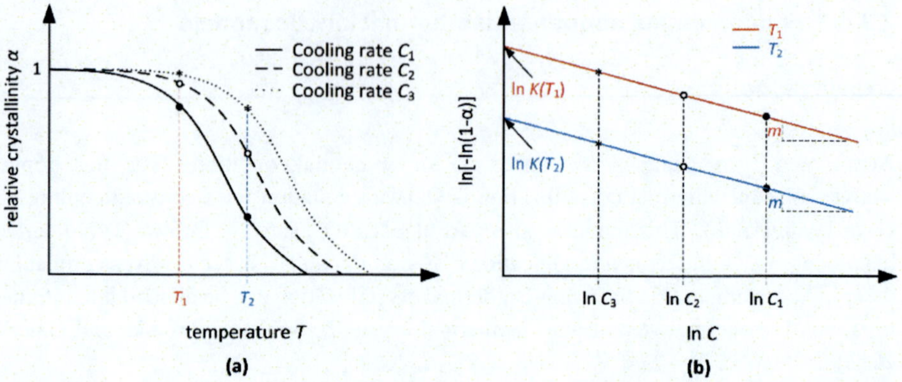

Figure 2.25 Ozawa model (a) temperature dependent relative crystallinities for different cooling rates C, (b) graphical determination of m and $K(T)$.

If second crystallization processes are negligible, a mastercurve for $K(T)$ can be generated by horizontally shifting the $\ln[-\ln(1-\alpha)]$ or $\log[-\ln(1-\alpha)]$ along the cooling rate axis by an amount of $m\ln(C)$ or $m\log(C)$. This provides a simple way of predicting non-isothermal transformation kinetics for a known cooling rate C, by substituting $K(T)$ from the mastercurve, and experimentally determined exponent m (Long et al., 1995).

The validity of the model can be checked by the linear relation in the $\ln[-\ln(1-\alpha)]$ vs. $\ln(C)$ or $\log[-\ln(1-\alpha)]$ vs. $\log(C)$-plot. Ozawa's theory is based on quasi-isothermal crystallization and supposes a constant crystallization rate, which is not true for non-isothermal crystallization. Crystallization is a dynamic process and hence the crystallization rate is a function of time and cooling rate (Abareshi et al., 2014). Literature showed that Ozawa model is not valid if the cooling rates vary in a wide range and that the method is highly sensitive to the variations between primary and secondary crystallization processes (Keridou et al., 2019; Abareshi et al., 2014).

Jeziorny-Avrami method

Jeziorny (1978) modified Avrami's equation (2.22) to describe the non-isothermal kinetics by introducing the reduced crystallization rate constant k_c according to eq. (2.27), wherein C corresponds to the cooling rate. The larger k_c, the larger the crystallization rate (Abareshi et al., 2014).

$$\ln k_c = \ln k/C \quad \text{resp.} \quad \log k_c = \log k/C \tag{2.27}$$

In non-isothermal crystallization, the temperature T can be related to crystallization time t by $t = (T_0 - T)/C$, where T_0 is the temperature at which the crystallization starts. If the relation of $\ln[-\ln(1-\alpha)]$ against $\ln(t)$ for the investigated cooling rates C is not linear, the model is not appropriate (Abareshi et al., 2014). Unfortunately, the direct application of the Avrami equation to the evaluation of non-isothermal crystallization corresponds to a mathematically fitting in which the parameters lose their physical meaning. For example, values of the exponent become usually higher than four (Keridou et al., 2019).

Mo-Liu method

Liu et al. (1997) proposed to combine the Avrami and Ozawa models. This results in eq. (2.28) respectively eq. (2.29) for a certain relative degree of crystallization α. Again, the temperature T can be converted to crystallization time t by $t = (T_0 - T)/C$, wherein T_0 is the temperature at which the crystallization starts and C is the cooling rate.

$$\ln[-\ln(1-\alpha)] = \ln(k) + n\ln(t) = \ln(K(T)) - m\ln(C) \tag{2.28}$$

$$\ln(C) = \ln(C_\alpha(T)) - b\ln(t) \tag{2.29}$$

$C_\alpha(T)$ is the required cooling rate to reach a certain crystallization degree at unit crystallization time and can be calculated with eq. (2.30).

$$C_\alpha(T) = [K(T)/k]^{1/m} \tag{2.30}$$

b_{nm} is the ratio of the Avrami coefficient to the Ozawa coefficient:

$$b_{\mathrm{nm}} = \frac{n}{m}. \tag{2.31}$$

As shown in Figure 2.26 (b), b_{nm} and $\ln(C_\alpha(T))$ are determined from the slope and intercept of the $\ln(C)$ against $\ln(t)$ plot, created for different relative degrees of crystallization α. $C_\alpha(T)$ increases with increasing relative degree of crystallinity α, which means that the relative degree of crystallinity in a unit time is higher when the cooling rate is high (Abareshi et al., 2014).

The validity of the Mo-Liu method is given, if a linear relation for all relative crystallization degrees is obtained. The main problem of this model is the non-physical sense of $C_\alpha(T)$ (Keridou et al., 2019).

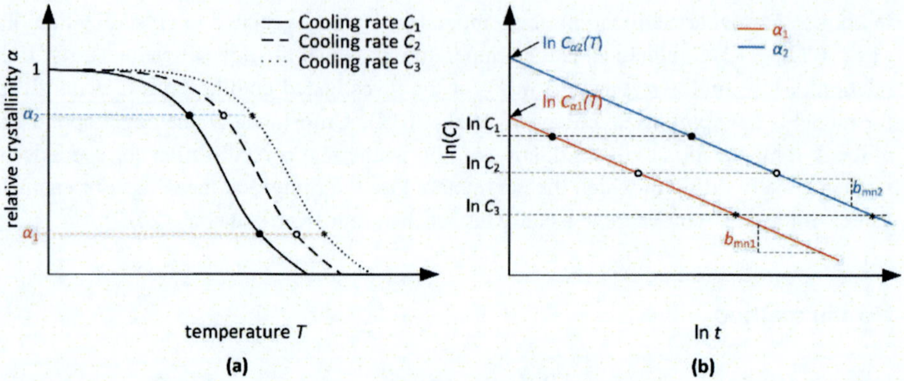

Figure 2.26 Mo-Liu model (a) temperature dependent relative crystallinities for different cooling rates C, (b) Determination of C_α and b_{nm} values for different relative crystallinities α).

2.2.4.3 Fundamental single step kinetics

Often crystallization processes are described with the fundamental single step kinetics. The differential form of the fundamental kinetic formula assumes, that the rate of the kinetic process can be expressed by the product of temperature dependent rate constant $K(T)$ and a kinetic model function $f(\alpha)$, see eq. (2.32) (Málek, 2000; Menezes et al., 2019). $f(\alpha)$ should be independent to procedural parameters such as sample mass, heating rate (non-isothermal) or temperature (isothermal) (Málek, 2000).

$$\frac{\mathrm{d}\alpha}{\mathrm{d}t} = K(T) \cdot f(\alpha) \tag{2.32}$$

Instead of the differential form in eq. (2.32), integral kinetics as shown in eq. (2.33) was formerly preferred for technical reasons (Menezes et al., 2019).

$$g(\alpha) = \int_0^\alpha \frac{\mathrm{d}\alpha}{f(\alpha)} = K(T) \cdot t \tag{2.33}$$

The temperature dependent rate constant $K(T)$ is assumed to follow the Arrhenius law, as shown in eq. (2.34). E_a is the activation energy, A is the pre-exponential factor, R is the universal gas constant and T is the absolute temperature. E_a and A are independent on temperature and the relative crystallinity α (Málek, 2000).

$$K(T) = A \cdot \exp\left(\frac{-E_\mathrm{a}}{R \cdot T}\right) \tag{2.34}$$

For *isothermal* conditions, eq. (2.35) and eq. (2.36) follow (Menezes et al., 2019).

$$\frac{\mathrm{d}\alpha}{\mathrm{d}t} = A \cdot \exp\left(\frac{-E_\mathrm{a}}{R \cdot T}\right) \cdot f(\alpha) \tag{2.35}$$

$$g(\alpha) = A \cdot \exp\left(\frac{-E_\mathrm{a}}{R \cdot T}\right) \cdot t \tag{2.36}$$

For *non-isothermal* crystallization reactions, the rate expressions must be transformed by $\mathrm{d}\alpha/\mathrm{d}T = \mathrm{d}\alpha/\mathrm{d}t \cdot \mathrm{d}t/\mathrm{d}T$, with the cooling rate $C = \mathrm{d}t/\mathrm{d}T$. Hence, for non-isothermal conditions, eq. (2.37) and eq. (2.38) are obtained, wherein $p(x) = \int_x^\infty \exp(-x)/x^2 \cdot \mathrm{d}x$ and $x = E_\mathrm{a}/(R \cdot T)$. $p(x)$ has no analytical solution, but there exist many approximations, which will be introduced later (Menezes et al., 2019).

$$\frac{\mathrm{d}\alpha}{\mathrm{d}T} = \frac{A}{C} \cdot \exp\left(\frac{-E_\mathrm{a}}{R \cdot T}\right) \cdot f(\alpha) \tag{2.37}$$

$$g(\alpha) = \frac{A}{C} \int_0^T \exp\left(\frac{-E_\mathrm{a}}{R \cdot T}\right) \mathrm{d}T = \frac{A \cdot E_\mathrm{a}}{C \cdot R} \int_x^\infty \frac{\exp(-x)}{x^2} \mathrm{d}x = \frac{A \cdot E_\mathrm{a}}{C \cdot R} \cdot p(x) \tag{2.38}$$

Ozawa (1992) states that the fundamental single step equation holds for diffusion, chemical reactions consisting of a single elementary process and for crystallization growth from pre-existing nuclei.

If several kinetic processes occur simultaneously or overlap (*multistep solid-state reactions*), eq. (2.32) must be expanded to eq. (2.39). i indicates the ith reaction to the overall reaction, c is a weighting factor ($\sum c_i = 1$ and $\sum c_i \alpha_i = \alpha$). The resulting complexity is immediately apparent (Koga, 2013).

$$\frac{\mathrm{d}\alpha}{\mathrm{d}t} = \sum_{i=1}^{n} c_i \cdot A_i \cdot \exp\left(-\frac{E_{a,i}}{R \cdot T}\right) f_i(\alpha_i) \tag{2.39}$$

Overlapping nucleation and growth processes are considered as multistep solid-state reactions, since they are different kinetic processes with different activation energies (Koga, 2013). Hence the temperature dependence of the overall rate may not fulfill the basic single step equation in general (Vyazovkin and Sbirrazzuoli, 2003). Exceptions are the so called isokinetic transformations, in which the activation energies of the different processes are equally. Nevertheless, it is assumed, that the equation holds relatively well for a relatively narrow temperature interval related to a certain value of relative crystallinity α. Vyazovkin and Sbirrazzuoli (2003) criticized the expression *activation energy* in the case of melt crystallization and has stressed that it is the *effective activation energy* or *overall activation energy* and it accounts for all the multiple steps involved in the crystallization process (e.g. nucleation and growth).

2.2.4.4 Activation energy

The activation energy is the dynamic factor controlling the crystallization of polymers (Shehzad et al., 2014). For the determination of the activation energy E_a, isoconversional methods can be used. They can be applied to non-isothermal crystallizations for evaluating the dependency of the effective activation energy on relative crystallinity and temperature (Vyazovkin and Sbirrazzuoli, 2003).
If E_a is largely independent of the relative crystallinity, the single step model can be used. Hence, isoconversional methods help to detect whether a single or complex reaction occurs during the non-isothermal process.
Isoconversional methods require more than one heating or cooling rate to obtain E_a vs. α values at progressive relative crystallinities. The isoconversional methods are hence also called multicurves methods (Shehzad et al., 2014). The activation energy is then determined graphically (Málek, 1992). The most common isoconversional methods are presented in the following. It should be noted, however, that the individual methods are presented here only very briefly. Detailed information, especially on the assumptions and boundary conditions, can be found in the original papers and in (Ozawa, 1992).

Ozawa-Flynn-Wall activation energy OFW

A common method to determine the activation energy is the Ozawa-Flynn-Wall OFW method, which is based on Doyle's linear approximation for $\log p(x)$ (see eq. (2.38)):

$$\log p(x) = a + bx, \tag{2.40}$$

with $a = -2.315$ and $b = -0.4567$ for $20 \leq x \leq 60$ (Ozawa, 1992; Koga, 2013).

Hence, eq. (2.41) follows from eq. (2.33). Differentiating and assuming a constant α leads to eq. (2.42), wherein E_a is the activation energy and R is the universal gas constant.

$$\begin{aligned}\log(C) &= \log\left(\frac{A \cdot E_a}{R \cdot g(\alpha)}\right) + \log(p(x)) \\ &= \log\left(\frac{A \cdot E_a}{R \cdot g(\alpha)}\right) - 2.315 - 0.4567 \cdot \frac{E_a}{R \cdot T}\end{aligned} \tag{2.41}$$

$$\left[\frac{\mathrm{d}\log(C)}{\mathrm{d}\frac{1}{T}}\right]_{\alpha} = -\frac{0.457}{R}E_{\mathrm{a}} \tag{2.42}$$

The plot of $\log(C)$ vs. $1/T$ for each α is a straight line with the slope $-0.457/\mathrm{R} \cdot E_{\mathrm{a}}$, (Ozawa, 1992; Koga, 2013; Ozawa and Koto, 1991).
In practice, this method is often applied only to the peak temperatures T_{p}, see eq.(2.43). In this case it is called peak shift method, Figure 2.27 (a).

$$\frac{\mathrm{d}\log(C)}{\mathrm{d}\frac{1}{T_{\mathrm{p}}}} = -\frac{0.457}{R}E_{\mathrm{a}} \tag{2.43}$$

Kissinger-Akahira-Sunose activation energy KIS

This method is based on Doyle's exponential approach for $p(x)$, shown in eq. (2.44), which is valid for $20 \leq x \leq 50$ (Ozawa, 1992).

$$\ln p(x) = -x - \ln(x^2) \tag{2.44}$$

Hence, eq. (2.45) and eq. (2.46) follow. R is the universal gas constant E_{a} is the activation energy at different cooling rates. The activation energy E_{a} is determined as slope of the plot $\ln(C/T^2)$ against $\ln(1/T)$ for different relative crystallinities (Abareshi et al., 2014; Kissinger et al., 1956).

$$\ln\left(\frac{C}{\frac{1}{T^2}}\right) = -\ln\left(\frac{E_{\mathrm{a}} \cdot g(\alpha)}{A \cdot R}\right) - \frac{E_{\mathrm{a}}}{R \cdot T} \tag{2.45}$$

$$\left[\frac{\mathrm{d}\ln(\frac{C}{T^2})}{\mathrm{d}(\frac{1}{T})}\right]_{\alpha} = -\frac{E_{\mathrm{a}}}{R} \tag{2.46}$$

This method is also predominantly used in the literature with the peak temperatures T_{p}, see eq. (2.47) and Figure 2.27 (b).

$$\frac{\mathrm{d}\ln(\frac{C}{T_{\mathrm{p}}^2})}{\mathrm{d}(\frac{1}{T_{\mathrm{p}}})} = -\frac{E_{\mathrm{a}}}{R} \tag{2.47}$$

Expanded Friedman activation energy

The method of Friedman (1964) is based on the fundamental kinetic equation in differential form (see eq. (2.35)). Taking the logarithm leads to eq. (2.48). With eq. (2.49) it can be seen that similar to the other methods, the activation energy E_a can be determined from the universal gas constant R and the slope of a certain plot (Friedman, 1964; Ozawa, 1971; Koga, 2013; Málek, 1992). The plot consists of $\ln(\mathrm{d}\alpha/\mathrm{d}t)$ against $1/T$ for fixed α and is called Friedman plot, Figure 2.27 (c).

$$\ln\frac{\mathrm{d}\alpha}{\mathrm{d}t} = \ln[A \cdot f(\alpha)] - \frac{E_a}{R \cdot T} \tag{2.48}$$

$$\left[\frac{\ln\left(\frac{\mathrm{d}\alpha}{\mathrm{d}t}\right)}{\frac{1}{T}}\right]_{\alpha} = -\frac{E_a}{R} \tag{2.49}$$

General remark

The methods of Ozawa-Flynn-Wall and Kissinger-Akahira-Sunose require taking the logarithm of C, which is why they cannot be directly applied to crystallization of polymer melts that occurs on cooling ($C<0$). This can be overcome by using the basic rate equation (expanded Friedman method).

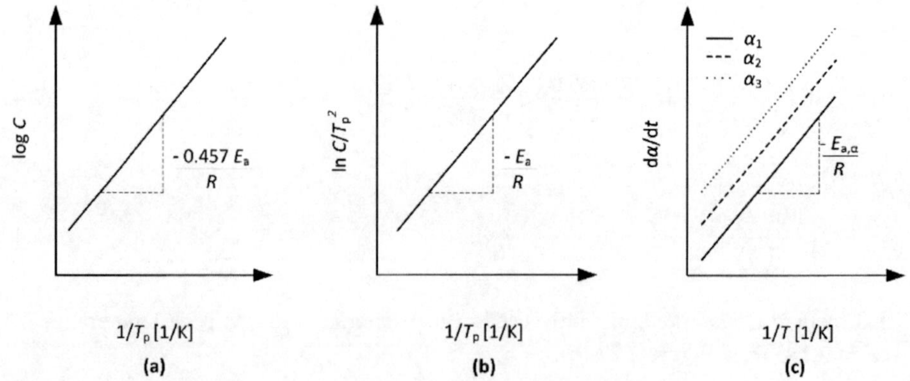

Figure 2.27 Determination of activation energy according to (a) Ozawa-Flynn-Wall, (b) Kissinger-Akahira-Sunose and (c) Friedman.

2.2.4.5 Kinetic model functions

Most methods to study the phase transformation involving nucleation and growth are based on the JMAK- *(Johnson-Mehl-Avrami-Kolmogorov)* or simply called *Avrami-rate equation*. As already described, this model was initially derived for isothermal conditions and is based on the assumptions of homogeneous nucleation or heterogeneous nucleation at randomly disperged particles, a temperature controlled and time-independent growth rate of the new phase, and a low anisotropy of the growing crystals (Málek, 2000). However, for the special cases that the transformation rate is independent on the thermal history and hence only dependent on the state variables of transformed fraction α and temperature T, the JMAK rate equation is also valid for non-isothermal conditions. This is the case for site-saturated nucleation, where nucleation is completed prior to crystal growth and isokinetic transformations, which means that the activation energies of the individual processes (here nucleation and growth) are equally (Málek, 2000; Farjas and Roura, 2006).

A slightly different expression of Avrami's equation (2.22) is given in eq. (2.50) (Málek, 2000). K is the crystallization rate constant (involves nucleation and growth rate parameters) at the isothermal temperature T and $k = K^n$ (compare eq. (2.22) with eq. (2.50)), so that K is independent on the Avrami exponent n.

$$1 - \alpha = \exp\left[-(K \cdot t)^n\right] \tag{2.50}$$

The crystallization rate in eq. (2.51) is obtained by differentiation with respect to time.

$$\frac{\mathrm{d}\alpha}{\mathrm{d}t} = K \cdot n \cdot (1 - \alpha) \cdot [-\ln(1 - \alpha)]^{1-1/n} \tag{2.51}$$

Using eq. (2.32), the kinetic function of eq. (2.51) equals:

$$f(\alpha) = n \cdot (1 - \alpha) \cdot [-\ln(1 - \alpha)]^{1-1/n}. \tag{2.52}$$

As already mentioned before, Ozawa's method is the most popular to describe non-isothermal crystallization. If one modifies Ozawa's equation (eq. 2.24) a little bit to eq. (2.53), the same kinetic function (see eq. (2.52)) is obtained.

$$1 - \alpha(T) = \exp\left[\left(-\frac{K(T)}{C}\right)^n\right] \tag{2.53}$$

The validity of the JMAK rate equation and the determination of the kinetic parameters, can be done in different ways (Málek, 2000). The most popular methods are the constructions of linearity plots, as already previously introduced in section 2.2.4.1 and section 2.2.4.2. Málek (2000) pointed out, that the linearity checks are performed with double logarithmic functions, which are generally not very sensitive to changes in their argument and hence not very reliable.
A better and more reliable method than the linearity checks, was developed by Ozawa and Malek by the introduction of $y(\alpha)$- and $z(\alpha)$-functions (Málek, 2000). On the basis of the constant activation energy, Ozawa derived an important kinetic equation by employing the time integral of the exponential term in the Arrhenius equation, see eq. (2.54). Θ is the generalized time, which can be interpreted as the hypothetical reaction time, calculated by extrapolating the reaction time at real temperatures to infinite temperature according to the Arrhenius equation (Koga, 2013; Málek, 2000). Θ can be calculated with eq. (2.55).

$$g(\alpha) = A \int_0^t \exp\left(\frac{-E_\mathrm{a}}{R \cdot T}\right) \cdot \mathrm{d}s = A \cdot \Theta \tag{2.54}$$

$$\Theta = \int_0^t \exp\left(\frac{-E_\mathrm{a}}{R \cdot T}\right) \cdot \mathrm{d}s \tag{2.55}$$

Derivation of eq. (2.55) after t leads to:

$$\frac{\mathrm{d}\Theta}{\mathrm{d}t} = \exp\left(\frac{-E_\mathrm{a}}{R \cdot T}\right). \tag{2.56}$$

Combining this with eq. (2.32) results in eq. (2.57), which is analogously to eq. 2.32, with t replaced by Θ and $K(T)$ replaced by A (Koga, 2013).

$$y(\alpha) = \frac{\mathrm{d}\alpha}{\mathrm{d}\Theta} = Af(\alpha) \tag{2.57}$$

Later, Malek derived an alternative equation, in which the pre-exponential factor is canceled (Koga, 2013):

$$z(\alpha) = \frac{\mathrm{d}\alpha}{\mathrm{d}\Theta} \cdot \Theta = f(\alpha)g(\alpha). \tag{2.58}$$

Once the activation energy has been determined, it is possible to find the kinetic model that best describes the measured data (Málek, 1992). Málek showed that for this purpose, it is useful to evaluate the two functions $y(\alpha)$ (eq. (2.57)) and $z(\alpha)$ (eq. (2.58)), which are qualitatively shown in Figure 2.28.

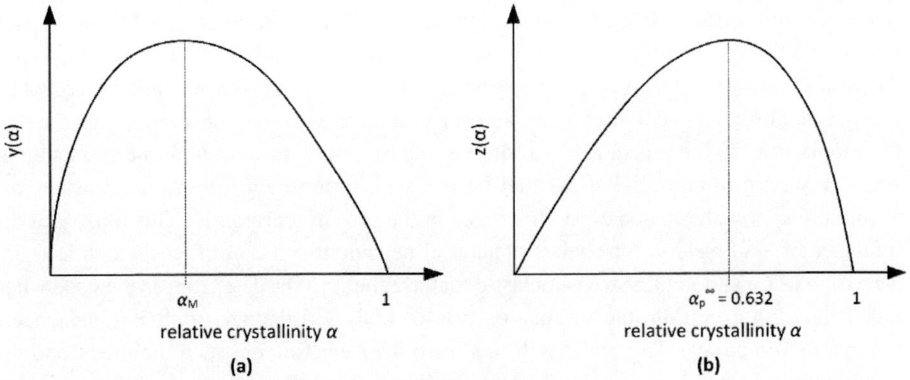

Figure 2.28 (a) $y(\alpha)$ function for $n > 1$, (b) $z(\alpha)$ function for JMAK model.

The peak positions α_M of $y(\alpha)$ and $\alpha_P{}^\infty$ of $z(\alpha)$ ($\alpha_M < \alpha_P{}^\infty$) are characteristic for the respective kinetic model functions. The value of the kinetic exponent n can be determined from the position of α_M (Málek, 1992, 2000):

- For $n \leq 1$: $\alpha_M = 0$.
- For $n > 1$: $\alpha_M = 1 - \exp(n^{-1} - 1)$.

$\alpha_P{}^\infty = 0.632$ is a characteristic value for the JMAK model. If $\alpha_P{}^\infty < 0.632$, then the conditions to apply JMAK model are not fulfilled (Málek, 1992, 2000). This may be due to an increasing complexity, which results e.g. from the influence of surface nucleation, from an influence of liberated crystallization heat or from autocatalytic behaviour (Málek, 2000). Here, the empirical rate equation defined by *Sestak and Bergrren* (SB-Model) has proved to be very powerful. More information about that model can be obtained in (Málek, 1992; Šesták and Berggren, 1971).

2.3 Viscoelasticity

Due to their long molecular chain structure (and the time consuming rotation and unwinding effects, see section 2.1), polymeric materials show a viscoelastic material behaviour. As the name suggests, elastic (functional relationship between stress and strain) and viscous (functional relationship between stress and strain rate) material behaviour are combined here. This results in a material behaviour that is not only dependent on the load level, but also on the load rate, load duration and load history. A higher load rate leads for example to a higher stiffness than a lower load rate. In addition, hysteresis occurs in the stress-strain curves during loading and unloading.

If a viscoelastic body is suddenly loaded with a constant stress (creep test), a part of the

strain occurs immediately and a part in the course of time due to creep. The latter may approach a limit value for thermosets or increase without bound for thermoplastics. When the load is removed (creep recovery), part of the strain springs back immediately while another part only recovers after a certain time and there even may be a remaining part for thermoplastics. In the relaxation test, the viscoelastic body is subjected to a spontaneous strain, resulting in an initial stress that decays with time to zero for thermoplastics and eventually approaches a non-zero limit value in the case of thermosets. This illustrates the influence of load history. Viscoelastic material behaviour can be further divided into linear viscoelastic and nonlinear viscoelastic material behaviour. The latter is the case when either the elastic portion, the viscous portion or both portions are nonlinear and occurs at large deformations. The linearity limits form the transition between the linear and the nonlinear viscoelastic material behaviour (Schwarzl, 2013; Brinson et al., 2008).
In the following, only the relaxation behaviour is considered. There is first a brief introduction to the linear viscoelastic material behaviour, which is valid for small strains and is the basis for chapter 5. The Boltzmann superposition principle and the generalized Maxwell model including the Prony series are introduced. This is followed by a section on linearity limits, which are experimentally investigated in chapter 6. Finally, the nonlinear viscoelasticity is treated, which is applied in the case of large strains, and the Schapery model is introduced. The understanding of nonlinear viscoelastic behaviour is of great importance in chapter 7. Further information about viscoelasticity can for example be found in (Brinson et al., 2008; Schwarzl, 2013).

2.3.1 Linear viscoelasticity - generalized Maxwell model

Since the primary loading of the interlayers in intact laminated safety glass is due to shear, the relaxation test is considered in the simple shear condition. The ratio of the time-dependent stress response $\tau(t)$ to the applied constant strain γ_0 results in a relaxation function $G(t)$ (eq. (2.59)) that also decays with time, see Figure 2.29. In the case of thermoplastics, the stress and hence the relaxation function approaches zero after a sufficient amount of time t, while for thermosets a non-zero limit $G_\infty = \tau_\infty/\gamma_0$ may be reached. Hence, the relaxation function $G(t)$ can be split into a time invariant equilibrium part G_∞ and a transient part $\Delta G(t)$.

$$G(t) = \frac{\tau(t)}{\gamma_0} = G_\infty + \Delta G(t) \tag{2.59}$$

The generalized Maxwell model in Figure 2.30 represents a common rheological model of stress relaxation $G(t)$ in the linear-viscoelastic range. It consists of N parallel spring-damper elements (= Maxwell elements), which describe the transient part of eq. (2.59), and a supplementary parallel linear elastic spring, which describes the equilibrium part

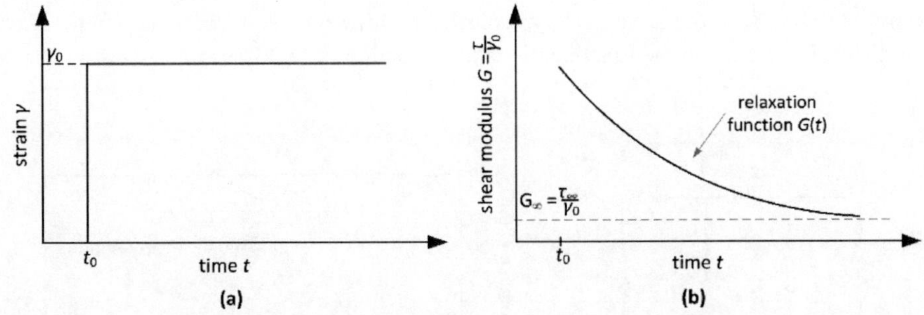

Figure 2.29 Viscoelastic material behaviour (a) sudden constant strain input, (b) decreasing stress output and relaxation function

of eq. (2.59). The mathematical description of the generalized Maxwell model is called Prony series, see eq. (2.60), in which the Prony parameters G_∞ resp. $G_0 = G_\infty + \sum G_i$, G_i resp. $g_i = G_i/G_0$ and $\tau_i = \eta_i/G_i$ have to be determined experimentally. The following boundary conditions must be taken into account: τ_i, G_∞ and G_i resp. g_i must be positive numbers $\tau_i \geq 0$, $G_\infty \geq 0$, $G_i \geq 0$, $g_i \geq 0$). G_0 is a strictly positive number $G_0 > 0$. The sum of the g_i has to be smaller than 1 ($\sum g_i \leq 1$).
Alternatively, fractional viscoelastic material models are used in literature (Di Paola et al., 2021; Kraus, 2018; Thiele et al., 2022). The biggest advantage of those models is the lower number of material parameters. However, since the generalized Maxwell model is implemented in some commercial FEM software and incorporated in some standards or general building approvals, e.g.(NEN 2608; Z-70.3-230), the consideration is limited to this model.

$$\begin{aligned} G(t) &= G_\infty + \sum_{i=1}^{N} G_i \cdot \exp -\frac{t}{\tau_i} \\ &= G_0 \cdot \left[1 - \sum_{i=1}^{N} g_i \cdot \left(1 - \exp -\frac{t}{\tau_i}\right)\right] \end{aligned} \tag{2.60}$$

Conventional test methods for determining the time-dependent stiffness of laminated safety glass interlayers are creep and relaxation tests on the laminates (or on the pure material), which can also record temperature dependency, when they are executed at different temperatures as e.g. done in (Schuler, 2003). However, those tests need high experimental efforts, especially with regard to the testing times. In addition, when the tests are conducted on laminates, large climate chambers are necessary in order to investigate the relaxation behaviour at different temperatures. For these reasons, Dynamic-Mechanical-Thermal-Analysis (DMTA) are usually carried out, in which small-scaled samples of the pure interlayer material are tested dynamically at different frequencies and temperatures. Further information on the functionality and evaluation of DMTA is introduced in chapter 3.

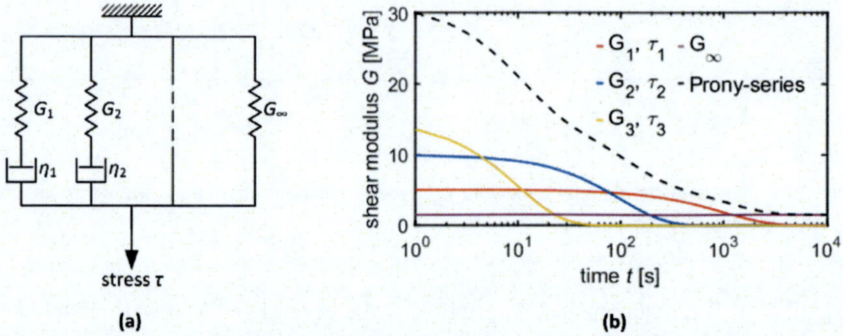

Figure 2.30 Generalized Maxwell model (a) rheological model, (b) composition of the Prony series with G_1 = 5 MPa, τ_1 = 1000 s, G_2 = 10 MPa, τ_2 = 100 s, G_3 = 15 MPa, τ_3 = 10 s and G_∞ = 1.5 MPa.

Figure 2.30 (b) demonstrates the effects of the different parameters in the Prony series. G_0 represents the initial stiffness at time $t = 0$ and G_∞ the stiffness at time $t \to \infty$, (=the stiffness after relaxation of all Maxwell elements). The relaxation time τ_i of the ith Maxwell element marks the time t at which the relaxation process of Maxwell element i has been completed to 63%. Figure 2.30 shows, that the contribution of each Maxwell element to the overall stiffness mainly decays within about one time decade, measured in a logarithmic scale. In order to represent the relaxation function $G(t)$ accurately, it has proven successful to adapt the total number of Maxwell elements N to the number of investigated time decades.

Using the generalized Maxwell model (and Prony series), the stress output can be calculated with:

$$\begin{aligned} \tau(t) &= \gamma_0 \cdot \left[G_\infty + \sum_{i=1}^{N} G_i \cdot \exp - \frac{t}{\tau_i} \right] \\ &= \gamma_0 \cdot G_0 \cdot \left[1 - \sum_{i=1}^{N} g_i \cdot \left(1 - \exp - \frac{t}{\tau_i} \right) \right]. \end{aligned} \tag{2.61}$$

In the linear viscoelastic region, the relaxation function $G(t)$ is independent on the applied strain magnitude γ_0. In fact, stress and strain behave linear at every time t, which can be visualized in isochronous stress-strain diagrams as exemplary shown in Figure 2.31 (a). This leads to linear scalability (Figure 2.31 (b)), which states that twice as much applied strain results in twice as much stress response. In addition, the *Boltzmann superposition principle (BSP)* is valid in the linear viscoelastic region, which can be used to calculate the stress output after an arbitrary strain input (Schwarzl, 2013). According to this principle, the material response to arbitrary loading processes can be calculated by the summation of responses obtained under infinitesimal constant loading increments $\Delta s \to 0$, see Figure 2.31 and eq. (2.62), wherein G is the stress relaxation function and s is a time with known strain history.

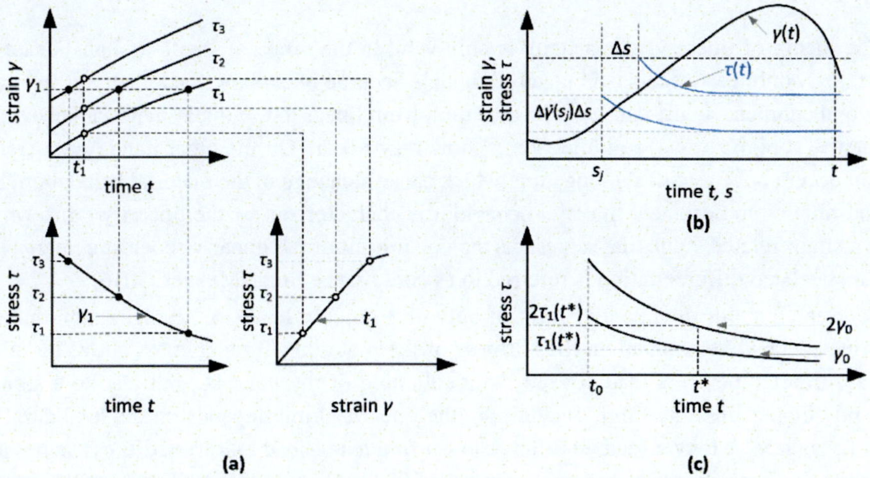

Figure 2.31 Linear viscoelasticity (a) isochronous stress-strain diagram, (b) Boltzmann superposition principle, (c) linear scalability.

$$
\begin{aligned}
\tau(t) &= \int_0^t G(t-s)\frac{\mathrm{d}\gamma(s)}{\mathrm{d}s}\mathrm{d}s \\
&= \int_0^t \left[G_\infty + \sum_{i=1}^{N} G_i \cdot \exp -\frac{t-s}{\tau_i}\right]\frac{\mathrm{d}\gamma(s)}{\mathrm{d}s}\mathrm{d}s \\
&= G_\infty \cdot \gamma(t) + \int_0^t \left[\sum_{i=1}^{N} G_i \cdot \exp -\frac{t-s}{\tau_i}\right]\frac{\mathrm{d}\gamma(s)}{\mathrm{d}s}\mathrm{d}s \\
&= G_\infty \cdot \gamma(t) + \int_0^t \Delta G(t-s)\frac{\mathrm{d}\gamma(s)}{\mathrm{d}s}\mathrm{d}s
\end{aligned}
\tag{2.62}
$$

At this point, it should be pointed out that the equations presented refer only to deviatoric strains. According to theory, relaxation also occurs under volumetric strains. This is expressed with a time-dependent compressibility. However, it has been shown in the literature that this relaxation is much smaller than the deviatoric one. Therefore, and since the time-dependent compressibility or the time-dependent Poisson's ratio (which is required to determine the compressibility from the Young's or shear modulus) is difficult to determine experimentally, the compressibility is often considered as time invariant and is calculated from the Young´s or shear modulus at very long times ($t \to \infty$) and the Poisson's ratio, which in turn is often assumed to be ≈ 0.5 (Stommel et al., 2018; Andreozzi et al., 2014).

2.3.2 Linearity limits

The theory of linear viscoelasticity is only valid in the range of small strains. For larger strains, nonlinear viscoelastic material models become necessary. Sackmann (2008) cites two phenomena as the cause of the transition from linear to nonlinear material behaviour: Firstly, at high strains, solidification or flow may occur. On the other hand, microcracks can occur above a critical elongation, which cause a change in the material behaviour. The delimitation to nonlinear material behaviour is characterized by the linearity limit, which is a strain-related value that represents the upper limit of the linear viscoelastic range. The stress-related differentiation is referred to as compliance limit (Mezger, 2016).

A research about the linearity limit of polymers quickly leads to the realization that the values depend very much on the stiffness of the material. Temperatures above the glass transition temperature lead not only to a softening of the material, but also to a significantly higher linearity limit. In addition, the linearity limit depends on the load duration or frequency. A longer load duration or lower frequency, lead to lower stiffness values and hence the linearity limit increases (Jansson, 1974; Mezger, 2016). The standard for the determination of dynamic-mechanical properties of plastics, however, only recommends a single frequency value of 1 Hz (EN ISO 6721-1) in order to determine the linearity limit. Often only rough indications of the linearity limit are given in literature. Schneider et al. (2016); Grellmann and Seidler (2014); Domininghaus (2013) and Batzer (1985) state for example that the linearity limit for polymers in the energy-elastic range is approximately 1 % and can reach up to 100 % in the entropy-elastic range. Yannas (1972) investigated the linearity limits of amorphous polymers more closely in uniaxial creep and relaxation tests. In experiments on polycarbonate, he showed that the linearity limits determined after $t = 100$ s are between 0.8 % and 1.25 % in the energy-elastic state. In addition, he has collected the results of various publications from the years 1944 to 1965, in which relaxation tests were carried out on natural rubber and polyisobutylene. Here, the determined linearity limits are 25 % to 150 % in the entropy-elastic range and 10 % in the range of the glass transition temperature. The values for the linearity and compliance limits were graphically displayed as a function of the difference between the examination temperature and the glass transition temperature ($T - T_g$), as shown in Figure 2.32 exemplary for the linearity limits. Schwarzl (2013) also compiled a summary of these investigations, including the graphic evaluations. Further linearity limits from literature are summarized in Table 2.1.

It should be noted that the indicated linearity limits refer to uniaxial tensile strain ε_{lin}. In some of the literature, it is not clear how the stated linearity limit ε_{lin} was calculated and whether it is an engineering strain ε_0, true strain ε_t or different expression. A different expression would e.g. be $1/3 \cdot (\lambda - 1/\lambda^2)$ with $\lambda = 1 + \varepsilon_{tech}$ (Yannas, 1972; Staverman and Schwarzl, 1956). This can be derived from Neo-Hooks hyperelastic material model for uniaxial tension: $\sigma_0 = E_0/3 \cdot (\lambda - 1/\lambda^2)$, wherein E_0 is the initial tensile modulus. For small strains $\varepsilon_0 \leq 1\%$, the difference between the three mentioned strain measures is negligible (Yannas, 1972; Bertilsson and Jansson, 1975).

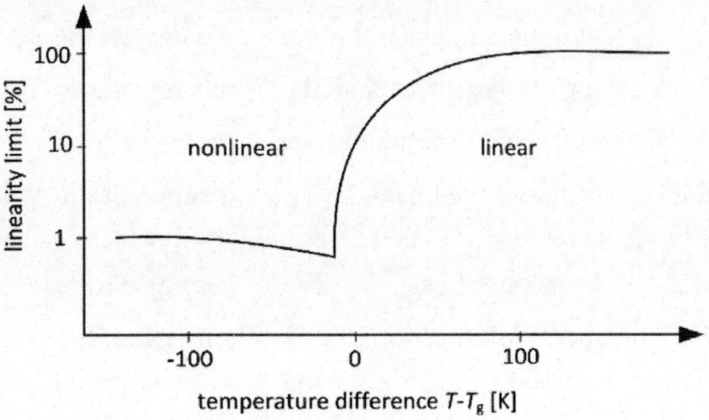

Figure 2.32 Linearity limit, based on Yannas (1972).

The linearity limits can be determined in quasi static relaxation or creep tests as well as in DMTA amplitude sweeps (see chapter 3). There are various ways of distinguishing between linear and nonlinear material behaviour. For the evaluation of quasi static tests, Papanicolaou and Zaoutsos (2019) mention the verification of the validity of Boltzmann's superposition principle and the linear relationship between stress and strain, which corresponds to a constant stiffness at any point in time (isochronous stress strain diagram). If one of the two conditions is violated, the material behaviour is no longer linear. Stress-strain-diagrams can also be generated out of DMTA amplitude sweeps. However, it is common to directly evaluate the storage modulus-strain-curve, see further chapter 3. For the determination of the linearity limit, a percentage deviation from the linear viscoelastic material behaviour must be specified. However, there are different specifications in the literature for this, which are briefly presented in chapter 3.

It can be concluded that the linearity limit is dependent on the type of polymer, temperature distance to the glass transition temperature $(T - T_g)$, load duration (or applied frequency) and the exact definition of linearity limit. According to Schwarzl (2013), different linearity regions are possible for simple shear, uniaxial strain and isotropic compression. However, Schwarzl (2013) assumes that the linearity limits for simple shear γ_{lin} are in the same order of magnitude.

Table 2.1 Linearity Limits.

Reference	Material	Temperature range	Linearity limit [%]
Mezger (2016)	polymers in general	energy elastic	0.1[a]
Retting (1972)	high polymeric materials	energy elastic	0.1 - 0.5
Schuler (2003)	amorphous thermoplastics	energy elastic	0.1 - 0.5
Staverman and Schwarzl (1956)	different polymers	energy elastic	0.1 - 1.0 [b]
Ensslen (2005)	thermoplastics	energy elastic	0.5
Bertilsson and Jansson (1975)	thermoplastics	energy elastic	0.5 - 1.0[c]
Jansson (1974)	polymers	energy elastic	0.7 - 1.0
Yannas (1972)	polycarbonate	energy elastic	0.8 - 1.25[d]
Schmachtenberg (1985)	amorphous materials	energy elastic	1.0
Yannas (1972)	polyisobutylene	glass transition	10[e]
Staverman and Schwarzl (1956)	rubber	entropy elastic	10 - 100[b]
Brinson et al. (2008))	rubbery materials	entropy elastic	20 - 50
Yannas (1972)	polyisobutylene, natural rubber	entropy elastic	25 - 150[e]

[a] Limit is given in shear mode γ_{lin},
[b] measure of tensile strain: $\lambda - 1/\lambda^2$,
[c] measure of tensile strain: $\varepsilon_0 = \Delta l/l_0$,
[d] measure of tensile strain: $1/3 \cdot (\lambda - 1/\lambda^2)$ or $\varepsilon_{true} = \ln(1 - \varepsilon_0)$. For small strains about 1%, the difference between those two strain measures is negligible (Yannas, 1972),
[e] measure of tensile strain: $1/3 \cdot (\lambda - 1/\lambda^2)$.

2.3.3 Nonlinear viscoelasticity - Schapery model

After glass fracture, the interlayer locally delaminates from the glass and undergoes large strains exceeding the linear viscoelastic range, see further e.g. (Delincé et al., 2008). Since tension is prevailing in the interlayer between the glass fragments (Figure 1.4), the tensile mode is considered in this section. In the nonlinear viscoelastic region, the relaxation function $E(t,\varepsilon)$ is not only dependent on loading duration t, but also on the prevailing strain ε (or stress σ) level. As a consequence, the Boltzmann superposition principle (BSP) and the linear scalability do not hold anymore.

Different material models to depict nonlinear viscoelastic material behaviour can be found in literature. (Brinson et al., 2008) and (Naumann, 2012) provide good overviews, which are briefly summarized here.
The simplest models are based on parabolic, exponential, logarithmic or hyperbolic approach creep-functions, etc. An example is Findley's nonlinear creep power law. The few parameters of such functions can be determined quite easily. However, those models only simulate a single step creep or relaxation test.
In order to be able to map arbitrary load histories, integral approaches were developed based on Boltzmann's superposition principle (BSP). Leadermann proposed 1948 a single integral approach (similar to BSP), where the nonlinear effects are incorporated in the stress resp. strain measure. He assumed that the dependencies on time and stress resp. strain level can be factorized and are not linked to each other. In Schapery's single integral approach in 1964, the creep resp. relaxation function was extended to an additional stress-level resp. strain-level dependency. Hence, contrary to Leadermann, time and stress resp. strain dependency are interconnected. This can be imagined in such a way, that there is a stress resp. strain dependent scaling of the time axis, which is then called *effective time* and has similarities to the time-temperatures-superposition principle, which is introduced in chapter 2.4. Additionally, Schapery introduces further nonlinearity factors to the creep resp. relaxation function. Knauss and Emri (1981) formulated a single integral approach wherein the nonlinear parameters are associated to the free volume. This offers the advantage that the approach becomes more versatile, because other factors can be considered besides the stress or strain level, which have an influence on the free volume v_f. These include for example stress-induced dilatation, moisture or other effects.
Other nonlinear viscoelastic models are the multiple-integral approach of Rivlin and Green, the deformation model of Schmachtenberg (1985), the model of Bergström and Boyce (1998), the constitutive equations of Pipkin and Rogers (1968) and the quasi linear viscoelastic theory proposed by Fung and Skalak (1981).

In the following, the Schapery model is examined in more detail. This model has the advantage that it can be used for linear and nonlinear viscoelastic material behaviour. For this purpose, the BSP in eq. (2.62) is supplemented by four different nonlinearity factors ($h_\infty(\varepsilon)$, $h_1(\varepsilon)$, $h_2(\varepsilon)$ and $a_\varepsilon(\varepsilon)$), which makes this model to the most flexible of all presented single integral models. The linear viscoelastic parameters remain mainly unchanged compared to eq. (2.62), only a conversion from shear to tension has to be performed. Hence, eq. (2.63) is obtained. E_∞ is the equilibrium modulus, $\Delta E(t)$ is the transient part of the linear viscoelastic tensile modulus at a given time t. If all nonlinearity factors equal 1 ($a_\varepsilon = h_\infty = h_1 = h_2 = 1$), BSP in eq. (2.62) is obtained. If $h_\infty = h_1 = h_2 = 1$, Knauss and Emri's free volume dependent model is obtained. Schapery (1969a) states that this model is often applicable for fibers.

$$\begin{aligned}
\sigma(t,\varepsilon) &= \int_0^t E(\varepsilon, t-s)\frac{\mathrm{d}\varepsilon}{\mathrm{d}s}\mathrm{d}s \\
&= h_\infty(\varepsilon)E_\infty\varepsilon(t) + h_1(\varepsilon)\int_0^t \Delta E(\psi-\psi')\left[\frac{\mathrm{d}[h_2(\varepsilon)\cdot\varepsilon(s)]}{\mathrm{d}s}\right]\mathrm{d}s \\
\psi &= \int_0^t \frac{\mathrm{d}t}{a_\varepsilon(t)} \\
\psi' &= \int_0^s \frac{\mathrm{d}s}{a_\varepsilon(s)}
\end{aligned} \tag{2.63}$$

If perfectly parallel relaxation curves emerge in relaxation tests with different strain levels, overlaying curves arise after solely vertically shifting the log-log relaxation curves. Hence, the relaxation modulus is factorized into a product of a strain level dependent function $h(\varepsilon)$ and a time-dependent function $E(t)$. This approach is also often studied in the literature due to its simplicity, see further e.g. (Hooper et al., 2012a; Titomanlio et al., 1980), and, according to Schapery (1969a), has been used successfully, especially for elastomers resp. unfilled polymers in their rubbery state.

For a single step relaxation function with a constant strain ε_0, eq. (2.63) can be reduced to eq. (2.64), wherein $h_{1,2}(\varepsilon) = h_1(\varepsilon)\cdot h_2(\varepsilon)$, or to eq. (2.65) with $\Delta E \neq \Delta E^*$ and $h^*_{1,2}(\varepsilon) \neq h_{1,2}(\varepsilon)$. Further, the Prony series can be incorporated to describe the linear viscoelastic parameters.

$$\begin{aligned}
\sigma(t,\varepsilon_0) &= \varepsilon_0\cdot\left[h_\infty(\varepsilon)E_\infty + h_1(\varepsilon)\cdot h_2(\varepsilon)\cdot\Delta E\left(\frac{t}{a_\varepsilon}\right)\right] \\
&= \varepsilon_0\cdot\left[h_\infty(\varepsilon)E_\infty + h_{1,2}(\varepsilon)\cdot\Delta E\left(\frac{t}{a_\varepsilon}\right)\right] \\
&= \varepsilon_0\cdot\left[h_\infty(\varepsilon)E_\infty + h_{1,2}(\varepsilon)\cdot\left(\sum_{i=1}^N E_i\cdot\exp-\frac{t}{a_\varepsilon\cdot\tau_i}\right)\right]
\end{aligned} \tag{2.64}$$

$$\begin{aligned}\sigma(t,\varepsilon_0) &= \varepsilon_0 \cdot \left[h_0(\varepsilon)E_0 - h^*_{1,2}(\varepsilon)\cdot \Delta E^*(\frac{t}{a_\varepsilon})\right] \\ &= \varepsilon_0 \cdot \left[h_0(\varepsilon)E_0 - h^*_{1,2}(\varepsilon)\cdot \left(\sum_{i=1}^{N} E_i \cdot (1-\exp -\frac{t}{a_\varepsilon \cdot \tau_i})\right)\right] \\ &= \varepsilon_0 \cdot \left[h_0(\varepsilon)E_0 - h^*_{1,2}(\varepsilon)\cdot E_0 \cdot \left(\sum_{i=1}^{N} e_i \cdot (1-\exp -\frac{t}{a_\varepsilon \cdot \tau_i})\right)\right]\end{aligned} \tag{2.65}$$

Schapery (1969a) states that variations in the nonlinearity factors h_∞, h_1 and h_2 are due to third and higher order strain effects in the Helmholtz free energy, while changes in a_ε are due to similar strong strain influences in both entropy production and free energy. Further information can be found in (Schapery, 1966, 1969a,b).

The nonlinearity factor a_ε has a similar effect than the time-temperature-superposition shift factor a_T (see chapter 2.4.1) and has to be a strictly positive number ($a_\varepsilon > 0$). These displacement factors cause a change in the relaxation times and thus a horizontal displacement of the relaxation curve. $h_\infty(\varepsilon)$ resp. $h_0(\varepsilon)$ modify the time invariant part of the relaxation function E_∞ resp. E_0 and $h_{1,2}(\varepsilon)$ resp. $h^*_{1,2}(\varepsilon)$ affect the transient part of the relaxation function $\Delta E(t)$ or $\Delta E^*(t)$. The impact of the different nonlinearity factors in eq. (2.64) and eq. (2.65) is exemplary shown in Figure 2.33. Note that the y-axes are in linear scale for Figure 2.33 (a) and (c) for a better understanding of the h-factors. In linear y-scale, h_∞ resp. h_0 lead to a vertical shift of the Prony series, without changing the shape of the curve. $h^*_{1,2}$ resp. $h_{1,2}$ are influencing the height of the time-dependent increment / decrement in stiffness and hence are stretching or compressing the curve in y direction. In logarithmic scales, h_∞ resp. h_0 and $h^*_{1,2}$ resp. $h_{1,2}$ all lead to a change in the course of the relaxation curve. A solely vertical shift of the curve without change in shape is reached with $h_\infty = h^*_{1,2}$ resp. $h_0 = h_{1,2}$. Figure 2.33 also illustrates that $h_\infty \neq h_0$ and $h^*_{1,2} \neq h_{1,2}$.

Schapery (1969a) originally used a two-step strain input at several strain magnitudes (resp. a two step stress input, e.g. creep and creep-recovery, at several stress magnitudes) to uniquely determine all the model parameters. He states that for single-step loading it is not possible to determine the parameters h_1 and h_2 individually. Instead, only $h_{1,2}$ or $h^*_{1,2}$ can be found. The linear model parameters were obtained by evaluating the linear viscoelastic test results, and the nonlinear parameters were obtained by the test results in the nonlinear viscoelastic range. Schapery exemplary used a power functions (rather than a Prony series) to describe the relaxation function, but states that his formulations are not restricted to that approximation. (Note: When the power law is valid, only $h_{1,2}/a^n_\varepsilon$ or $h^*_{1,2}/a^n_\varepsilon$ wherein n is the power law exponent, can be determined in single-step input. It is not possible to provide unique values for the individual nonlinearity parameters.)

Later, Strganac et al. (1995) and Golden et al. (1996, 1999) used Dynamic-Mechanical-Thermal-Analysis (DMTA) measurement data in the linear range in combination with re-

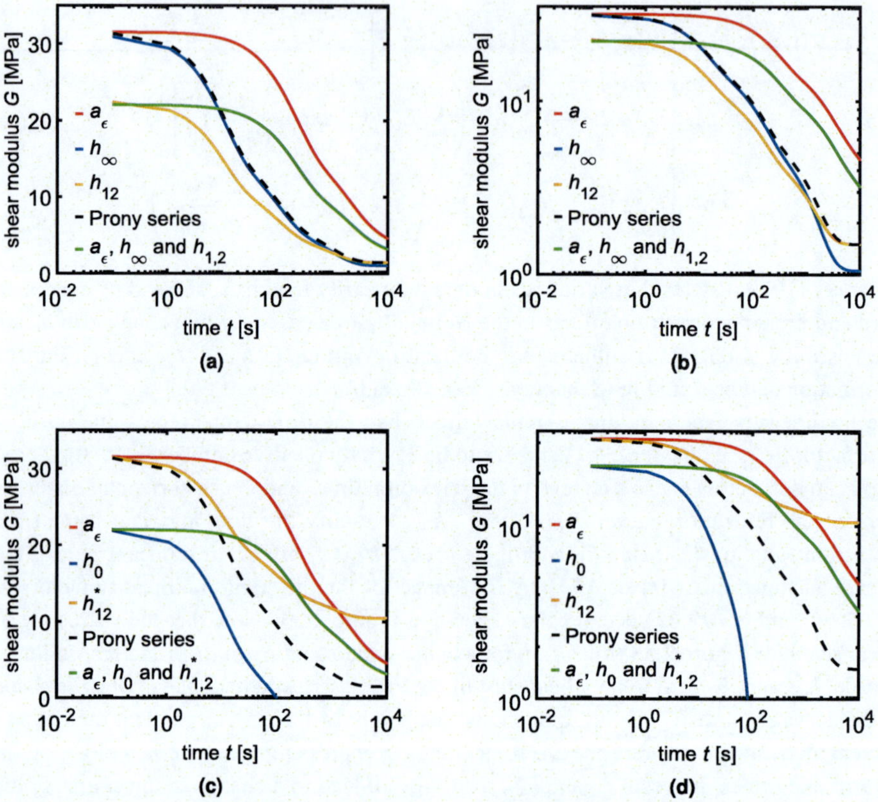

Figure 2.33 Effect of nonlinearity parameters in Schapery model, (a) a_ε = 20; h_∞ = 0.7 and $h_{1,2}$ = 0.7 with a linear y-scale, (b) a_ε = 20; h_∞ = 0.7 and $h_{1,2}$ = 0.7 with logarithmic y-scale (c) a_ε = 20; h_0 = 0.7 and $h^*_{1,2}$ = 0.7 with a linear y-scale, (d) a_ε = 20; h_0 = 0.7 and $h^*_{1,2}$ = 0.7 with logarithmic y-scale.

laxation/creep tests or DMTA measurements in nonlinear range by oscillating around different preload levels to determine the model parameters. Again, the power law is mostly used to describe the linear viscoelastic creep or relaxation function.

An additional note has to be made on σ and ε. Since large strains are considered now, a distinction has to be made between engineering values σ_0 and ε_0 and true values σ_t and ε_t. Engineering values refer to the initial, and the true sizes to the instantaneous geometries (see further chapter 3). Schapery (1969a) used engineering values in his paper. He states that this choice satisfies the condition that $\sigma \cdot d\varepsilon$ must be equal to an increment of work per unit initial volume, according to the underlying thermodynamic theory (Schapery (1966),

Schapery (1969b)). He also states, that by simply changing the notation other stress and strain pairs which satisfy the virtual work condition can be used.

2.4 Time-superposition principles

Engineering structures are often designed for a service life of 20-50 years. However, it is obvious that creep or relaxation tests on polymers can hardly be performed over such a long period of time. Time-superposition principles allow relatively short experiments to be performed, which can then be used to predict the long-time behaviour (Brinson et al., 2008). The best known example is the time-temperature-superposition principle, in which the individual creep or relaxation tests are carried out at different isothermal temperatures (Schwarzl, 2013). Apart from a time-temperature equivalence, there are also others, which can be helpful in the characterization of polymeric materials. These include, for example, time-crystallization, time-stress / strain, time-physical age or time-moisture superposition principles. In addition, different effects can be combined, e.g. time-temperature-structural relaxation, time-temperature-crystallinity or time-temperature-stress / strain-superposition (Brinson et al., 2008). In the following, the time-temperature-superposition principle, the time-crystallinity-superposition principle and the time-stress / strain-superposition principle will be discussed. General information on the time-physical age-superposition can be found e.g. in (Struik, 1987; Brinson et al., 2008; Naumann, 2012) and (Joshi, 2014). Kraus (2018) and Sackmann (2008) investigated the effect of physical aging for PVB interlayers. Information on the time-moisture-superposition principle for PVB can be found in (Botz, 2020). General information about the a combined time-temperature-structural relaxation-superposition principle can be found in (Narayanaswamy, 1971).

2.4.1 Time-temperature-superposition

It was mentioned before (see chapter 2.1), that the material behaviour of viscoelastic polymeric interlayers not only depends on the load duration but also on the prevailing temperature. Following Schwarzl (2013), there exists a thermodynamical correlation between temperature and time as relaxation processes are based on thermally activatable molecular movements and rearrangement processes. Hence, data taken at one temperature can be superimposed on data measured at another temperature by using a *time-temperature-superposition principle TTSP* (ISO 18437-6; Brinson et al., 2008; Schwarzl, 2013). To do so, the single isothermal relaxation functions (usually represented as $\log G$ vs. $\log t$) are horizontally shifted on the logarithmic time axis with a temperature dependent logarithmic shift factor $\log a_{\mathrm{T}}$ until a smooth *mastercurve* is created for a specific reference temperature T_{ref}. This mathematically corresponds to the multiplication of the measured times t with the temperature dependent shift factors a_{T}, see eq. (2.66).

$$\log a_{\mathrm{T}}(T, T_{\mathrm{ref}}) = \log t - \log t_{\mathrm{ref}} = \log \frac{t}{t_{\mathrm{ref}}}$$
$$a_{\mathrm{T}}(T, T_{\mathrm{ref}}) = \frac{t}{t_{\mathrm{ref}}} \tag{2.66}$$

According to Rouse Jr (1953), in addition to the horizontal shift a_{T}, a slight vertical shift b_{T}, see eq. 2.67 and Figure 2.34 (a), should be considered due to the density differences between the reference temperature T_{ref} and the actual temperature T. In many cases, however, this is so small that it can be neglected. Only for very large entropy regions the vertical shift factor has an influence (Dealy and Plazek, 2009; Brinson et al., 2008; Schwarzl, 2013). The product of the vertical shift factor and the shear relaxation function $G(t,T)$ is also known as the *reduced shear modulus* G_r, eq. (2.68). The relaxation mastercurve at T_{ref} is obtained by plotting $\log G_{\mathrm{r}} = \log(b_{\mathrm{T}} G(t))$ vs. $\log(t/a_{\mathrm{T}})$, see Figure 2.34 (b).

$$b_{\mathrm{T}} = \frac{G(T_{\mathrm{ref}}, \frac{t}{a_{\mathrm{T}}})}{G(T,t)} = \frac{\rho_{\mathrm{ref}} \cdot T_{\mathrm{ref}}}{\rho \cdot T} \tag{2.67}$$

$$G_{\mathrm{r}}(t,T) = b_{\mathrm{T}} \cdot G(T,t) = G(T_{\mathrm{ref}}, \frac{t}{a_{\mathrm{T}}}) \tag{2.68}$$

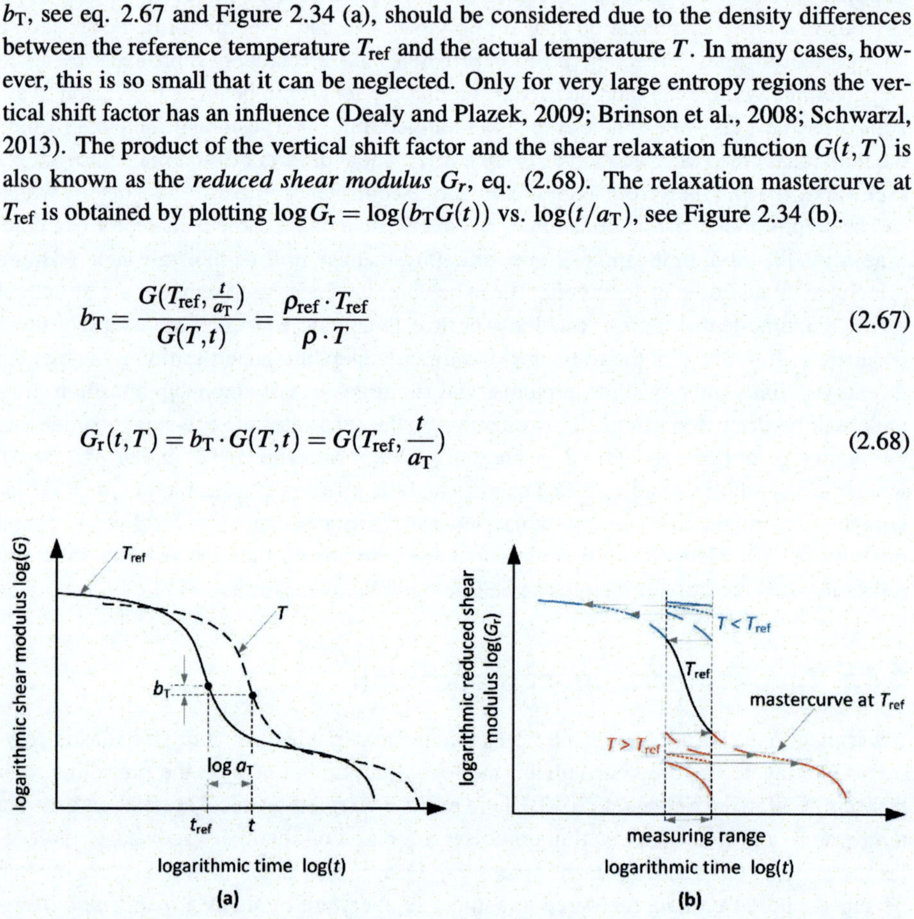

Figure 2.34 Time-temperature-superposition principle (TTSP) for thermorheologically simple materials (a) horizontal and vertical shift factors, (b) creation of mastercurve.

Once the mastercurve is created for a specific temperature T_{ref}, the time-dependent material behaviour can be approximated by a viscoelastic model, e.g. the generalized Maxwell model and mathematically described by e.g. a Prony series (see chapter 2.3). To determine the material behaviour at another temperature than the reference temperature T_{ref}, TTSP allows to determine the temperature-dependent shift factors and to shift the mastercurve / Prony series to temperatures $T \neq T_{\text{ref}}$ of interest according to eq. (2.69).

$$G(T,t) = b_{\text{T}} \cdot G(T_{\text{ref}}, a_{\text{T}} \cdot t_{\text{ref}}) \tag{2.69}$$

TTSP-Models

In literature different functional forms for the dependency of the horizontal shift factors a_{T} on temperature T exist, such as the *William-Landel-Ferry* (WLF) (Williams et al., 1955), the *Arrhenius equation* (Schwarzl, 2013; Brinson et al., 2008), *polynomials* (Z-70.3-236; Z-70.3-256) or the *arc-tangent* function (Woicke et al., 2004). In general, it should be noted that the TTSP cannot be used beyond their application limits, e.g. for temperatures outside the examined temperature range or the data cannot be extrapolated to an infinite load duration.

In the region of the glass transition and the entropy-elastic range, the molecular movement extend from small side chain movements to full rotations and position changes. Commonly, the empirically derived *William-Landel-Ferry (WLF)* shift function (eq. (2.70)), is used as functional form for the shift factors. The two material parameters C_1 and C_2 in eq. (2.70) have to be determined experimentally, wherein T is the considered temperature and T_{ref} is the reference temperature in [°C]. It is worth mentioning, that the parameters C_1 and C_2 are in value strongly dependent on the choice of T_{ref}. Eq. (2.70) represents a hyperbola in a $\log a_{\text{T}}$ vs. T plot, whose radius of curvature is determined by $C_1 \cdot C_2$. The parameters C_1 and C_2 represent the distance between the hyperbolic asymptotes and the reference temperature T_{ref} (Schwarzl, 2013).

$$\log a_{\text{T}}(T, T_{\text{ref}}) = -\frac{C_1(T - T_{\text{ref}})}{C_2 + T - T_{\text{ref}}} \tag{2.70}$$

Although eq. (2.70) was found empirically, it can be derived from the concept of free volume of a liquid (Brinson et al., 2008). Doolittle (1951) formulated eq. (2.71) to describe the viscosity of a liquid, wherein v is the total specific volume, v_{f} is the free volume and $f_{\text{v}} = v_{\text{f}}/v$ is the fractional free volume. At temperatures above the glass transition temperature T_{g}, the fractional free volume f_{v} is defined as $f_{\text{v}} = f_{\text{g}} + \Delta f_{\text{T}}$, wherein $f_{\text{g}} = 0.025$ and $\Delta f_{\text{T}} = \alpha_{\text{v}}(T - T_{\text{g}})$.

$$\begin{aligned}\ln\eta(T) &= \ln a + b\frac{v - v_f}{v_f} = \ln a + b\left(\frac{1}{f_v} - 1\right)\\ &= \ln a + b\left[\frac{1}{f_g + \alpha_v(T - T_g)} - 1\right]\end{aligned} \tag{2.71}$$

Converting eq. (2.71) to a base 10 logarithm, the shift factor in eq. (2.72) is obtained between the viscosity at an arbitrary temperature T and the glass transition temperature T_g.

$$\begin{aligned}\log\frac{\eta(T)}{\eta(T_g)} &= \frac{b}{2.303}\left[\frac{1}{f_g + \alpha_v(T - T_g)} - \frac{1}{f_g}\right] = -\frac{b}{2.303 f_g}\left[\frac{T - T_g}{\frac{f_g}{\alpha_v} + T - T_g}\right]\\ &\approx -\frac{C_1(T - T_g)}{C_2 + T - T_g} = \log a_T\end{aligned} \tag{2.72}$$

For temperatures below the glass transition region (energy-elastic range), it is assumed that a certain activation energy E_a is necessary to initiate side chain movements with a certain velocity (eq. (2.73)) and hence to start the relaxation process characterized by the relaxation time τ. The temperature dependent shift factors a_T can then be calculated according to eq. (2.74), which is often referred to as *Arrhenius* equation. E_a is the activation energy, which has to be determined experimentally, R the universal gas constant, T the considered temperature and T_{ref} the reference temperature in [K].

$$\ln\tau(T) = \ln A - \frac{E_a}{R \cdot T} \tag{2.73}$$

$$\begin{aligned}\log a_T(T, T_{ref}) &= \log\frac{\tau(T)}{\tau(T_{ref})} = -\frac{E_a}{2.303 \cdot R}\left(\frac{1}{T} - \frac{1}{T_{ref}}\right)\\ &= -0.434\frac{E_a}{R}\left(\frac{1}{T} - \frac{1}{T_{ref}}\right)\end{aligned} \tag{2.74}$$

If the mastercurve was created from isothermal curves at temperatures from both the energy-elastic and entropy-elastic range, a discontinuity at $T \approx T_g$ is evident in the $\log a_T$ vs. T plot. The shift factors at temperatures above this bend can be adjusted with a WLF and those below with the Arrhenius function. This is exemplary shown in Figure 2.35.
To avoid combining WLF and Arrhenius functions to approximate the temperature-dependent shift factors, alternatively a *polynomial* dependency (eq. (2.75)) can be used. Depending on the degree N of the polynomial, $N+1$ material parameters $P_i = 0, \ldots, N$ have

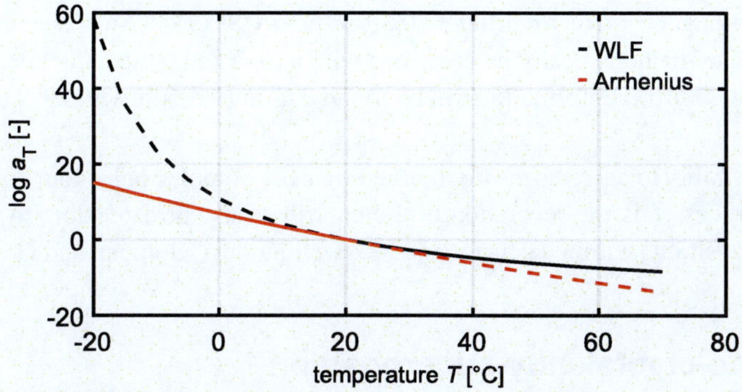

Figure 2.35 Shape of WLF and Arrhenius function. The WLF parameters are exemplary chosen to C_1 = 17 and C_2 = 51.6°C. For the Arrhenius equation $-0.434 \cdot E_a/R$ = 2800 K was used. The reference temperature was chosen to T_{ref} = 20°C. The solid lines mark the range of application, the dashed lines show the course of the respective function outside the application area.

to be determined experimentally. Here again, the value of the parameters P_i significantly dependent on the choice of T_{ref}.

$$\log a_T(T, T_{ref}) = P_0 + \sum_{i=1}^{N} P_i \cdot T^i \tag{2.75}$$

Alternatively to the polynomial function, Woicke et al. (2004) used a *arctan-function* to describe the shift parameter-temperature dependency over a temperature range that includes both, the energy elastic and the entropy elastic range.

Materials for which the above described TTSP is applicable for the entire relaxation curve, behave *thermorheologically simple* (ISO 18437-6; Schwarzl, 2013). This means that all the relaxation times of a relaxation test show the same dependency on temperature and a_T has the same value for the whole viscoelastic function. Graphically spoken, the relaxation curves have the exact same shape at different temperatures.

Thermorheologically complex material behaviour

Polymers can generally possess multiple distinct molecular transitions or crystalline phases with their own characteristic temperature dependencies. In such cases, *thermorheologically complex* material behaviour may occur, which means that the relaxation times τ at a certain temperature T_{ref} are not simply related to those at another temperature T by a constant ratio a_T, e.g. $a_{T_i} = \tau_{T_{ref}} / \tau_{T_i}$ is not a constant value for different i. Instead, the individual displacement factors can depend on time t as well as temperature T. Graphically spoken, in addition to a horizontal shift, the shape of the relaxation curves stretches or compresses along the x-axis for different temperatures (ISO 18437-6; Fesko and Tschoegl, 1971). This can mathematically be described with a time and temperature dependent shift factor $a_T(T,t)$. Further information can be found e.g. in (Urzhumtsev, 1974).

Another possibility to describe the thermorheologically complex behaviour due to different molecular processes is the separation of the contributions corresponding to the different molecular transitions within a generalized Maxwell model (Klompen and Govaert, 1999).

2.4.2 Time-crystallinity-superposition

According to ISO 18437-6, polymers with crystalline phases may behave thermorheologically complex. Different approaches for modelling thermorheological complexity due to crystallinity effects exist in literature.

In the 1950s, some authors e.g. Tobolsky and McLoughlin (1955); Nagamatsu et al. (1958) found that for polymers with crystalline structures, in addition to the temperature-dependent horizontal shift a_{T}, an important vertical shift along the y-axis becomes necessary in order to create a mastercurve from isothermal relaxation tests at different temperatures, see Figure 2.36. Nagamatsu et al. (1958) refers to eq. (2.76). This formula implies the vertical shift factor $b_{T,\chi}$ (eq. (2.77)), which is not only dependent on temperature T but also on crystallinity χ. The second term corresponds to a vertical shift due to temperature differences (density differences are neglected) and the last two terms correspond to a vertical shift due to crystallinity differences. G_r is the reduced modulus, χ is the absolute degree of crystallinity, χ_{ref} is the absolute degree of crystallinity at this reference temperature, T is the temperature of measurement and T_{ref} is the reference temperature in [K].

$$\log G_{\mathrm{r}} = \log G + \log \frac{T_{\mathrm{ref}}}{T} - \log\left(\frac{\chi}{1-\chi}\right) + \log\left(\frac{\chi_{\mathrm{ref}}}{1-\chi_{\mathrm{ref}}}\right) \tag{2.76}$$

$$b_{\mathrm{T},\chi} = \frac{G_{\mathrm{r}}}{G} = \frac{T_{\mathrm{ref}}}{T} \cdot \left(\frac{\chi_{\mathrm{ref}}}{1-\chi_{\mathrm{ref}}} \frac{1-\chi}{\chi}\right) \tag{2.77}$$

Urzhumtsev (1974) extended equation (2.77) by the vertical displacement fraction caused by the influence of a change in the molecular chain mobility. Unfortunately, the experimental evaluation of the vertical displacement factor including the shift due to density changes, crystallinity changes and mobility changes involves a number of technical difficulties. Hence Urzhumtsev (1974) suggests to progressively shift the $G(T,t)$ curves along

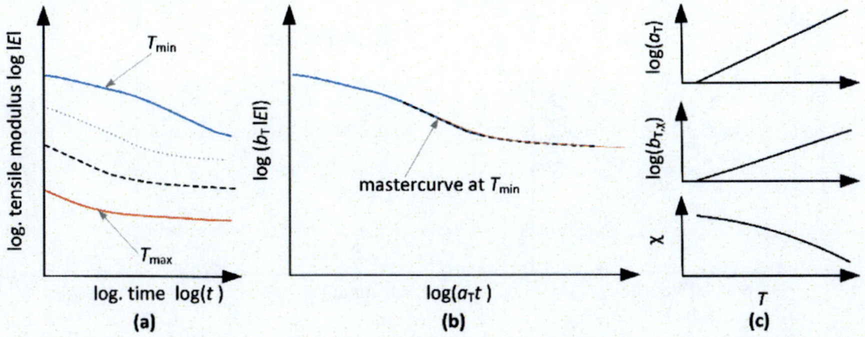

Figure 2.36 Time-crystallinity-superposition principle according to Nagamatsu et al. (1958). (a) Tensile relaxation functions at different temperatures T, (b) application of the TCSP to create a mastercurve at $T_{ref} = T_{min}$, (c) horizontal shift factor a_T, vertical shift factor $b_{T,\chi}$ and crystallinity χ in dependency of temperature.

the ordinate axis until for the overlapping part of the curves a_T does not depend on the time t.

Lamberti et al. (2007) performed isothermal DSC and rheological viscosity measurements at various temperatures and found that both the kinetic crystallization data and the rheological data (hardening function $\Gamma = \eta^*/\eta^*_{ref}$, also called normalized rheological function, wherein η^* is the complex viscosity) form linear relationships in a plot of log(t) vs. T. He showed that a unique relationship between the relative crystallinity α and hardening Γ exists, which means that the value of the hardening function is only dependent on the total amount of crystallinity, regardless how the crystallinity was produced (small amount of large crystallites, or large amount of small crystallites).

Analogously to the time-temperature-superposition principle (horizontal shift factor a_T), he introduced the *time-crystallinity-superposition principle TCSP* (horizontal shift factor a_χ) and created a mastercurve for the viscosity (more precisely complex viscosity $|\eta^*|$). The procedure is exemplary shown in Figure 2.37. He suggested eq. (2.78) with a reference relative degree of crystallinity of $\alpha_{ref} = 0$, a_1 and a_2 fitting parameters as functional form of the horizontal shift factor. This equation is referred to as *Lamberti-Peters-Titomanlio (LPT)*.

$$\log a_\chi(\alpha, \alpha_{ref}) = a_1 \cdot \alpha^{a_2} \tag{2.78}$$

In contrast to Lamberti et al. (2007), Roozemond et al. (2012) proposed a two-step shifting procedure, which consists of a horizontal shift factor a_χ and a vertical shift factor b_χ in order to create mastercurves for the shear modulus (more precisely storage modulus G' and loss modulus G'', see chapter 3) and viscosity (more precisely complex viscosity $|\eta^*|$)

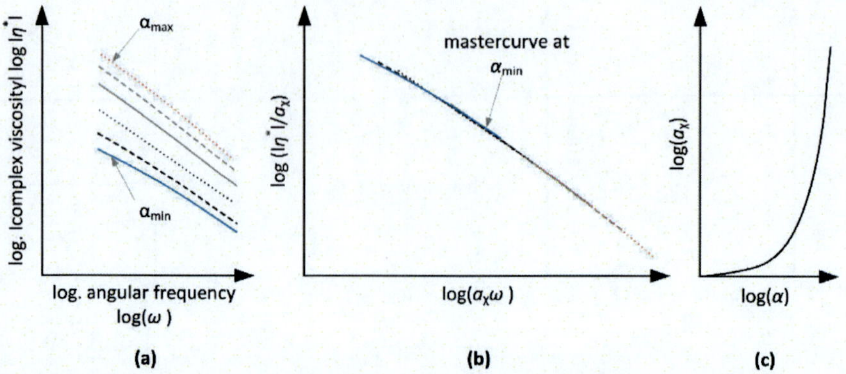

Figure 2.37 Time-crystallinity-superposition principle according to Lamberti et al. (2007). (a) Complex viscosity in dependency of the angular frequency for different relative degrees of crystallinities α at a constant temperature, (b) application of the TCSP to create a mastercurve at α_{ref}, note: $|\eta^*| = |G^*|/\omega$, (c) exemplary course of shift factors in dependency of relative crystallinity.

from isothermal rheological data at different degrees of relative crystallinity α. For the functional form of the horizontal shift factor, Roozemond et al. (2012) refers to (Lamberti et al., 2007) eq. (2.78) with $\alpha_{\text{ref}} = 0$, a_1 and a_2. For the vertical shift factor, he suggests eq. (2.79), wherein b_1 and b_2 are fitting parameters..

$$\log b_\chi(\alpha, \alpha_{\text{ref}}) = b_1 \cdot \alpha^2 - b_2 \cdot \alpha \tag{2.79}$$

Pantani et al. (2015) established simultaneous morphological and rheological measurements in order to investigate the effect of crystallinity on the viscoelastic parameters. Pantani et al. (2015) confirms in his literature review that the viscosity increases with increasing relative crystallinity and that this effect is normally described by a horizontal shift factor (like Lamberti et al. (2007) did). However, Pantani et al. (2015) says that the increase of the viscoelastic shear modulus (more precisely G' and G'', see chapter 3) by crystallinity is different from that of the viscosity. In fact, similar to Roozemond et al. (2012), two different shift factors are considered. The first one affects the modulus values (vertical shift) and the second one the relaxation times (horizontal shift). Contrary to Roozemond et al. (2012), Pantani et al. (2015) vertically shifted each Maxwell-mode individually.

Time-temperature-crystallinity-superposition TTCSP

According to Carotenuto et al. (2019a,b) and Minale et al. (2019), TTSP and TCSP can be superimposed to a *time-temperature-crystallinity-superposition TTCSP*, see eq. (2.80).

They have recently successfully applied this time-temperature-crystallinity-superposition principle on EVA-samples. Based on rheological temperature-frequency sweeps (where the sample was reheated before each new temperature level, thus zeroing the physical age), they generated shear modulus (more precisely storage modulus G', loss modulus G'', see chapter 3) and viscosity master curves (more precisely complex viscosity $|\eta^*|$). By means of rheological tests in form of temperature sweeps (cooling and heating cycle), and time sweeps at different temperatures, they were able to estimate the degree of crystallinity for each data point in the Dynamic-Mechanical-Thermal-Analysis (DMTA). To create the mastercurves, in a first step the temperature-dependent shift factor a_{T} was determined. Only the data points with relative crystallinities close to 0 or 1 were considered and shifted horizontally along the frequency-axis. In the second step the crystallinity-dependent shift factor was determined, by minimizing the horizontal distance between data having similar storage modulus values.

$$\log a_{\mathrm{T},\chi}(T,\alpha;T_{\mathrm{ref}},\alpha_{\mathrm{ref}}) = \log a_{\mathrm{T}} + \log a_{\chi} \tag{2.80}$$

According to the author of this thesis, the TTCSP is a special form of the TTSP for thermorheologically complex materials. Since the degree of crystallization is dependent on temperature and time, a crystallization degree-dependent shift factor is equivalent to a time- and temperature-dependent shift factor.

2.4.3 Time-stress / strain-superposition

In fractured laminated glass, the polymeric interlayer undergoes large deformations, so that nonlinear viscoelastic material models become necessary (see section 2.3.3). Schapery (1969a) introduced a strain dependent a_{ε} (resp. stress dependent a_{σ}) horizontal shift factor for the transient part of the material response, which is also referred to as *time-strain / stress-superposition principle TSSP*. Analogously to the TTSP, the TSSP reduces the experimental effort. The basic idea of the TSSP in strain controlled measurements (relaxation tests) is based on the agreement of the (transient) stresses for different times and strain levels. Accordingly, the (transient) stress at a specific strain level and test duration corresponds to the same (transient) stress at a different strain level and test duration. For stress controlled measurements (creep tests), the TSSP applies analogously (Brinson et al., 2008). The construction of a mastercurves from creep tests at different stress levels is shown exemplary in Figure 2.38.
Most literature in the nonlinear viscoelastic range deals with creep tests or stress controlled Dynamic-Mechanical-Thermal-Analysis (DMTA), so the stress-time displacement factor a_{σ} is used. Rand et al. (1996) performed creep tests at different stress levels and different temperatures. The higher the applied stress, the higher the transient part of the compliance. He created mastercurves for the lowest investigated stress level. The time-stress shift fac-

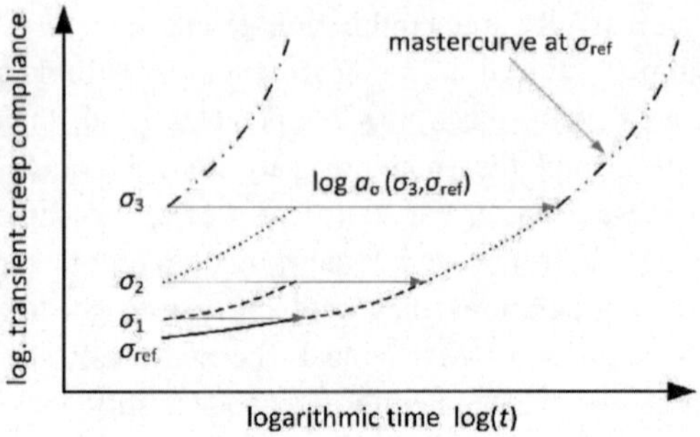

Figure 2.38 Time-stress-superposition, based on (Brinson et al., 2008). The creep tests were performed at a constant temperature T.

tor decreases with increasing stress levels, hence a higher stress level is shifted to higher load durations (since t/a_σ). The time-stress shift factors used to create the mastercurves at different temperatures differ from each other. Strganac et al. (1995) has investigated the nonlinear viscoelastic behaviour in oscillatory tests at different preloads (stress levels) and temperatures. In order to compare the results of different temperatures, the pre-stresses and the time-stress shift factors were expressed as a percentage of the temperature-dependent yield stress σ_y. The smallest stress ratio served as a reference for the modulus mastercurve (more precisely, the storage modulus mastercurve). Here the time-stress shift factor was 1. The higher the stress ratio, the greater the measured modulus. The course of the logarithmic time-stress shift factor $\log a_\sigma$ to the stress ratio σ/σ_y is linear. Klompen and Govaert (1999) used eq. (2.81) as functional form for the time-stress-shift factor, wherein σ_l is a yield stress dependent nonlinearity parameter. Golden et al. (1999) performed Dynamic-Mechanical-Thermal-Analysis to investigate the nonlinear viscoelastic material behaviour. The with increasing stress decreasing logarithmic stress-strain shift factor is approximated with a third order polynomial function, (eq. (2.82)).

$$a_\sigma = \frac{\sigma/\sigma_l}{\sinh(\sigma/\sigma_l)} \tag{2.81}$$

$$\log a_\sigma = P_3 \cdot \sigma^3 + P_2 \cdot \sigma^2 + P_1 \cdot \sigma + P_0 \tag{2.82}$$

Gacem et al. (2008) used time-dependent time-stress shift factors to predict the long term creep behaviour. The additional stress dependency stretches or compresses the curve along the x-axis. He performed creep tests with different stress levels and created a mastercurve for the highest stress level. He used eq. (2.83) to mathematically describe the time-stress shift factor, wherein a and b are stress independent constants and $f(\sigma)$ a stress dependent function.

$$a_{\sigma,\mathrm{t}} = f(\sigma) \cdot \left[1 - \frac{2}{\pi} \cdot \arctan\left(\exp a\frac{t-b}{t}\right)\right] \tag{2.83}$$

Time-temperature-stress-superposition TTSSP

Similar to eq. (2.80), a combined shift factor for temperature and stress / strain effects can be obtained via eq. (2.84) (Brinson et al., 2008). As shown in Figure 2.39, this enables to combine test data from different temperatures and strain levels into one mastercurve. One possible way, would be to first perform creep tests at constant stress and different temperatures and then create a mastercurve at a reference temperature (TTSP) (Figure 2.39 (a) → (c)). This process can then be repeated for other constant stress levels, resulting in a large number of mastercurves for T_{ref} and the different stress levels (Figure 2.39 (c) → (b)). A combined mastercurve is then created by subsequently applying the TSSP. The TTSP and TSSP can then be used to shift this combined mastercurve to different temperatures and stress levels (Figure 2.39 (b) → (d)). Similarly, it can be started with creep tests with different stresses and several constant temperature levels. By applying the TSSP, mastercurves are generated for the different temperature levels at a reference stress σ_{ref}, which in turn can serve as input for a subsequent TTSP. This corresponds to the path (b) → (d) → (a) → (c) in Figure 2.39.

$$\begin{aligned} \log a_{\mathrm{T},\varepsilon}(T,\varepsilon;T_{\mathrm{ref}},\varepsilon_{\mathrm{ref}}) &= \log a_{\mathrm{T}} + \log a_{\varepsilon} \\ \log a_{\mathrm{T},\sigma}(T,\sigma;T_{\mathrm{ref}},\sigma_{\mathrm{ref}}) &= \log a_{\mathrm{T}} + \log a_{\sigma} \end{aligned} \tag{2.84}$$

Kolařík and Pegoretti (2006) explained the validity of the TSSP with the free volume theory by adding a stress / strain dependent term to the expression of the fractional free volume F_{v}, which results in eq. (2.85), wherein $\Delta f_{\varepsilon} = (1-2\mu)\cdot\varepsilon(t)$ with μ corresponding to the Poisson's ratio. According to this, the free volume grows with increasing strain, which causes the relaxation process to progress more quickly. Hence, a higher strain has a similar effect than a higher temperature. This contradicts the results of Strganac et al. (1995).

$$f_{\mathrm{v}}(T,\varepsilon(t)) = f_{\mathrm{g}} + \Delta f_{\mathrm{T}} + \Delta f_{\varepsilon} \tag{2.85}$$

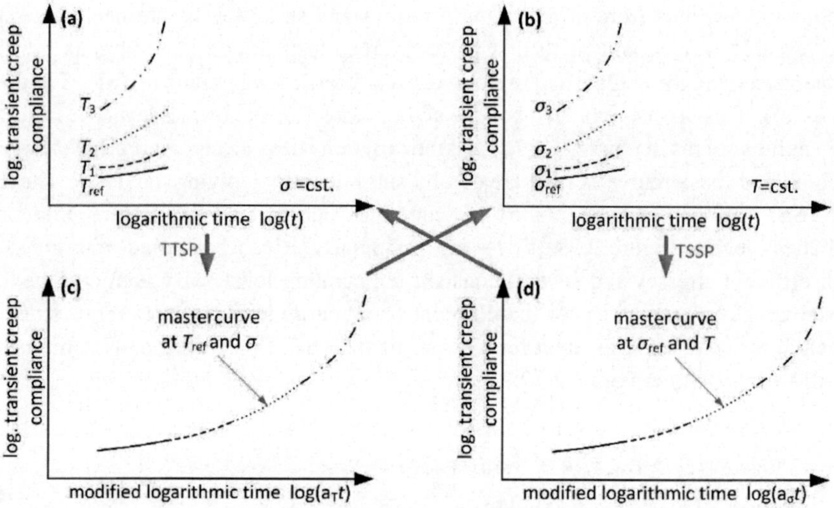

Figure 2.39 Time-temperature-stress-superposition, based on (Brinson et al., 2008). (a) Transient creep compliance at different temperatures T but constant stress level σ, (b) transient creep compliance at constant temperatures T but different stress levels σ, (c) application of TTSP for mastercurve creation at T_{ref} and σ, (d) application of TSSP for mastercurve creation at σ_{ref} and T.

3 General description of the experimental methods

This chapter first provides an overview of the materials studied. Subsequently, the general mode of operation, evaluation methodology and the specific experimental setup of the various experimental methods applied are presented. Finally, an overview table is provided showing which experiments were carried out on which materials. The detailed information on the test parameters and the results are given in chapters 4-7.

3.1 Overview of tested materials

The tests were carried out on the following materials:

- SG6000, also known as SentryGlas® XTRA™ and hereafter referred to as **SG**. This is the latest ionoplast (available since 2019) from the company Kuraray. The aim of developing the new ionoplast was to bring the lamination process closer to that of PVB without losing the enhanced structural properties of ionoplastics. Compared to the predecessor interlayer SG5000, the adhesion to the air side of the glass has been improved so that extra primers are no longer needed during lamination and it is no longer necessary to pay attention to the orientation of the glass panes. In addition, haze formation is less sensitive to the cooling rate in the autoclave, so that a high optical quality can be achieved even at a lower cooling rate. The shear relaxation modulus values are significantly higher than those of structural PVB, but lower than those of the predecessor ionoplast SG5000 (Schneider, 2019).
- EV200 DAYLIGHT, hereafter referred to as **cEVA**. This is a crosslinked ethylene vinyl acetate from the company Glaast with 28 wt%VAC and a gel rate of 94%. In addition, noncrosslinked EV200 DAYLIGHT, hereafter referred to as **EVA** was investigated for comparison with cEVA.
- Saflex® RB from Eastman, hereafter referred to as **Standard PVB**.
- Saflex® DG from Eastman, hereafter referred to as **Structural PVB**.
- Saflex® QS from Eastman, hereafter referred to as **Acoustic PVB**. The acoustic interlayer is a trilayer, whereby the very soft core layer is enclosed by stiffer cover layers.

M. Schuster, *Characterization of Laminated Safety Glass Interlayers*,
Mechanik, Werkstoffe und Konstruktion im Bauwesen 66,
https://doi.org/10.1007/978-3-658-39821-7_3

Sample preparation and storage

When characterizing the laminated glass interlayers, care must be taken to ensure that they are in a state similar to the installed state during the tests. SG and PVB, for example, are hygroscopic materials, which means that they readily bind moisture from the environment until the chemical potentials of the water in the atmosphere and the water in the interlayer are equal and equilibrium is reached. This changes the interlayer moisture content set at the manufacturing plant, which in turn affects the stiffness behaviour and adhesion. To ensure that the set interlayer moisture content does not change, the manufacturers give recommendations to the lamination sites for storing the interlayers, which will be presented here briefly.
For PVB rolls, a cool storage temperature with $T < 10°C$ is recommended. However, this is due to the problem that the individual layers of the roll adhere to each other at higher temperatures. Additional non-adherent layers in the roll eliminate this problem, so that in this case an average storage temperature of up to 30°C is possible. In case of open interlayer rolls or packages, the relative air humidity should be selected so that the interlayer moisture does not change the value set in the factory. For example, with a set interlayer humidity of 0.43% (usual for standard and structural PVB), a storage temperature of 20°C and a relative humidity of approx. 25%rH would meet the requirements (Eastman Chemical Company, 2003; Kuraray Trosifol World of Interlayers, 2012).
For SG, which is produced as sheets or as rolls, it is recommended to reseal the package airtight after removing the required amount of material in order to have a dry storage. The sealed packages can then be stored in room climate. In addition, the manufacturer recommends not to use interlayers with interlayer moisture higher than 0.2% (Kuraray Trosifol World of Interlayers, 2019).
EVA is not so sensitive to humidity, and can be stored at temperatures up to 30°C and relative humidities up to 80%rH, (Folienwerk Wolfen GmbH, 2020). However, EVA develops three-dimensional chemical crosslinking during the lamination process (EVA becomes cEVA). If the mechanical behaviour of the interlayer is to be tested in the installed state, interlayer that has already been crosslinked must be examined. For this purpose, the material should pass through the lamination process (so that the typical degree of crosslinking of approx. 95% occurs). An additional non-adhesive film can be placed between the interlayer and the glass sheets.
In addition, EVA (and cEVA) and SG exhibit semicrystalline regions. The number and size of the crystallites depends on the temperature history, which is why it is also recommended to consult the interlayer manufacturer if the melting temperature is exceeded (e.g. due to high temperatures required for sample installation).

Since only small test specimens were used for the purpose of the present work, all materials were delivered in approx. DIN A3 size. For each material, the sample preparation as well

as the sample storage was coordinated with the interlayer manufacturers.
The SG interlayers have gone through the lamination process before they were shipped airtight and protected from sunlight from the manufacturer to the TU Darmstadt (Figure 3.1 (a)). After sample preparation, the specimens were stored for several days (minimum. 5 days) at room temperature in sealed glass containers with dry beads (Figure 3.1 (b) and (c)). The *SteinerChemie* dry beads consist in addition to the high-moisture-binding silica gel, also of a color indicator that shows the moisture content by changing color from orange to green. By checking the visual appearance, it was possible to ensure dry storage and hence that the equilibrium moisture content is maintained within the interlayer. The duration of 5 days was recommended by the manufacturer in order to redevelop the secondary crystallites that melt for example during the lamination process or during the sample installation.

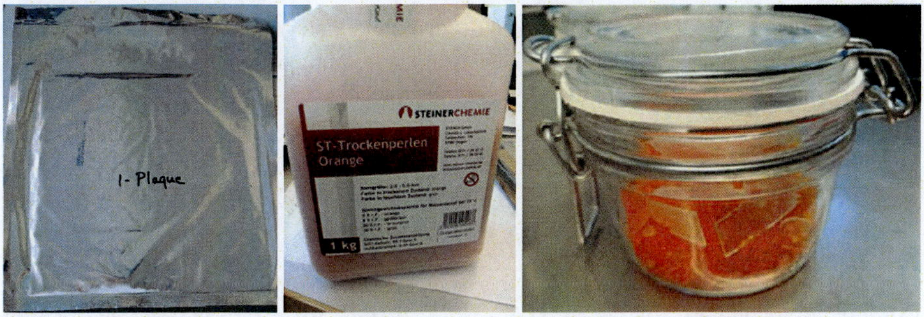

(a) (b) (c)

Figure 3.1 (a) Sealed bag with SG interlayers (b) SteinerChemie drying beads, (c) sample storage in sealed glass containers filled with highly moisture-binding silica gel.

As the EVA and cEVA interlayers are less sensitive to humidity, they were simply handed over to the TU Darmstadt in an envelope without further protection. Nevertheless, the samples were also stored in drying beads (at least 48 h) before being tested in order to guarantee reproducibility of the tests.

The PVB interlayers have not undergone the lamination process in advance. They were shipped light protected in aluminum bags that were closed with tapes (having a moisture barrier) to the TU Darmstadt. Preliminary tests have shown that considerable shrinkage of the interlayers therefore occurs at elevated temperatures (Figure 3.2).
For this reason, after consultation with the manufacturer, the interlayers were heated up to 100°C and left in the oven at this temperature for one hour before the test specimens were produced. Subsequent sample storage prior to the start of testing again corresponded to several days (at least 48 h) of dry storage in drying beads, unless otherwise specified. SG

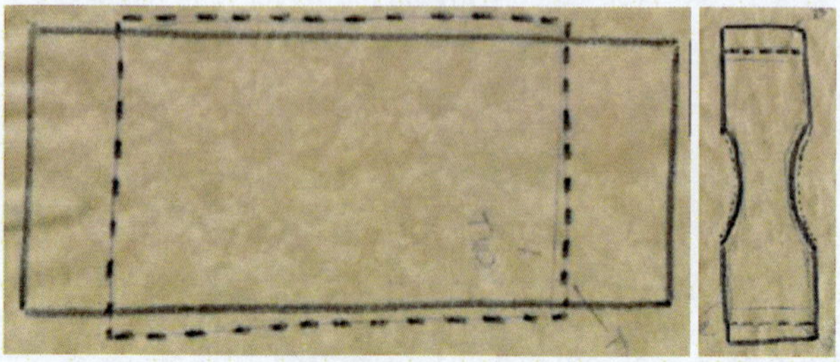

(a) (b)

Figure 3.2 Solid line corresponds to the initial geometry of PVB, dashed line corresponds to the geometry after heating to 100 °C (a) shrinkage of a rectangular sample, (b) shrinkage of uniaxial tensile rod.

and PVB interlayer remnants that were not needed immediately upon arrival at the institute were vacuum packed so that the equilibrium moisture content remained unchanged and samples could be taken from the interlayer again at a later time.

3.2 Differential Scanning Calorimetry

3.2.1 Functionality

Differential Scanning Calorimetry (DSC) is a technique for measuring the heat flow of samples as a function of temperature (in the case of non-isothermal measurement) or time (in the case of isothermal DSC-measurement). It allows the analysis of almost all energetic effects that occur in a solid or liquid during thermal treatment (NETZSCH-Gerätebau GmbH). This includes physical transformations, such as glass transition, melting and crystallization or chemical reactions, such as polymerization, cross-linking and vulcanization (EN ISO 11357-1; NETZSCH-Gerätebau GmbH).

In DSC measurements, the sample is subjected to a controlled temperature program together with a reference crucible in the same heater, see Figure 3.3. Due to the unequal heat capacities of the two crucibles, a temperature difference arises from which the difference between the heat flow (per time unit transferred heat quantity: $\Phi = \mathrm{d}Q/\mathrm{d}t$, equals the rate of enthalpy: $\mathrm{d}H/\mathrm{d}t$ assuming isobaric conditions) of the crucible with sample and the reference crucible is derived in dependence on temperature or time (EN ISO 11357-1). In literature, the values given for the heat flow $\Phi = \mathrm{d}Q/\mathrm{d}t$ or the enthalpy rate $\mathrm{d}H/\mathrm{d}t$ are often values normalized with the sample mass, which leads to the unit [J/g]. This is adopted in the present work. An exemplary DSC-curve, also called thermogram, is shown in Figure 3.5. In the case of non-isothermal measurements, a heating rate or cooling rate can be defined. The measured enthalpy behaves proportionally to the set temperature rate, which means that high rates result in larger enthalpy values. This is used to make small effects more visible in the thermogram. Small rates, on the other hand, are advantageous for separating effects that are close to each other (EN ISO 11357-2).

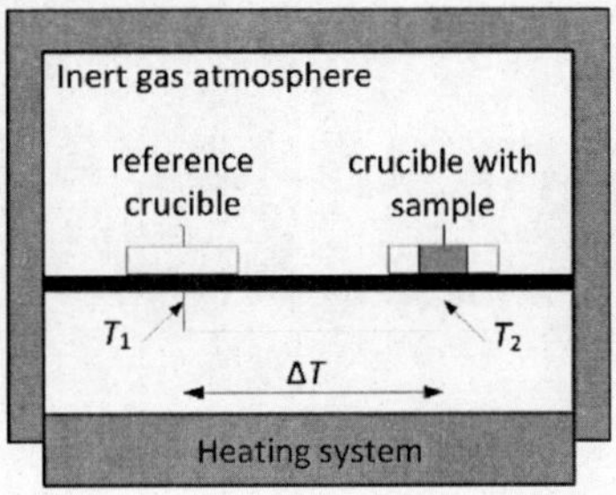

Figure 3.3 DSC functionality.

3.2.2 Test set-up

The DSC experiments were conducted in a DWDK Netzsch Type 214 Polyma DSC, which is shown in Figure 3.4 (a), and / or on a Mettler Toledo DSC. The used aluminum crucibles represented in Figure 3.4 (b) have a diameter of 5 mm. The samples were cut from the interlayer foils, which have been stored at defined conditions beforehand (see section 3.1). The samples masses were between 4 - 10 mg. A small hole was made in the lids of the crucibles to allow pressure equalization with the environment. The samples were placed in the crucibles so that they had good contact with the bottom of the crucible. Once this was ensured, the lids were cold-welded to the crucibles, weighed with a precision balance and placed in the DSC (right crucible in Figure 3.4 (b)). The empty reference crucible (with cold-welded lid) was also weighed and placed in the DSC (left crucible in Figure 3.4 (b)). The measurements were carried out in a nitrogen atmosphere.

3.2.3 Interpretation of the results

As shown in Figure 3.5 (a), different processes can be recognised by the course of the DSC-curve. In the following, the phenomena relevant to the materials studied are briefly presented.

Glass transition

As shown in Figure 3.5 (a), the glass transition can be recognized by a step in the DSC-curve. The step height correlates with the specific heat capacity. If the step is small in

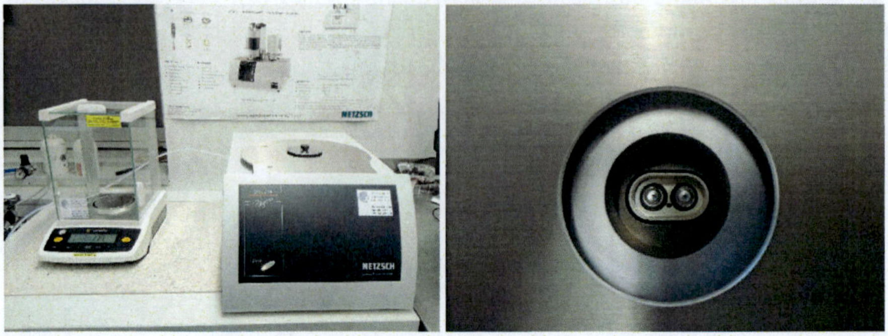

(a) (b)

Figure 3.4 Test set up DSC, (a) precision balance on the left and DWDK Netzsch Type 214 Polyma DSC on the right, (b) crucibles in the DSC.

a semicrystalline polymer, this means that the amorphous phase is small compared to the crystalline phase (Hammer, 2010).
The glass transition temperature T_g can be determined by different methods. In addition to the half-step method, there is also the turning-point and equal-surface method. Further information on this can be found in EN ISO 11357-2 and DIN 51810-2.
Since the glass transition is a kinetic phenomenon (see chapter 2.1.4), the measured value depends on the temperature rate, the thermo-mechanical history of the sample and the evaluation conditions (Hammer, 2010).

Enthaply relaxation

During the first heating process, an endothermic peak superimposed on the glass transition step is often observed in the DSC curve, which influences the evaluation of the glass transition. This effect is called enthalpy relaxation, and is caused by physical aging of the sample below T_g (see further chapter 2.1.4). The longer a sample is stored below the glass transition, the higher is the effect of physical aging and the greater is the enthalpy relaxation peak (see Figure 3.5 (b)). This gives important information about the thermal and mechanical history of a sample (storage temperature, storage time, cooling rate etc.), (Hammer, 2010).
The enthalpy relaxation peak can be eliminated, by first slightly heating the sample over the glass transition temperature T_g and then shock-cooling it before heating it for a second time.

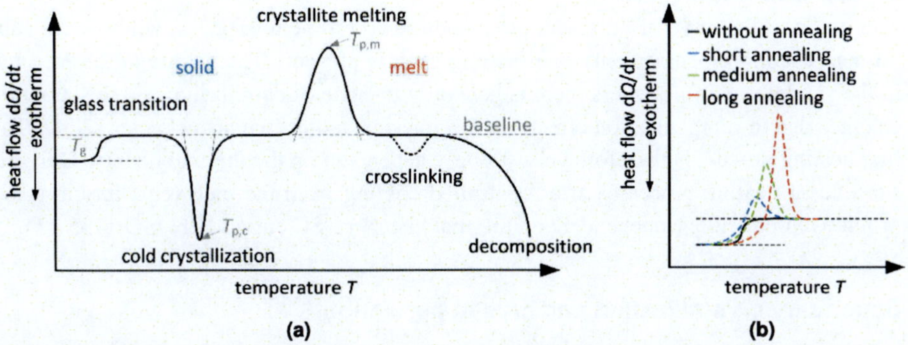

Figure 3.5 (a) thermogram showing glass transition, cold crystallization, crystallite melting, crosslinking and decomposition. The characteristic temperatures are indicated. Index g: glass transition, index c: crystallization, index m: melting, index p: peak, (b) enthalpy relaxation after different annealing times below T_g.

Crystallization

Crystallization is an exothermic process, which means that the heat flow in the test specimens crucible is lower than in the reference crucible. This corresponds to a conversion with heat release, which can be identified by an exothermic peak in the DSC curve. It usually occurs during cooling of a semicrystalline sample starting from the melt (Schuster et al., 2018).

The crystallization peak is characterized by different temperatures: the peak temperature $T_{p,c}$, the offset and onset temperatures, which are determined at the intersection points of the baseline and the tangent lines applied at the inflection point of the peak, and the temperatures at which the crystallization process starts and ends, which are defined as the temperatures at which first deviations of the curve from the baseline occur. Since crystallization is controlled by kinetic processes, the exothermic peak is dependent on the testing parameters, e.g. the cooling rate.

Melting

The melting of crystallites is identified by an endothermic peak in the DSC heating curve. Here, the heat flow in the test specimen crucible is greater than that in the reference crucible. Analogously to the crystallization peak, the melting peak is characterized by different temperatures, e.g. the peak temperature $T_{p,m}$, the onset and offset temperatures and the temperatures at which the melting process starts and ends. The melting process is a thermal transition (thermodynamically controlled). Hence, the endothermic peak is inde-

pendent on the heating rate.
The first heating curve of polymers often shows several melting peaks, while in the subsequent heating processes only one melting peak is present. This is due to the fact that in the solid phase of polymers there is generally no state of equilibrium and the structure changes due to reorganization effects during physical aging (see also chapter 2.1.4). The first heating process is therefore very strongly dependent on the thermal history. Only the subsequent heating processes after controlled cooling from the melt (equilibrium state) provide reproducible (independent on thermal history) DSC curves (EN ISO 11357-1).

Secondary crystallization and crystal perfection

Secondary crystallization or cold crystallization is recognized by an exothermic peak in the heating curve. It is caused when the sample had no time to crystallize in the previous cooling process. Since mobility increases above the glass transition temperature, small crystallites form at relatively low temperatures above the glass transition temperature, see also section 2.2. Crystal perfection or recrystallization from smaller to larger crystallites is difficult to detect with DSC, because endothermic melting and exothermic crystallization processes take place simultaneously, (Hammer, 2010).

Degree of crystallinity

The crystallinity of a semicrystalline sample is determined by the ratio of the current enthalpy of fusion to the enthalpy of fusion of the fully crystallized material, (EN ISO 11357-5; Schick, 2009). The obtained crystallinity is based on mass fraction (Mo and Zhang, 1995; Wakabayashi and Register, 2006; Marx and Cooper, 1974).
To determine the enthalpy of fusion, the specific heat flow $\Phi = \mathrm{d}Q/\mathrm{d}t$ (or the rate of enthalpy $\mathrm{d}H/\mathrm{d}t$) of the melt peak must be integrated, whereby the integration limits are defined by the intersection points with the peak baseline t_0 and t_{end} or T_0 and T_{end}:

$$\Delta H_{\mathrm{m}} = \int_{t_0}^{t_{\mathrm{end}}} \Phi \,\mathrm{d}t \tag{3.1}$$

$$\Delta H_{\mathrm{m}} = \frac{1}{C} \int_{T_0}^{T_{\mathrm{end}}} \Phi \,\mathrm{d}T \tag{3.2}$$

However, the construction of the peak baseline is not a simple task for polymers, as can be seen for example in Figure 3.6, in which different baselines (extension of the thermogram from the melt, horizontal baseline, linear baseline, sigmoidal baseline) have been examined for an EVA-based sample (Schick, 2009; Ledergerber et al., 2016).

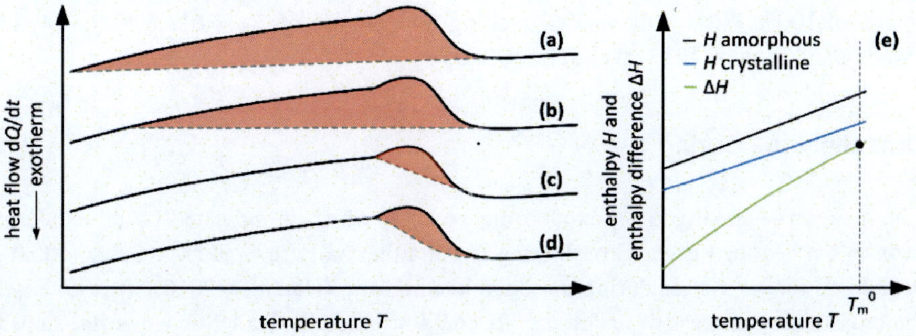

Figure 3.6 (a)-(d) Area (red patches) between DSC curve and baseline for different baseline definitions. (a) Extrapolation from the melt, (b) horizontal, (c) linear, (d) sigmoidal. (e) Enthalpy of amorphous state, crystalline state and difference between both (note: axis of ΔH differs from axis for H). The heat of fusion is determined at the equilibrium melting temperature T_{m}^{0}.

The melting region is often very broad and multiple melting peaks may be observed. In addition, many semicrystalline polymers contain a rigid amorphous phase (RAF) (see, chapter 2.2) whose devitrification must also be taken into account when establishing the baseline. Some authors, e.g. Schick (2009); Mathot and Pijpers (1983) and Reading et al. (2001), therefore recommend to create the baseline through extrapolation from the melt data. This recommendation will be followed during the evaluation of the test results.

Once the melting enthalpy ΔH_{m} has been determined, it must be divided by the heat of fusion of the pure crystallites $\Delta H_{\mathrm{m},100\%}$, see eq. (3.3). $\Delta H_{\mathrm{m},100\%}$ is commonly given for the equilibrium melting temperature.

$$\chi_{\mathrm{m}} = \frac{\Delta H_{\mathrm{m}}}{\Delta H_{\mathrm{m},100\%}} \tag{3.3}$$

However, this may be significantly higher than the temperature of the observed melting or crystallization peak, which is why the temperature dependency of the heat of fusion of the pure crystallites should be taken into account (see chapter 2.2). This can for example be done by using the enthalpy functions for amorphous and crystalline material provided in some data banks, e.g ATHAS data bank (Wunderlich, 2005), see Figure 3.6 (b), or from the heat capacity difference between crystal and melt, (Mathot and Pijpers, 1983; Schick, 2009).

EVA-based and ionoplastic interlayers contain PE-crystallites. Due to the lack of data, the literature value $\Delta H_{\mathrm{m},100\%\mathrm{PE}} = 293$ J/g is used for the heat of fusion of the pure PE-crystallites (Schubnell, 2001; Blaine, 2002). This value corresponds to the enthalpy of fusion of 100% PE at a theoretical melting temperature of T_m = 414.6 K (≈ 141.6°C) (Mathot and Pijpers, 1983; Wunderlich, 2005).

Crosslinking / Curing

Chemical cross-linking is an exothermic process, which is indicated by an exothermic peak in the heating curve. Crosslinking is not a thermally reversible process, so that the peak is no longer visible during a second heating run, (Hammer, 2010). It is noted, that the crosslinking process of uncured EVA could be studied with DSC. However, since the EVA in laminated glass is crosslinked, this curing process is not further investigated here.

Crystallization kinetics

The isothermal and / or non-isothermal DSC data can be used to determine the crystallization kinetics described in chapter 2.2. Málek (1992, 2000) assumes that the rate of the kinetic process $d\alpha/dt$ is proportional to the measured heat flow (which corresponds to the enthalpy rate for isobaric conditions) normalized per sample mass, Φ [W/g] see eq. (3.4). ΔH_c corresponds to the total enthalpy change associated with the crystallization process and can be calculated analogously to eq. (3.1) for isothermal and eq. (3.2) for non-isothermal conditions.

$$\frac{d\alpha}{dt} = \frac{\Phi}{\Delta H_c} \tag{3.4}$$

The fractional conversion / relative crystallinity is obtained by integrating eq. (3.4) and results in eq. (3.5) for isothermal conditions. For non-isothermal crystallization reactions, the rate expressions must be transformed by $d\alpha/dT = d\alpha/dt \cdot dt/dT$, with the cooling rate $C = dt/dT$. Hence, for non-isothermal conditions, eq. (3.6) is obtained (Menezes et al., 2019; Papageorgiou et al., 2010).

$$\alpha = \frac{1}{\Delta H_c} \int_0^t \Phi \, dt \tag{3.5}$$

$$\alpha = \frac{1}{\Delta H_c \cdot C} \int_{T_0}^{T} \Phi \, dT \tag{3.6}$$

The specific heat flow Φ equals eq. (3.7), which is obtained by combining eq. (3.4), eq. (2.32) and eq. (2.34). T_0 is defined as the temperature at which $\Phi(T_0) = 0$.

$$\Phi = \Delta H_c \cdot A \cdot \exp \frac{-E_a}{R \cdot T} \cdot f(\alpha) \tag{3.7}$$

The relative degree of crystallization determined in this way is then used to determine crystallization kinetics parameters according to different methods (see chapter 2.2, e.g. Avrami, Evan, Ozawa, Jeziorny-Avrami, Mo-Liu etc). If the verification of the $y(\alpha)$ and $z(\alpha)$ plots is desired, eq.(2.57) and eq. (2.58) can be expressed by eq.(3.8) for isothermal conditions and eq. (3.9) for nonisothermal conditions.

$$y(\alpha) = \Phi \quad \text{and} \quad z(\alpha) = \Phi \cdot t \tag{3.8}$$

$$y(\alpha) = \Phi \cdot \exp - \frac{E_a}{R \cdot T} \quad \text{and} \quad z(\alpha) = \Phi \cdot T^2 \tag{3.9}$$

3.3 Dynamic-Mechanical-Thermal-Analysis

In Dynamic-Mechanical-Thermal-Analysis (DMTA), the test sample is loaded by sinusoidal oscillations. A viscoelastic body shows a delayed response with a phase shift of $0° \leq \delta \leq 90°$, (Figure 3.7 (a)). The tests can be carried out both stress and strain controlled (Brinson et al., 2008; Schwarzl, 2013; Kraus et al., 2017). In the frequency domain, the Prony series results from the Fourier-transformation of eq. (2.60) to eq. (3.10). The real part $\mathrm{Re}(G^*)$ is called storage modulus $G'(\omega)$ and can be calculated via $G'(\omega) = |G^*(\omega)| \cdot \sin \delta$. The imaginary part $\mathrm{Im}(G^*)$ is called loss modulus $G''(\omega)$ and can be calculated via $G''(\omega) = |G^*(\omega)| \cdot \cos \delta$. The relation between the complex, storage and loss modulus is illustrated in Figure 3.7 (b) (Schwarzl, 2013).

$$\begin{aligned} G^*(\omega) &= G_\infty + \sum_{i=1}^{N} G_i \frac{\omega^2 \tau_i^2}{1+\omega^2 \tau_i^2} + i \sum_{i=1}^{N} G_i \frac{\omega \tau_i}{1+\omega^2 \tau_i^2} \\ &= G_0 \left(1 - \sum_{i=1}^{N} g_i + \sum_{i=1}^{N} g_i \frac{\omega^2 \tau_i^2}{1+\omega^2 \tau_i^2} + i \sum_{i=1}^{N} g_i \frac{\omega \tau_i}{1+\omega^2 \tau_i^2} \right) \end{aligned} \tag{3.10}$$

From both, the complex modulus $G^*(\omega) = G'(\omega) + \mathrm{i}G''(\omega)$ can be formed, which is a complex number. The amount of the complex modulus $|G^*(\omega)|$ is calculated from the ratio of the stress amplitude to the strain amplitude, $|G^*(\omega)| = \tau_0/\gamma_0$. The loss factor $\tan \delta = G''/G'$ indicates the ratio of the loss and storage modulus and is therefore a measure of the internal damping. Alternatively to the dependency on the angular frequency ω, the complex, storage and loss modulus can also be expressed as a function of frequency f: $|G^*(f)|$, $G'(f)$, $G''(f)$. The Prony parameters G_∞, G_0, G_i, g_i and τ_i are identical to those in the time domain (eq. (2.60)). In addition, there are some approximate equations to calculate $G(t)$ using $G'(\omega)$ and $G''(\omega)$. The simplest approximation corresponds to $G(t) \approx G'(1/\omega)$. Further approximations can be taken from e.g. (Schwarzl, 2013).

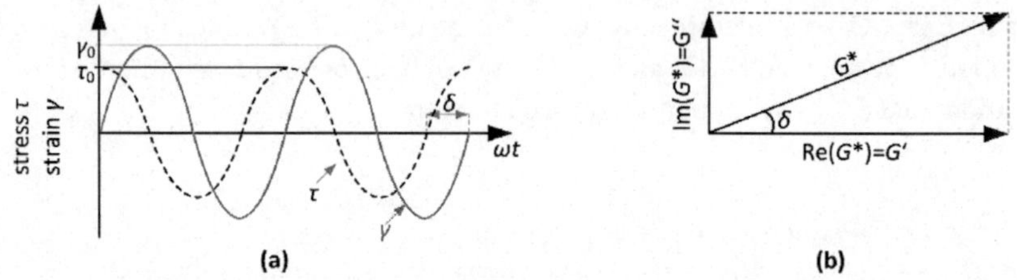

Figure 3.7 DMTA (a) functionality. γ_0 and τ_0 correspond to the strain and stress amplitude. δ indicates the phase shift, (b) definition of complex modulus G^*, storage modulus G' and loss modulus G''.

3.3.1 Different sweeps

DMTA are usually performed in such a way that two of the adjustable parameters (frequency, excitation amplitude and temperature) remain constant while the third varies. In some cases it can also be interesting to leave all three adjustable parameters constant and evaluate the change of the mechanical properties over time (NETZSCH-Gerätebau GmbH; EN ISO 6721-1). Amplitude sweeps, temperature sweeps, temperature-frequency sweeps, frequency-temperature sweeps, and time sweeps are presented in the following.

Amplitude sweeps

In amplitude sweeps (DMTA-AS) the test takes place at a constant temperature and at a constant frequency. The amplitude of the applied stress or strain is varied (Figure 3.8 (a)). DMTA-AS are usually performed in order to chose an excitation amplitude in the linear viscoelastic range for the temperature-frequency sweeps. DMTA-AS can be used to determine the linearity limits.

Temperature sweeps

In temperature sweeps (DMTA-TS), the temperature is varied with a defined heating or cooling rate. The frequency and excitation amplitude remain constant, (Figure 3.8 (b)). DMTA-TS are used to determine transition temperatures, e.g. the glass transition temperature T_g, the melting temperature T_m or the crystallization temperature T_c.

Temperature-frequency sweeps

Temperature-frequency sweeps (DMTA-TFS) consist of frequency sweeps performed at different isothermal temperatures with a constant excitation amplitude, (Figure 3.8 (c)). TFS are performed to determine the parameters of the Prony series and the TTSP, which characterize the time- and temperature-dependent material behaviour.

Frequency-temperature sweeps

In frequency-temperature sweeps (DMTA-FTS) a series of temperature sweeps is performed at different frequencies with a constant excitation amplitude, (Figure 3.8 (d)). Analogously to the DMTA-TFS, this measurements can be used to determine the parameters of the Prony series and the TTSP.

Time sweeps

During a time sweep (DMTA-tS), the frequency, excitation amplitude and temperature remain constant. DMTA-tS are performed in order to investigate kinetic processes at specific temperatures and frequencies.

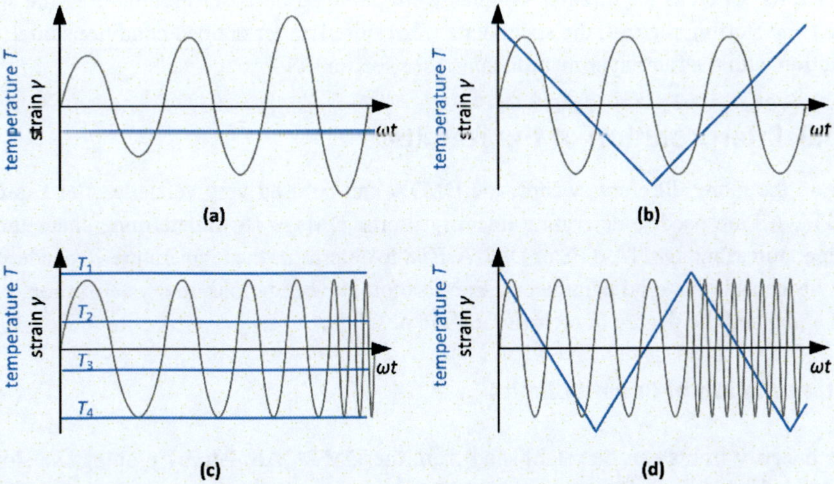

Figure 3.8 Different sweeps (a) amplitude sweep at a constant temperature, (b) cooling and heating temperature sweep, (c) temperature-frequency sweep, (d) frequency-temperature sweep.

3.3.2 Test set-up

The DMTA tests were performed on an Anton Paar MCR 302 rheometer. Depending on the sample type, one of the three following measuring systems was used:

- Solid state fixture (SRF), see Figure 3.9 (a). Here the tested samples are of rectangular shape ($\approx$ 50 mm x 10 mm, $\approx$ 1 mm thick).
- Plate-plate (PP) configuration, see Figure 3.9 (b). Here the tested sample is of circular shape (8 mm diameter, $\approx$ 1 mm thick).
- Textured plate-plate configuration (tPP). In contrast to the PP measuring system, the upper plate here has a textured surface. The sample is of circular shape (8 mm diameter, $\approx$ 1mm thick).

In all three cases, the lower part of the measuring system is fixed, while the upper part of the measuring system can oscillate. The temperature in the test chamber is controlled by a convection oven (CTD 450). Depending on the desired temperature range, this can be connected to a compressed air system or a nitrogen system. In the compressed air system, the air is brought to the desired temperature by a liquid connected with a thermostat (viscotherm 2) and then fed to the convection oven. If the nitrogen system is used, the tank with liquid nitrogen is connected to an evapuration unit (EVU200). This is then connected to the convection oven.
The rectangular samples (in the case of SRF measuring system) were cut from the interlayer foils, whereas the circular samples were punched out. Before mounting the specimens and starting the test, the specimens were stored under defined conditions after consultation with the interlayer manufacturer, see section 3.1.

3.3.3 Interpretation of the results

Due to the many different sweeps, the DMTA can be used very versatile. For example, DMTA-AS are used to determine linearity limits, DMTA-TS to determine characteristic temperatures, and DMTA-TFS or DMTA-FTS to determine the temperature-dependent linear viscoelastic material parameters. Furthermore, investigations on crystallization, physical aging and the degree of crosslinking are possible.

Determination of linearity limits

The linearity limits can be established from the DMTA-AS data by plotting the storage modulus G'-strain amplitude γ-curve and evaluating it graphically. A maximum deviation from the stiffness in the linear viscoelastic area has to be defined. Additionally it has to be defined, whether the deviation refers to the first measured value (smallest deformation) or to the average value of the measuring points in the linear viscoelastic range. In ideal tests

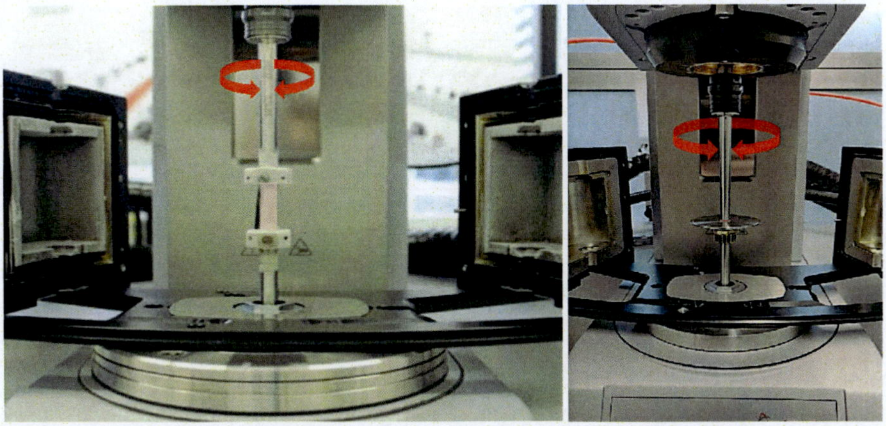

Figure 3.9 Test set up DMTA, (a) solid state fixture, (b) plate-plate configuration.

this makes no difference, since the linear viscoelastic modulus is constant. In reality, however, small measurement fluctuations occur, which can lead to different results depending on the evaluation method. While Ehrenstein (2011) defines the linearity limit as the strain level at which a deviation of 1% from the first measured value is reached, Bertilsson and Jansson (1975) write that the linearity limit occurs at 1% deviation from the average value in the linear viscoelastic range. Yannas (1972) evaluated the linearity limits in its work with a deviation of 3%. Mezger (2016) also writes that the deviation of the storage modulus in the linear viscoelastic area necessary to determine the linearity limit can be set at 3%, 5% or even 10%. Mezger (2016) recommends a deviation of 5% for polymers. In the standard for determination of the complex shear modulus and phase angle for bitumen and bituminous materials, the linearity limit is defined at a deviation of the storage or loss module of 5% from the initial value (EN 14770). In the second part of the standard for the testing of lubricants the limit of the linear viscoelastic range is determined if G' deviates by 10% from the respective plateau value (DIN 51810-2). In the subsequent studies on the linearity limit, a deviation of 3% from the mean value of the linear viscoelastic range G'_{LVE} was specified for determining the linearity limit γ_{lin}. The procedure is shown exemplary in Figure 3.10.

It should be noted that after exceeding the linearity limit, the measured modulus values do not have the same physical meaning as those measured at small strains. The stress response at high strains is no longer sinusoidal. Rather, a superposition of several harmonic oscillations is created, which can be analyzed by means of Fourier transformation. However, the rheometer does not recognize this change and generates modulus values in

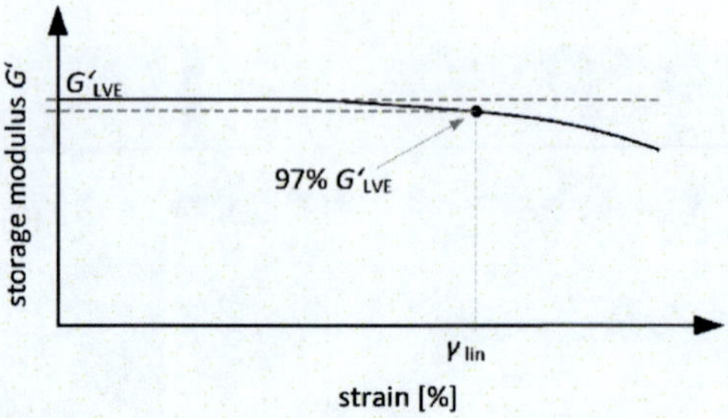

Figure 3.10 Determination of the linearity limits.

the same way as in the linear case by neglecting the higher harmonic contributions. The measured values, hence, are no more the true storage and loss modulus (Hyun et al., 2002).

Determination of characteristic temperatures

The glass transition temperature T_g can be determined with various approaches, which deliver significantly different results. Hence, it makes more sense to specify a glass transition range instead of one specific temperature. Figure 3.11 (a) shows the evaluation approaches standardized according to ISO 6721-11. Accordingly, T_g can be determined either as the maximum of the loss modulus $G''(T)$ or the loss factor $\tan\delta = G''/G'$, as a turning point or in the onset of the storage modulus $G'(T)$. Figure 3.11 (a) shows only the identification of the glass transition (α - relaxation), further relaxation processes can be determined analogously by, for example, the further local maxima of the loss factor $\tan\delta$ (Schuster et al., 2018).

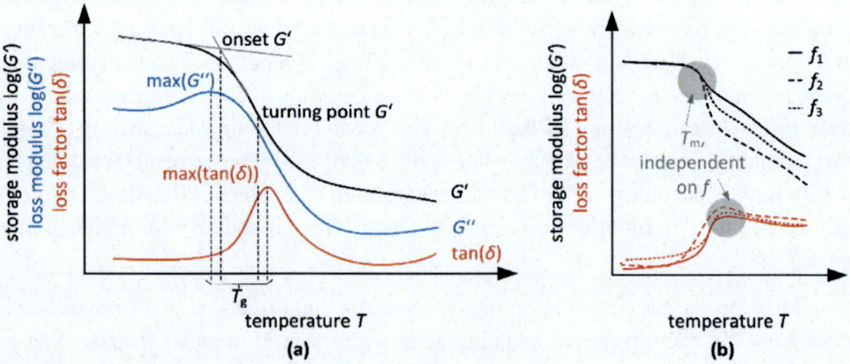

Figure 3.11 Determination of the glass transition temperature T_g (a) evaluation of data with one constant frequency, (b) evaluation of multi-frequency data.

Since the melting of crystallites is contrary to the glass transition temperature frequency-independent, variable-frequency DMTA allow the distinction of both phenomena (Wakabayashi and Register, 2006). This is shown in Figure 3.11 (b).

Determination of mastercurve, Prony parameters and TTSP parameters

The result of a DMTA-TFS are numerous isothermal frequency-dependent complex modulus (and storage modulus, loss modulus and loss factor) curves.
In order to create mastercurves for the complex modulus, storage modulus, loss modulus and loss factor at a reference temperature T_{ref}, the TTSP (section 2.4.1) is applied. It should be noted, however, that a division by a_{T} in the time domain corresponds to a multiplication by a_{T} in the frequency domain.
In order to determine the displacement factors a_{T} and b_{T}, three diagrams are first introduced. Those diagrams can be created with DMTA - temperature-frequency sweep data. They eliminate the effect of shifting individual curves and result in smooth curves (= temperature independent curves) if the simple TTSP holds and solely horizontal shifts a_{T} are necessary to create modulus-mastercurves (Stommel et al., 2018; Münstedt, 2016; Guedes, 2011; Dealy and Plazek, 2009). They include :

- *Van-Gurp Palmen Plot*, which depicts the loss angle δ or the loss factor $\tan\delta$ vs. the logarithmic amount of the complex modulus $\log|G^*|$.
- *Cole-Cole Plot*, which depicts the storage modulus G' vs. the loss modulus G''.
- *Wicket Plot*, which depicts the loss factor $\tan\delta$ vs. the storage modulus G'.

If the density function of the examined material is unknown (see eq. (2.67)), the vertical shift factor b_{T} can be determined independently on the horizontal shift factor a_{T} by creating a Van-Gurp Palmen plot (Van Gurp and Palmen, 1998). b_{T} corresponds to the horizontal shift required to get a smooth curve.
In order to determine a_{T} without making any assumption about b_{T}, the loss factor $\tan\delta$ can be used. Since, the loss factor in eq. (3.11) is a quantity that is not affected by b_{T} (Dealy and Plazek, 2009; Bae et al., 2011). If e.g. the van-Gurp-Palmen plot proves that $b_{\mathrm{T}} = 1$, the storage G' or complex modulus $|G^*|$ can be used to determine the horizontal shift factor a_{T}. The advantage of using G' or $|G^*|$ is that these functions (except for the entropy elastic range of elastomers) strictly increase with frequency.

$$\tan\delta = \frac{G''}{G'} = \frac{b_{\mathrm{T}}G''(a_{\mathrm{T}}\omega)}{b_{\mathrm{T}}G'(a_{\mathrm{T}}\omega)} \tag{3.11}$$

The horizontal shift of the individual $|G^*(f)|$ curves has been implemented in MATLAB®. For this purpose, initially only two isothermal curves (successive temperature steps) are considered, see Figure 3.12, wherein $|G^*{}_{T_i}(f)|$ serves as reference for $|G^*{}_{T_{(i+1)}}(f)|$. All measuring points within the overlapping range are created on the respective other curve. In Figure 3.12 this is shown exemplary for $|G^*{}_{f_{n-1}}|_{T_i}$. The shift factor is then determined with a least squares formulation of all the interrelated points within the overlapping area, see eq. (3.12), wherein $f_{\mathrm{ref},i}$ is the frequency of datapoint i on the reference curve in the overlapping range, f_i is the frequency of datapoint i on the other curve in the overlapping range. An alternative incremental shifting procedure is described by Bae et al. (2011).

$$\mathrm{lsq} = \sum_{i=1}^{N} (f_{\mathrm{ref},i} - a_{\mathrm{T}} f_i)^2 \tag{3.12}$$

The total shift factor for each curve relative to the global reference temperature T_{ref} curve is calculated via a multiplication of the incremental displacement factors a_{T}. Alternatively, the total logarithmic shift factors is obtained by a summation of the individual logarithmic displacement factors $\log a_{\mathrm{T}}$. Subsequently different TTSPs (see chapter 2.4.1) can be analyzed and the most suitable principle selected via least squares optimization in MATLAB®. It should be mentioned that the incremental methods are easy to implement but have the disadvantage that errors propagate incrementally. This might have a considerable influence on the frequency range / time range over which the mastercurve extends. Kraus (2018) showed a method to determine the shift factors globally.

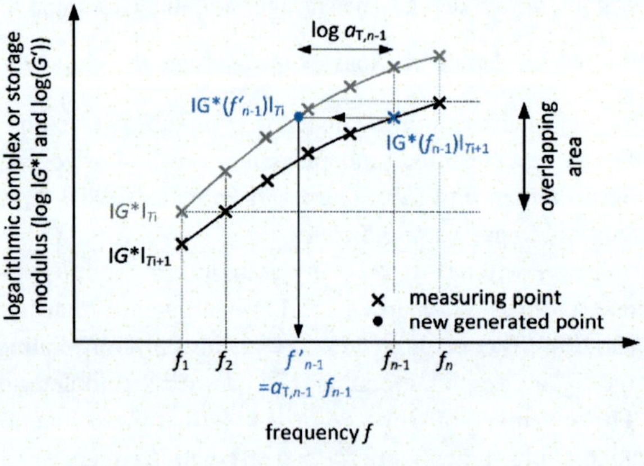

Figure 3.12 Mastercurving process for two successive temperature steps T_i and T_{i+1}.

In order to determine the Prony parameters, the measurement results for both the storage modulus G' and the loss modulus G'' are taken into account. g_i, G_∞ and τ_i of the generalized Maxwell model have to be identified in a way that they fit simultaneously the experimentally generated mastercurves of the storage and loss modulus. This represents a nonlinear optimization problem with $2N+1$ unknowns, where N is the number of spring-damper elements in the generalized Maxwell model. With the knowledge that one Maxwell element per measured frequency decade is sufficient to map the relaxation behaviour, the number of spring-damper elements N equals the number of examined frequency decays. Additionally, this allows to set a relaxation time τ_i to a certain positive value per decade, and hence to reduce the number of unknown parameters to $N+1$. In this thesis, the method developed in (Kuntsche, 2015) and (Kuntsche et al., 2015) was used, which is why it is only briefly presented here. Another procedure to determine the unknown parameters is described in (Kraus et al., 2017) and (Kraus, 2018). Kuntsche (2015) proposed a two-step procedure, which was implemented in MATLAB®. In a first step, a genetic algorithm is used to generate suitable start parameters, which are then transmitted in a global gradient based optimization tool. Since the values of the moduli spread over a wide range, the target function (see eq. (3.13), wherein $G'(\omega_j)$ and $G''(\omega_j)$ are the storage and loss modulus calculated with eq. (3.10), $\overline{G'}(\omega_j)$ and $\overline{G''}(\omega_j)$ are the storage and loss modulus out of experiment) is formed in the logarithmic region. This has the advantage that the high and low values have a similar influence on the error square. In addition, the loss modulus is weighted 10 times higher than the storage modulus in order to take the oscillation of the loss modulus into account (Göhler, 2010). If the Prony parameters are known, the shear modulus in the time domain is obtained by means of eq. (2.60).

$$f(g_1,...,g_N,G_\infty) = \sum_{j=1}^{J}\left(\left[\log G'(\omega_j) - \log\overline{G'}(\omega_j)\right]^2 + 10\left[\log G''(\omega_j) - \log\overline{G''}(\omega_j)\right]^2\right) \tag{3.13}$$

Investigation of crystallinity

Carotenuto and Minale investigated the nonisothermal crystallization kinetics with DMTA-TS. They state that the relative crystallinity α equals eq. (3.14), wherein G' is the measured storage modulus, G'_m is the storage modulus of the totally molten polymer, G'_s is the storage modulus at the end of the crystallization. This equation was also proposed by Khanna (1993), however there it was used to investigate isothermal crystallization.

$$\alpha = \frac{G' - G'_\mathrm{m}}{G'_\mathrm{s} - G'_\mathrm{m}} \tag{3.14}$$

Minale et al. (2019) estimated the storage modulus of the totally molten polymer G'_m and the storage modulus at the end of the crystallization G'_s from temperature sweeps, where the sample is first cooled from the melt with a specific cooling rate and frequency and then reheated to the melt. If the $G'(T)$ data of the cooling cycle and of the heating cycle are plotted in one diagram, a hysteresis in the melting or crystallization temperature range can be recognized. For G'_{m}, an Arrhenius equation (eq. 3.15) is fitted to that part of the cooling data, where the crystallization has not started yet. Analogously, for G'_{s}, a second Arrhenius equation (eq. 3.15) is fitted to that part of the heating data, where the melting has not started yet.

$$G'_{\mathrm{m}} = k_{\mathrm{m}} \cdot \exp \frac{E_{\mathrm{a,m}}}{RT}; \quad G'_{\mathrm{s}} = k_{\mathrm{s}} \cdot \exp \frac{E_{\mathrm{a,s}}}{RT} \tag{3.15}$$

In order to determine the Arrhenius parameters $(k_{\mathrm{m}}, E_{\mathrm{a,m}})$ for G'_{m}, in this work, the Arrhenius equation was transformed into a linearized form, see eq. (3.16). A linear regression of $\ln G'_{\mathrm{m}}$ vs. $1/T$ enables the determination of the searched parameters. For G'_{s} the procedure was analogous.

$$\ln G'_{\mathrm{m}} = \frac{E_{\mathrm{a,m}}}{R} \cdot \frac{1}{T} + \ln k_{\mathrm{m}} \tag{3.16}$$

Pogodina et al. (1999) proposed to define the relative crystallinity α based on the logarithmic storage modulus, see eq. (3.17).

$$\alpha = \frac{\log G' - \log G'_{\mathrm{m}}}{\log G'_{\mathrm{s}} - \log G'_{\mathrm{m}}} \tag{3.17}$$

Boutahar et al. (1998) suggested two other definitions. They recommend eq. (3.18) for suspensions with $\alpha < 0.4$. A suspension is a two-phase system consisting of a finely dispersed, non-soluble solid. The particles in the liquid have a size > 500 nm.

$$\alpha = 1 - \frac{G'_{\mathrm{m}}}{G'} \tag{3.18}$$

For colloids, they suggested eq. (3.19). A colloid is similar to a suspension, however the particle size is smaller (2 nm - 500 nm).

$$\alpha = \left(\frac{G' - G'_{\mathrm{m}}}{G'_{\mathrm{s}}} \right)^{\frac{1}{3}} \tag{3.19}$$

Eqs. (3.14), (3.17), (3.18) and (3.19) have been compared in (He et al., 2019) for isothermal crystallization at a specific temperature. The storage modulus of the totally molten polymer G'_{m} and the storage modulus at the end of the crystallization G'_{s} have been determined from time sweeps. For this purpose, the samples were first rapidly cooled down from the melt to the desired crystallization temperature. The time sweep was performed in a multi-frequency mode, resulting in many s-shaped storage modulus - time curves with different constant frequencies. G'_{m} corresponds to the value of G' at the beginning of the time sweep, and G'_{s} to the G' at infinite time (= which corresponds to a plateau value in the time sweeps).

Basu and Banerjee (1994) calculated the crystallinity of a polymer with DMTA temperature sweeps by applying eq. (3.20), wherein $\log E'_{T_{\mathrm{g}}} - \log E'_{T_{\mathrm{m}}}$ corresponds to the decrease of the logarithmic storage modulus between the glass transition temperature and the melting temperature. Basu and Banerjee (1994) assumed that during the glass transition, amorphous polymers exhibit a decrease in the storage modulus of $\Delta \log E' = 2.46$.

$$\chi = 1 - \frac{\log E'_{T_{\mathrm{g}}} - \log E'_{T_{\mathrm{m}}}}{2.46} \tag{3.20}$$

Based on the paper by Basu and Banerjee (1994), the temperature-dependent degree of crystallization χ_{T} is calculated in the present work by replacing $\log E'_{T_{\mathrm{g}}} - \log E'_{T_{\mathrm{m}}}$ in eq. (3.20) with $\log G'_{T_{\mathrm{g}}} - \log G'_{T}$ (between T_{g} and the temperature T of interest). The relative degree of crystallization α is subsequently determined by comparing the absolute degree of crystallization at T with the absolute degree of crystallization before the start of melting T_{m}.

Since in this thesis, the crystallinity is determined from non-isothermal data, all above mentioned methods are tested.

Investigation of physical aging, secondary crystallization, crystal perfection

Physical aging can be investigated in DMTA-ts at temperatures below T_{g}. Therefore the sample should first be heated to a temperature above T_{g} in order to delete the present (but unknown) physical aging effects, and subsequently be cooled down to the temperature of interest. This temperature is then held constant during the whole time sweep. The stiffness increase due to physical aging can be analyzed in storage modulus G'-time curves. Alternatively, DMTA-TS or DMTA-TFS can be performed on aged samples, which means that the tests are performed with samples that have been stored for different time durations at the temperature of interest after deleting the unknown physical age.
Similar investigations can be performed in order to analyze the effects of secondary crystallization or crystal perfection. The relevant temperature here is the melting T_{m} or crystallization T_{c} temperature.

Investigation of curing / chemical crosslinking

Chemical crosslinking can be detected in DMTA-TS, since during the process, the stiffness is increasing with higher temperatures. Chemical crosslinking also can be detected in DMTA-ts, where the sample is held for different times at a specific crosslinking temperature, since the stiffness is increasing with time. Chemical crosslinking is an non-reversible process, which means that in a subsequent TS, the material behaves stiffer than before crosslinking. Additionally, the heating and cooling curves are identical for an identical crosslinking degree (when absence of crystallites). However, since the EVA-based interlayers are crosslinked, no investigations on the curing process are performed here.

3.4 X-Ray Diffraction

3.4.1 Functionality

X-Ray Diffraction (XRD) is one of the most common methods for determining the degree of crystallization of semicrystalline polymers. In contrast to other methods, such as DSC or DMTA, where properties related to the degree of crystallization are measured, XRD directly measures the crystallinity with the three dimensional crystal structure (Lee et al., 1995).
Max van Laue discovered in 1912 that X-rays are diffracted at crystal lattices, when the radiation hits the crystal at a certain angle. William Henry and William Lawrence Bragg (father and son), explained this phenomena 1914 by assuming that crystals consists of different layers with plane distance d, which are occupied more or less dense by atoms, (Lechner et al., 2014; Suryanarayana and Norton, 2013).
When X-rays are incident on the atoms of a crystalline structure, they can interact with the electrons of the structure and cause them to oscillate. This re-radiates spherical waves (secondary waves), whose wavelengths and frequency are identical to the incident radiations (= coherent scattering). This can also be seen as a weak reflection of the X-rays at the individual lattice planes of the crystal, as shown in Figure 3.13 (a) (Suryanarayana and Norton, 2013).
Since the wavelength of X-rays (10^{-3} nm - 10 nm (Suryanarayana and Norton, 2013)) is in the same order than the distance between the crystal planes d (unknown), those weak reflections of various planes can interfere with each other, whereby an intensity maximum (increase of wave magnitude) is only reached when the path differences are integer multiples of the wavelength. This condition is described by Bragg's eq. (3.21), wherein λ is the wavelength of the incident beam and $z = 1, 2, 3, ...$ This is exemplary shown in Figure 3.13 (b). Θ is called Braggs or glazing angle (measured). For path differences that are integer multiples of half wavelengths, destructive interference occurs and both waves cancel each other out.

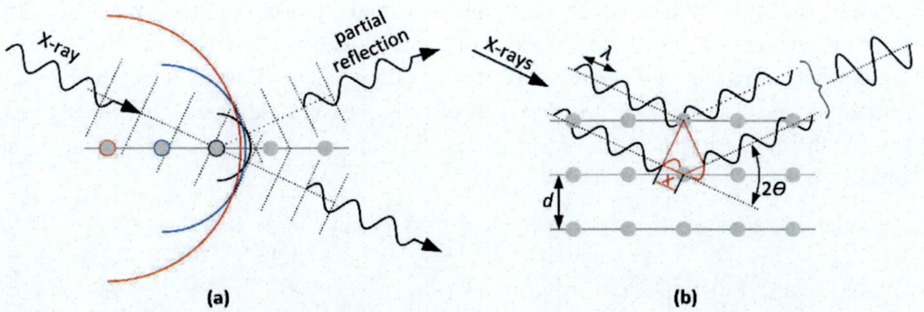

Figure 3.13 XRD functionality (a) X-rays interacting with the atoms of a crystal plane. Each atom hit by the X-ray is the source of a new spherical wave. Red: wave front of first hit atom, blue: wave front of second hit atom, black: wave front of third hit atom. This creates the impression that some of the X-rays are reflected at the grid plane and another part can penetrate to lower grid planes, (b) X-rays with identical wavelengths λ and phase hit a crystal lattice with plane distance d and are scattered off two different atoms within the lattice. The diffracted beams are in phase and the scattered waves interfere constructively. The path difference between the scattered waves equals $2x$, with $x = d \cdot \sin\Theta$.

$$2 \cdot d \sin\Theta = z \cdot \lambda \tag{3.21}$$

Crystals are three dimensional structures. Hence, multiple sets of equally spaced parallel planes exist. Those planes are typically defined by so called Miller indices (Suryanarayana and Norton, 2013). They represent vectors that indicate the orientation of the considered plane in the three dimensional coordinate system (x, y, z) by indicating for each direction the reciprocal of the intercept. The *Miller indices* $(1, 1, 2)$ exemplary represent a plane, that intercepts the coordinate system at $x = 1$, $y = 1$ and $z = 0.5$.
For each set of equally spaced parallel planes, there exists one angle 2Θ that fulfills Bragg's law and because the incident beam, the normal to the reflection plane and the diffracted beam are always co-planar, the crystal structure (e.g. size of elementary cell) can be determined from the diffraction pattern.

The most popular XRD technique is the *Debye-Scherer Method* (Figure 3.14 (a)), in which the sample to be analyzed is either pulverized and pressed into the form of a rod or analyzed as foil. The rod is placed in the middle of a circular cylindrical bent film (detector) and is vertically irradiated with monochromatic X-rays (Lechner et al., 2014). The X-rays diffracted at the same lattice planes (with the same interplanar distance d) lie on a cone whose axis coincides with the direction of the incident ray. Different planes generate diffraction cones with different angles. The intersection lines of the cone shells with the film result in the Debye-Scherrer diagrams. A slice through those concentric rings results in the typical diffractogram, with intensity on the y-axis and 2Θ on the x-axis. The peak maximums are called Bragg reflections (Suryanarayana and Norton, 2013). Modern powder diffraction instruments do not need photographic films to detect the cones of diffraction. Instead, point or line detectors directly measure the intensities. Since the X-ray and the cylindrical bent film resp. detector are on opposites sides of the sample, one speaks from transmission geometry.

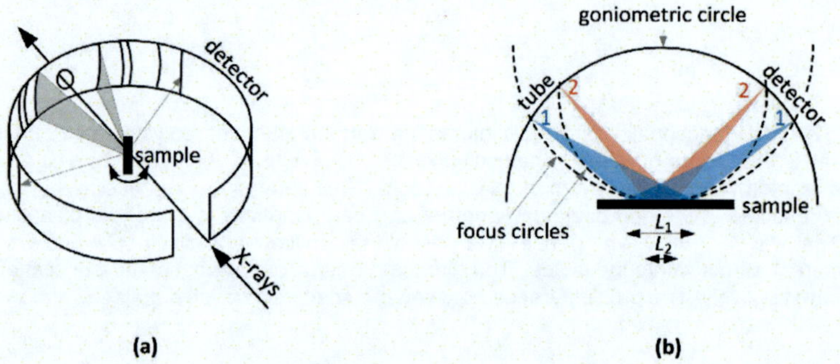

Figure 3.14 XRD configurations (a) Debye-Scherer, (b) Bragg-Brentano.

In the *Bragg-Brentano Parafocusing Geometry* (Figure 3.14 (b)), the X-ray tube, sample and detector are arranged in a focusing circle. The radius of the focusing circle decreases with increasing Θ (Spieß et al., 2005).
Different configurations are possible: In the Θ-2Θ mode, the X-ray source remains stationary, the sample is tilted at Θ and the detector rotates 2Θ around the sample. In the Θ-Θ geometry, the sample remains at a fixed angle and the X-ray source and the detector move at angle Θ (Bauch and Rosenkranz, 2017).
Since the X-ray and the detector are on the same side of the sample, one speaks from refection geometry.
It should be noted, that semicrystalline polymers have no real crystal lattice. However, the crystallites represent certain ordered regions so that XRD is applicable. The samples are not pulverized (Lechner et al., 2014).

3.4.2 Test set-up

The XRD experiments were conducted in a Bruker D2 PHASER (Bragg Brentano geometry), see Figure 3.15 (a), with CuKα radiation. Several layers of foil were placed in an amorphous sample carrier. The number of layers (≈ 6) was determined in a way that the top layer was flush with the specimen holder, see Figure 3.15 (b). The samples were cut from the interlayer foils, which have been stored at defined conditions beforehand.

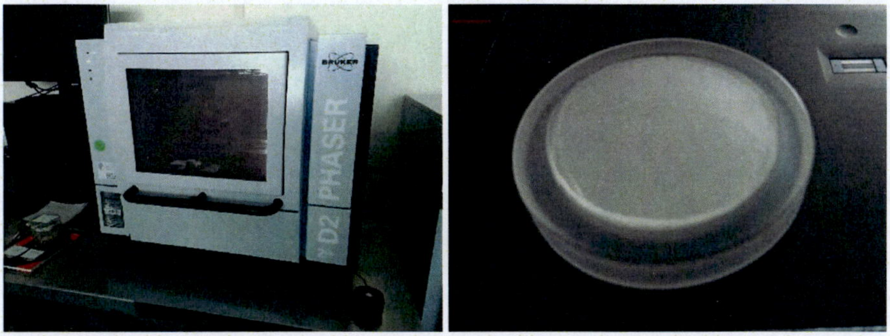

(a) (b)

Figure 3.15 (a) Test set up XRD, (b) sample in sample holder.

3.4.3 Interpretation of the results

XRD to polymers can be used to identify the polymer, the polymer state (highly crystalline, semicrystalline or amorphous), the crystallinity and the crystallite size. In the context of this work the diffractograms are used exclusively to determine the polymer state and the degree of crystallization in the interlayers.

Polymer state and polymer identification

Different polymer states have specific and characteristic appearance in the diffractogram. While highly crystalline polymers show sharp and well defined Bragg reflections, semicrystalline polymers show broader ones and amorphous polymers show very broad halos. To identify the polymer, the maximum values of the reflections have to be evaluated in terms of their position and intensity (Schwarzl, 2013).

Crystallinity

The crystallinity in terms of mass fraction χ can be determined by evaluating the reflections of the diffractograms with respect to their area, see eq. (3.22) (Elias, 2009; Lee et al., 1995; Mo and Zhang, 1995).

$$\chi = \frac{A_c}{A_c + K_x \cdot A_a} \tag{3.22}$$

wherein A_c and A_a are the integrated intensities scattered over a suitable angular interval by the crystalline and amorphous phases of the sample under investigation. K_x is a calibration constant, that corrects the signal intensity of the amorphous phase to the crystalline phase, since they have different densities and hence differences in scattering intensities (Stadler et al., 2009). However, the approximation of $K_x \approx 1$ is used here (Matthews et al., 1949; Kamath and Wakefield, 1965).

The shape of Bragg reflections can be described by different functional forms, see Figure 3.16. Common ones are the Gaussian function (eq. (3.23)), which has a rounded maximum and no tails and the Lorentzian function (eq. (3.24)), which is sharp at its maximum and has long tails (Figure 3.16 (a)). The characteristic values of the functions are shown in Figure 3.16 (b). y_0 corresponds to the offset baseline, x_c to the center, w to the width and A to the area. The width of the Gaussian function is calculated with $w = \mathrm{FWHM}/\sqrt{\ln 4}$, wherein FWHM is the full width at half maximum. The width of the Lorentzian function simply corresponds to FWHM. Overlapping reflections must be deconvoluted. For this purpose, the *Peak Fitting Module* in *Origin 8.5.1* (OriginLab®) is used.

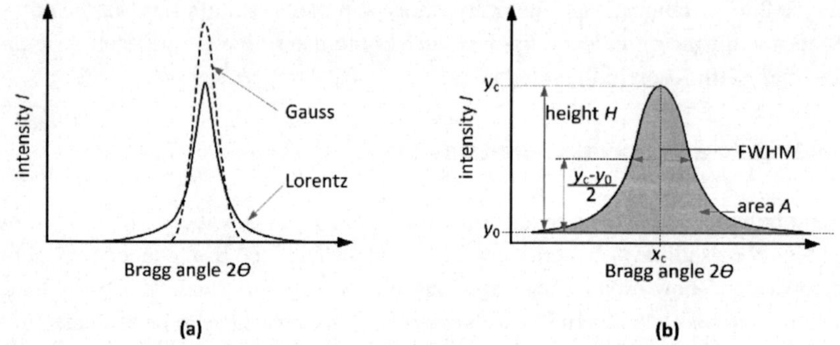

Figure 3.16 (a) Gaussian and Lorentzian peak approximations, (b) peak parameters.

$$y = y_0 + \frac{A}{w \cdot \sqrt{\pi/2}} \cdot \exp\left[-2 \cdot \left(\frac{x - x_c}{w}\right)^2\right] \tag{3.23}$$

$$y = y_0 + 2 \cdot \frac{A}{\pi} \cdot \left(\frac{w}{4 \cdot (x - x_c)^2 + w^2}\right) \tag{3.24}$$

3.5 Relaxation tests in uniaxial tensile mode

3.5.1 Functionality

In the uniaxial tensile relaxation test different specimen elongations (which correspond to different strain levels) are applied at a constant rate of 400 mm/min and then kept constant for a duration of 1 - 2 hours. Thereby the time-dependent stress reduction at approximate constant strain is investigated.

3.5.2 Test set-up

The uniaxial tensile relaxation tests were carried out on the Zwick Roell Z050 universal testing machine with hydraulic jaws. In order to control the climate during the tests, a climatic chamber was used to enclose the test area, see Figure 3.17 (a) and (b). The force was recorded using the load cell integrated in the Zwick (50 kN, 0.2% relative uncertainty, class 0.5 according to DIN EN ISO 7500-1-2016). Since slippage may occur, the strain of the specimen was measured during the test using digital image correlation. The Vic-Gauge-2D camera system from isi-sys GmbH was used for this purpose. The climatic chamber has only a narrow window, so only one camera could be used. It had to be aligned so that it was perpendicular to the specimen, see Figure 3.17 (a) and (b).

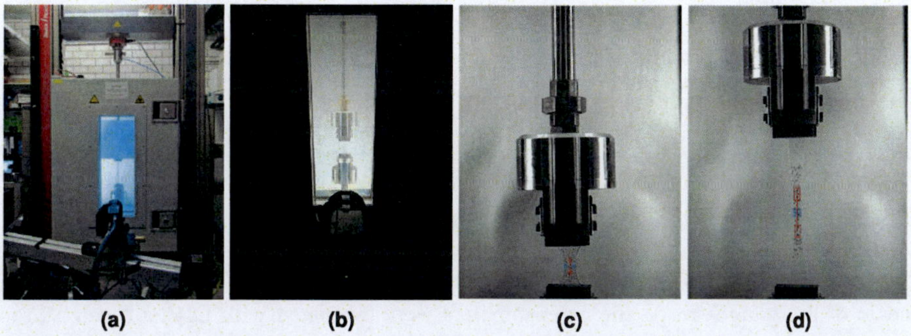

Figure 3.17 (a) Test set-up of the uniaxial relaxation test after installation of the sample, (b) test set-up during test (ambient light off), (c) sample with gauges before start of the test, (d) sample with gauges during the relaxation test.

Becker tensile rods, as depicted in Figure 3.18 (a), were used as sample geometry, Becker (2009). They were punched from the interlayer foils (after defined storage in consultation with the respective interlayer manufacturer; this included 1 hour at T = 100°C in order to prevent shrinkage). The Becker tensile rods were stored for approx. 12 h in the preconditioned climatic chamber. Immediately before the start of the test, the thickness of the specimens was measured at three different points using a digital micrometer and the grey value pattern was applied manually with a black pencil *Staedler permanent, Lumocolor, thickness S*. After clamping the sample into the hydraulic jaws, four virtual gauges (2 horizontally next to each other for measuring transverse strain and 2 vertically on top of each other for measuring longitudinal strain) were then placed on the grey value pattern using Vic-Gauge-2D software and tracked over the duration of the experiment, see Figure 3.17 (c) and (d). In order to obtain good lighting conditions, LED light chains were installed in the climatic chamber.

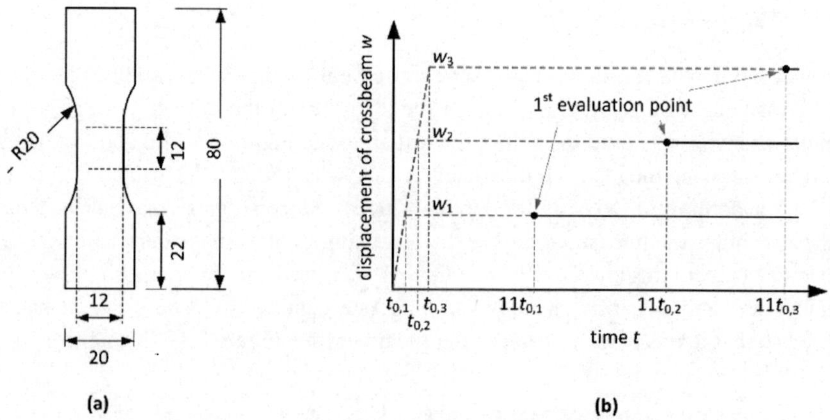

Figure 3.18 (a) Becker tensile rod, dimensions in [mm], (b) relaxation tests at different displacement levels. The first evaluation point is marked.

3.5.3 Interpretation of the results

The force decrement was measured during the relaxation tests. Additionally, the displacements of the gauges resp. the extensometers were evaluated with DIC software. From these data and the previously measured geometry, technical and true stresses as well as true strains (logarithmic strains, Hencky strains) and the Poisson´s ratio were evaluated. Especially for large deformations (i.e. when the linearity limits are exceeded), it is necessary to distinguish between true and technical stresses and strains. Technical values refer to the initial cross section, see eq. (3.25), wherein ε_0 and σ_0 are the technical strain and stress, Δl is the longitudinal change in length, l_0 is the initial length, F is the force and A_0 is the initial cross-section. As a result of increasing strains, however, there is reduction of the cross-sectional area. Hence, the true sizes (current cross section) should be considered, see eq. (3.26), wherein ε_t and σ_t are the true strain and stress, A is the current cross-section and μ is the Poisson's ratio, which can be obtained with eq. (3.27).

$$\varepsilon_0 = \frac{\Delta l}{l_0} \quad \sigma_0 = \frac{F}{A_0} \tag{3.25}$$

$$\varepsilon_t = \ln(1+\varepsilon_0) \quad \sigma_t = \frac{F}{A} = \sigma_0(1+\varepsilon_0)^{2\mu} \tag{3.26}$$

$$\mu = -\frac{\varepsilon_{t,y}}{\varepsilon_{t,x}} \tag{3.27}$$

The relaxation function $E(t)$ is calculated with the true stress $\sigma_t(t)$ and the true strain $\varepsilon_t(t) \approx$ constant. This corresponds to the secant modulus at a certain time t. It must be noted here, that the deformation is not applied suddenly, but at a rate of 400 mm/min. In order to neglect the influence of the load history, the withdrawn DIN 53441 recommends to choose a point in time for the beginning of the evaluation of relaxation experiments that corresponds to about ten times the time required to apply the deformation ($t \geq 10 \cdot t_0$). In (Schapery, 1969a) and (Brinson et al., 2008), it is suggested to evaluate the first data point at $t \geq 5 \cdot t_0$. In this work, the evaluation started at times $t = t_0 + 10 \cdot t_0$, as shown in Figure 3.18 (b)).

Nonlinear viscoelastic mastercurve from relaxation tests at constant strain but different temperatures, and Schapery's nonlinearity factors

It is assumed that the TTSP remains valid at large strains. If the tests are carried out with a constant strain at different temperatures, the TTSP (determined by means of DMTA in the linear viscoelastic range) is used to create a mastercurve at a reference temperature for the strain level under investigation.

The strain-dependent horizontal displacement factor is set to a_ε = 1. To determine the remaining nonlinearity factors h_0 and $h^*_{1,2}$ resp. h_∞ and $h_{1,2}$, the direct method of Brüller (1993) is applied.

If the linear viscoelastic material parameters are known, the time-invariant E_∞ resp. E_0 and the transient part ΔE resp. ΔE^* of the linear viscoelastic modulus can be evaluated for the n time points of the relaxation test in the nonlinear range and be stored in the nx2 matrix $\boldsymbol{E}_{\text{LVE}}$. $\boldsymbol{E}_{\text{NL}}$ is the vector of length n, which contains the values of the nonlinear relaxation test to be described. Subsequently, eq. (3.28) can be solved to $\boldsymbol{H}_{\text{NL}}$, which is the vector containing the two nonlinearity factors h_0 and $h^*_{1,2}$ resp. h_∞ and $h_{1,2}$.

$$\boldsymbol{E}^{\text{T}}_{\text{LVE}} \cdot \boldsymbol{E}_{\text{LVE}} \cdot \boldsymbol{H}_{\text{NL}} = \boldsymbol{E}^{\text{T}}_{\text{LVE}} \cdot \boldsymbol{E}_{\text{NL}} \tag{3.28}$$

If it is chosen to work with h_∞ and $h_{1,2}$ and using the Prony series for the linear viscoelastic material behaviour, $\boldsymbol{E}_{\mathbf{LVE}}$ and $\boldsymbol{H}_{\mathbf{NL}}$ correspond to:

$$\boldsymbol{H}_{\mathbf{NL}} = \begin{bmatrix} h_\infty \\ h_{1,2} \end{bmatrix}; \quad \boldsymbol{E}_{\mathbf{LVE}} = \begin{bmatrix} E_\infty & \Delta E(t_1) \\ E_\infty & \Delta E(t_2) \\ \vdots & \vdots \\ E_\infty & \Delta E(t_n) \end{bmatrix} = \begin{bmatrix} E_\infty & \sum_{i=1}^N E_i \cdot \exp - \frac{t_1}{a_\varepsilon \cdot \tau_i} \\ E_\infty & \sum_{i=1}^N E_i \cdot \exp - \frac{t_2}{a_\varepsilon \cdot \tau_i} \\ \vdots & \vdots \\ E_\infty & \sum_{i=1}^N E_i \cdot \exp - \frac{t_n}{a_\varepsilon \cdot \tau_i} \end{bmatrix}$$

If it is chosen to work with h_0 and $h^*_{1,2}$ and using the Prony series for the linear viscoelastic material behaviour, $\boldsymbol{E}_{\mathbf{LVE}}$ and $\boldsymbol{H}_{\mathbf{NL}}$ correspond to:

$$\boldsymbol{H}_{\mathbf{NL}} = \begin{bmatrix} h_0 \\ h^*_{1,2} \end{bmatrix}; \quad \boldsymbol{E}_{\mathbf{LVE}} = \begin{bmatrix} E_0 & \Delta E^*(t_1) \\ E_0 & \Delta E^*(t_2) \\ \vdots & \vdots \\ E_0 & \Delta E^*(t_n) \end{bmatrix} = \begin{bmatrix} E_0 & -\sum_{i=1}^N E_i \cdot (1 - \exp - \frac{t_1}{a_\varepsilon \cdot \tau_i}) \\ E_0 & -\sum_{i=1}^N E_i \cdot (1 - \exp - \frac{t_2}{a_\varepsilon \cdot \tau_i}) \\ \vdots & \vdots \\ E_0 & -\sum_{i=1}^N E_i \cdot (1 - \exp - \frac{t_n}{a_\varepsilon \cdot \tau_i}) \end{bmatrix}$$

For all cases $\boldsymbol{E}_{\mathbf{NL}}$ equals:

$$\boldsymbol{E}_{\mathbf{NL}} = \begin{bmatrix} E_{\mathrm{NL}}(t_1) \\ E_{\mathrm{NL}}(t_2) \\ \vdots \\ E_{\mathrm{NL}}(t_n) \end{bmatrix}$$

Nonlinear viscoelastic mastercurve from relaxation tests at different strains but constant temperature, and Schapery's nonlinearity factors

At constant temperature and different strain levels, it is checked whether a continuous mastercurve can be created by purely horizontal shifting the individual curves with a_ε. For this purpose, a similar procedure is used as for the TTSP in the linear viscoelastic range (Figure 3.12).

3.6 Overview of the performed tests

Table 3.1 summarizes the experiments carried out for each interlayer type. x indicates that the indicated test is evaluated in the following chapters. (x) indicates that the tests have been performed but the results are only shown in the appendix. More information on the tests (sample amount, test parameters) can be found in the respective chapter.

Table 3.1 Overview of performed tests.

Test	**Determination of...**	**EVA**	**cEVA**	**SG**	**Std. PVB**	**Struct. PVB**	**Ac. PVB**
XRD	crystallinity	x	x	x	(x)	/	/
DSC	crystallinity, characteristic temperatures, crystallization kinetics	x	x	x	(x)	/	/
DMTA-TS	characteristic temperatures	x^2	x^1	x^2	/	/	/
DMTA-TFS	Prony series in dependence of temperature and crystallinity	/	x^1	x^2	x	x	(x)
DMTA-FTS	Prony series in dependence of temperature and crystallinity	/	x^1	/	/	/	/
DMTA-tS	physical aging	/	/	x^2	/	/	/
DMTA-AS*	linearity limits	/	/	x^3	x^3	$(x)^3$	/
Uniaxial relaxation	nonlinear Schapery parameters	/	/	/	x	x	(x)

* over a wide range of temperatures
[1] SRF configuration
[2] PP configuration
[3] tPP configuration

4 Investigation of the semicrystalline structure of EVA and ionoplastic interlayers

This chapter refers to the materials SG, cEVA and EVA. According to theory (chapter 2.1.5), those materials exhibit partially crystalline regions. The aim of the following investigations is to verify the crystalline regions experimentally and to determine the initial degree of crystallization. In addition, the characteristic temperatures and enthalpies (glass transition temperature T_g, melting T_m and crystallization temperature T_c, melting ΔH_m and crystallization enthalpies ΔH_c), the crystallization kinetics as well as the stiffness behaviour (G') as a function of different temperature programs are to be specified.
In section A.2.4 of the appendix, it is experimentally shown, that PVB does not have crystalline phases.

4.1 Crystalline regions and initial crystallinity

In order to investigate whether the materials exhibit partially crystalline regions, Differential Scanning Calorimetry (DSC) tests were first carried out. To determine the initial degree of crystallization, the first heating run of the DSC was evaluated in more detail. In addition, the initial degrees of crystallization were determined by means of X-Ray Diffraction (XRD) experiments. The evaluation methods are described in chapter 3.

4.1.1 Differential Scanning Calorimetry

This chapter shows the results of one DSC measurement (consisting of multiple temperature cycles) per material. The DSC experiments for SG and cEVA were conducted in a *DWDK Netzsch Type 214 Polyma* DSC. The DSC experiment for EVA was conducted on a *Mettler Toledo* DSC. Three different temperature rates (+-20 K/min, +-10 K/min, +-3 K/min for SG resp. +-2 K/min for cEVA and EVA) have been investigated in a temperature range of [-60:140]°C, [-60:100]°C and [-60:120]°C respectively. In addition, at the end of the temperature program, the SG sample was left in the DSC at a constant temperature of 23°C for 24 h and finally heated up to 50°C. The sample mass was 4.1 mg for

M. Schuster, *Characterization of Laminated Safety Glass Interlayers*,
Mechanik, Werkstoffe und Konstruktion im Bauwesen 66,
https://doi.org/10.1007/978-3-658-39821-7_4

SG, 7.0 mg for cEVA and 9.6 mg for EVA respectively. The test parameters are summarized in Table A.1 in the appendix.

General observations

The different temperature programs are shown in Figure 4.1 and the results in Figure 4.2. First of all, one notices that the first cooling and heating process differs from the shape of all other cooling or heating curves. This can be explained with secondary crystallization, due to physical aging and will be examined in detail later. In the range of low temperatures, especially at the higher temperature rates, a step can be seen which characterizes the glass transition area. In addition, endothermic peaks in the high temperature range can be seen in the heating curves. These are assigned to the melting of crystalline phases. In the cooling curves exothermic peaks are visible, which characterize the crystallization process. The peak positions show a clear dependence on the cooling rate. It is also noticeable that the crystallization peak temperatures are below the melting peak temperatures. The characteristic temperatures and enthalpies will be discussed in detail later. It is noteworthy that no crosslinking process is observed in the uncrosslinked EVA.

Investigation of the first heating process

By determining the enthalpy of fusion ΔH_{m} in the first heating process (sample state "as received"), the initial crystallinity χ of the samples can be determined with eq. (3.3). For this purpose, a baseline was created in MATLAB®, which was extrapolated from the melt until an intersection with the DSC curve was obtained. For SG, the melting enthalpy (marked with a gray patch in Figure 4.3 (a)) was determined to $\Delta H_{\mathrm{m}} = 27.9$ J/g, which corresponds to an initial crystallinity of $\chi = 9.5\%$. For cEVA (Figure 4.3 (b)) the melting enthalpy was determined to $\Delta H_{\mathrm{m}} = 52.3$ J/g, which corresponds to an initial crystallinity of $\chi = 17.8\%$ and for EVA (Figure 4.3 (c)) the melting enthalpy was determined to $\Delta H_{\mathrm{m}} = 49.4$ J/g, which corresponds to an initial crystallinity of $\chi = 16.9\%$. In all three cases, $\Delta H_{\mathrm{m,100\%PE}} = 293$ J/g was used.

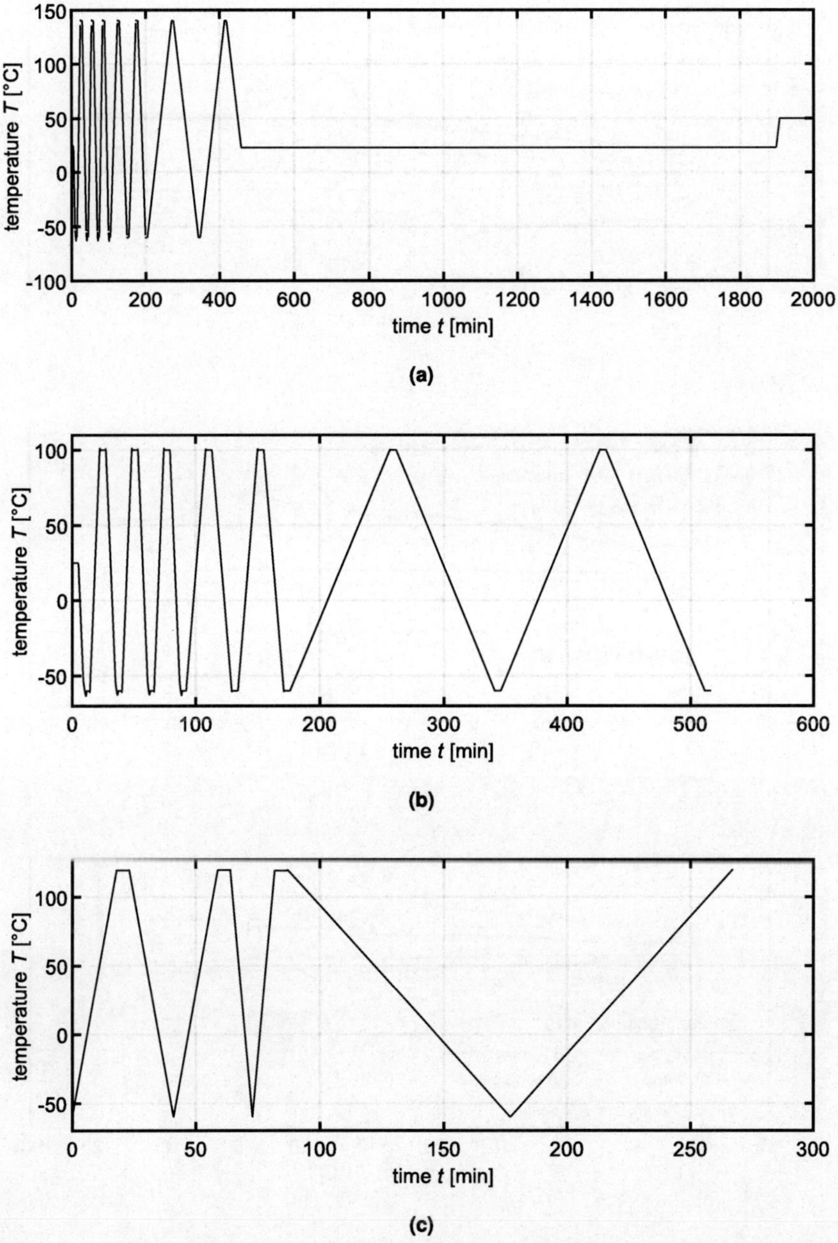

Figure 4.1 DSC temperature programs (a) SG: erasing thermal prehistory with 1 cycle 20 K/min, 2 cycles +-20 K/min [-60:140]°C, 2 cycles +-10 K/min [-60:140]°C, 2 cycles +-3 K/min [-60:140]°C, 24 h at 23°C, +3 K/min [23:50]°C (b) cEVA: erasing thermal prehistory with 1 cycle 20 K/min, 2 cycles +-20 K/min [-60:100]°C, 2 cycles +-10 K/min [-60:100]°C, 2 cycles +-2 K/min [-60:100]°C, (c) EVA: erasing thermal prehistory with 1 cycle 10 K/min, 1 cycle +-10 K/min [-60:120]°C, 1 cycle +-20 K/min [-60:120]°C, 1 cycle +-2 K/min [-60:120]°C.

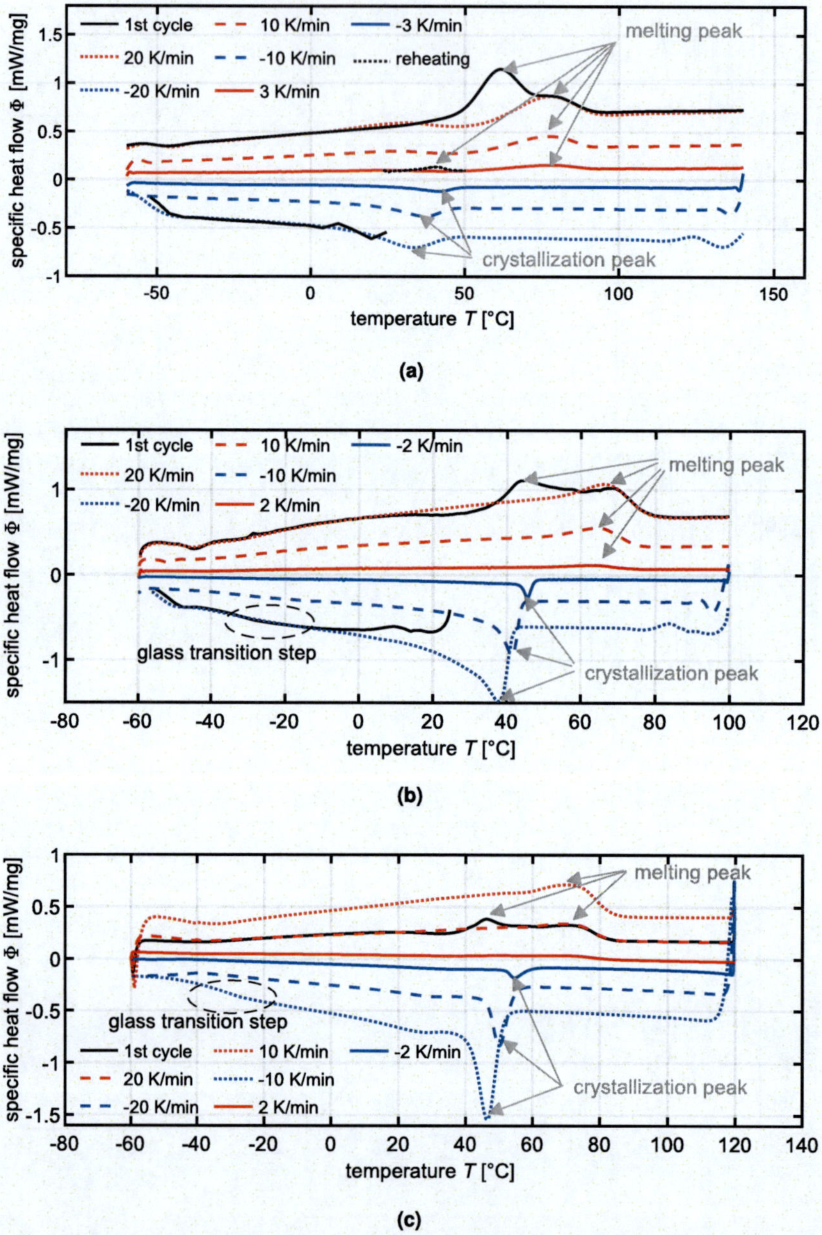

Figure 4.2 DSC results, exothermic direction is downwards. The red lines mark the results gathered during heating and the blue lines the results gathered during cooling. For several cycles with the same temperature rate, the mean value curve is shown. (a) SG, (c) cEVA, (c) EVA.

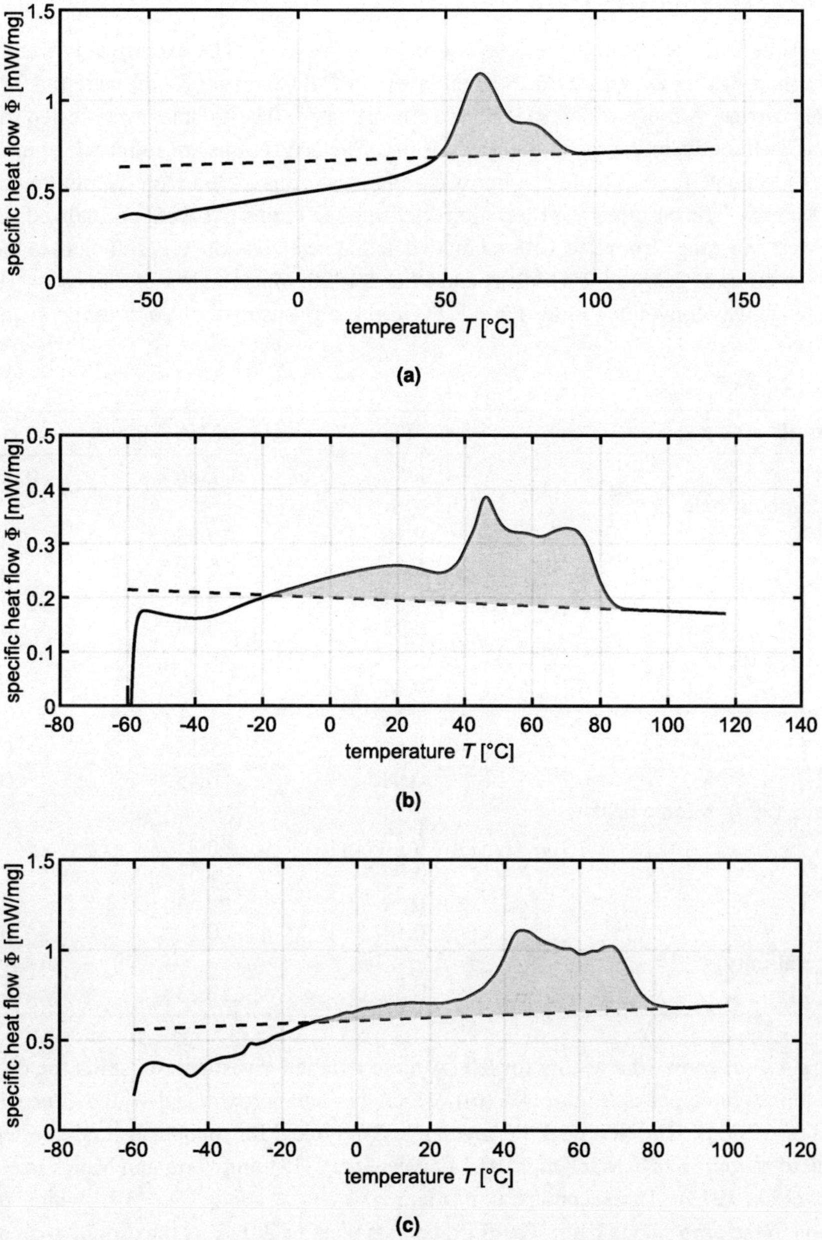

Figure 4.3 DSC results from the first heating scan, exothermic direction is downwards. (a) SG: ΔH_m = 27.9 J/g, χ = 9.5%, (b) cEVA: ΔH_m = 52.3 J/g, χ = 17.8%, (c) EVA: ΔH_m = 49.4 J/g, χ = 16.9%.

4.1.2 X-Ray Diffraction

The results from one test per interlayer type are presented. The experiments were conducted in a *Bruker D2 PHASER*. Several layers of foil (diameter 3 cm) were placed in a sample carrier. A range of $2\Theta = 5\text{-}70°$ was investigated. The samples were tested "as received", without deleting the thermal prehistory. The test set-ups are summarized in Table A.2. To evaluate the XRD plots, a linear baseline was constructed with the minima of the XRD curves. The baseline was then subtracted and the intensities were normalized so that they were in a range from 0 to 1. For a more detailed investigation, it is sufficient to look at an angle range of $2\Theta = 5\text{-}30°$ (Murthy and Minor, 1990; MacKnight and Earnest Jr, 1981). This range was convoluted in *Origin 8.5.1* (OriginLab®) using the *Peak Fitting module*.

Table 4.1 XRD results.

Sample	SG	cEVA	EVA
y_0 [-]	0.0086	-0.0386	-0.0343
Amorphous halo			
x_c [°]	19.16	19.83	19.23
w [°]	5.21	6.19	6.46
A [°]	5.97	6.66	6.20
H [-]	0.92	0.69	0.61
Bragg (110)			
x_c [°]	19.74	21.44	20.95
w [°]	1.15	1.61	1.20
A [°]	0.12	1.15	0.98
H [-]	0.08	0.45	0.52
Bragg (200) or ionic peak			
x_c [°]	6.27	23.57	23.10
w [°]	3.04	1.51	1.37
A [°]	0.54	0.34	0.28
H [-]	0.14	0.14	0.13
Crystallinity			
χ [%]	9.86	18.23	16.92

Figure 4.4 (a) shows the results for SG. A more detailed investigation, splits the diffractogram into three peaks (Figure 4.4 (b)), which can be approximated with a *Gauss* function (eq. (3.23)). The first peak (at low angles) is typical for ionoplasts, and is often referred to as *ionic peak* (MacKnight and Earnest Jr, 1981; Longworth and Vaughan, 1968; Marx et al., 1973). The second peak has its maximum at about $2\Theta = 19°$, which corresponds to the amorphous halo. The third and smallest peak has its maximum at an angle of $2\Theta = 19.75°$ and is associated with the Bragg (110) reflection, which corresponds to one of the two PE-crystals typical Bragg reflections.

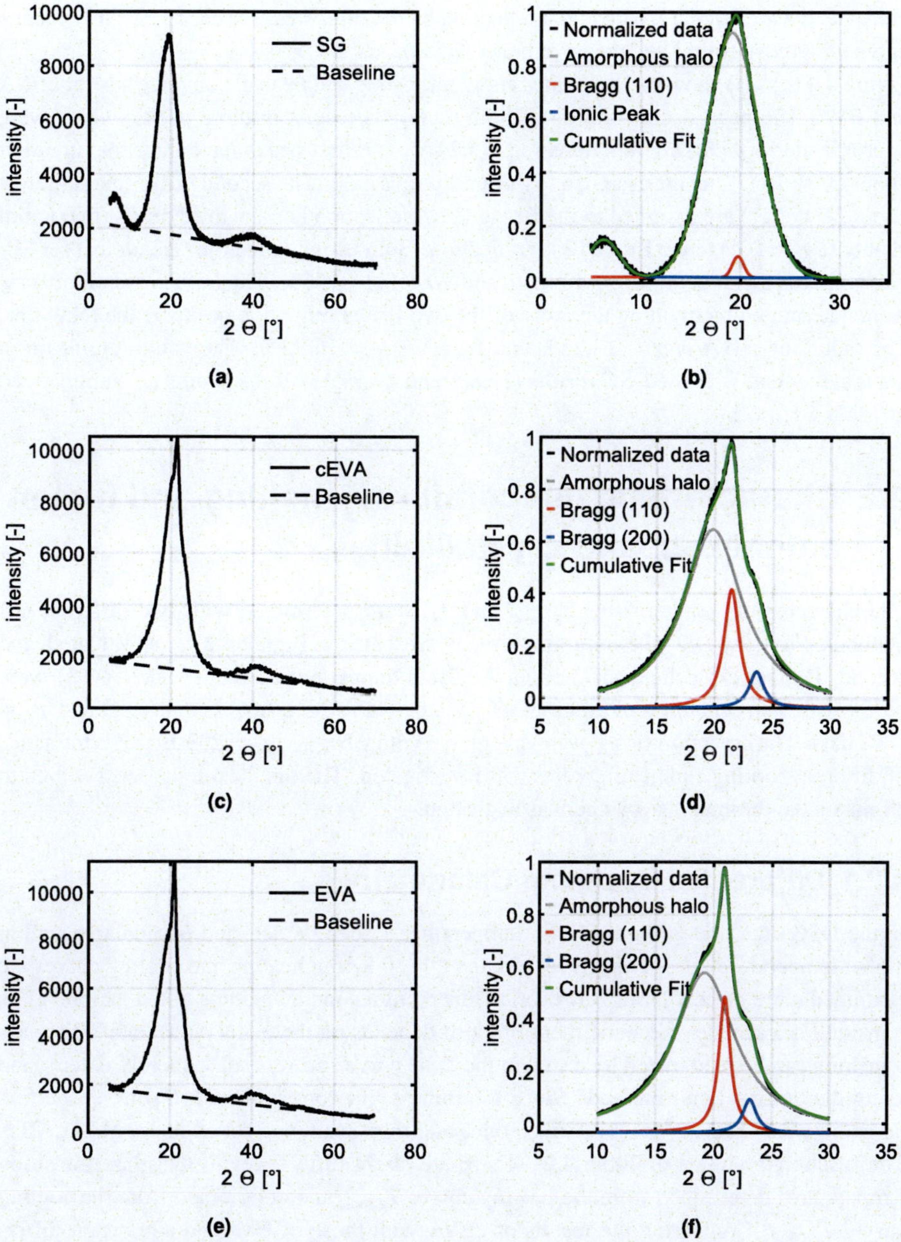

Figure 4.4 XRD diffractograms and peak deconvolutions. The peak positions, peak function parameters, peak areas and crystallinities are summarized in Table 4.1, (a) and (b) SG, (c) and (d) cEVA, (e) and (f) EVA.

The initial crystallinity of $\chi = 9.86\%$ was calculated by dividing both the area of the ionic peak and the area of the Bragg (110) reflection by the total area, see eq. (3.22). The results agree with those from DSC measurements (9.5%).
Figure 4.4 (c) - (f) show the XRD measurement results and peak deconvolutions for cEVA and EVA. The peak deconvolution of both materials results in three peaks, which were approximated by *Lorentz* functions (eq. (3.24)). The first peaks have their maximum at about $2\Theta \approx 19.5°$ and are assigned to the amorphous halo. The second peaks (maximum at about $2\Theta \approx 21°$) are assigned to the Bragg (110) reflection and the third peaks (maximum at about $2\Theta \approx 23°$) to the Bragg (200) reflection. Both Bragg reflections are typical for PE-like crystals (Beyer et al., 1984; Kamath and Wakefield, 1965). The initial crystallinities χ were determined by dividing the areas of the two Bragg reflection peaks by the total area. The value for cEVA is $\chi = 18.23\%$ and for EVA $\chi = 16.92\%$. The values are again in good agreement with the DSC results (17.8% and 16.9%). All the results are summarized in Table 4.1.

4.2 Characteristic temperatures, melting and fusion enthalpies, and crystallinities

The characteristic temperatures T_g, T_m and T_c, were determined with two different test methods. First, the DSC thermograms already presented in Figure 4.2 were evaluated, and second, Dynamic-Mechanical-Thermal-Analysis temperature sweeps (DMTA-TS) were performed. The enthalpies of melting ΔH_m and crystallization ΔH_c were determined using DSC data. The crystallinities χ were calculated with $\Delta H_{m,100\%PE} = 293$ J/g. The influence of different cooling and heating rates was investigated. The first heating process served to eliminate the thermal history and was neglected.

4.2.1 Differential Scanning Calorimetry

In the DSC data, the glass transition temperature T_g can be detected particularly well at higher temperature rates. For this reason, the $C = 20$ K/min heating and cooling curves are examined more closely here. However, at this point it should again be noted that the glass transition is a kinetic effect and therefore also depends on the temperature rate. The glass transition can be recognized by a step in the DSC curve and was subsequently determined using the turning point method. Since a turning point corresponds to a peak in the first derivative, the DSC curve was derived once and then examined for peaks in MATLAB®. The results are shown in Table 4.2. There are slight differences in the glass transition temperatures determined from the heating curves $T_{g,heat}$ and determined from the cooling curves $T_{g,cool}$. Comparing the results of cEVA with those of EVA, no significant differences can be seen. The glass transition temperature thus appears to be independent of the degree of crosslinking and equals $T_{g,cool} \approx$ -22°C in the cooling and $T_{g,heat} \approx$ -27°C in the heating curve. For SG, the determination of the glass transition is not obvious. In the cooling curves, the inflection point seems to be superimposed by the crystallization peak. Hence $T_{g,cool} \approx 20°C$ can only be estimated. In the heating curve it is estimated to $T_{g,heat} \approx 15.7°C$.

To further investigate the melting and crystallization behaviour, the baselines must first be determined. For this purpose, the DSC curves starting from the melt were extrapolated into the low temperature range (by creating a linear regression above the melt or crystallization peak with MATLAB®). Looking at Figure 4.5, it becomes clear that for all three materials, both the melting range in the heating curves and the crystallization range in the cooling curves are very wide and that melting starts directly after the glass transition and crystallization takes place until the glass transition is reached. This indicates that the investigated materials have a wide size distribution of crystallites. These wide ranges are marked by the peak temperatures T_p, whereby the baseline was first subtracted from the DSC curve to find the peak. To determine the melting ΔH_m and crystallization enthalpies ΔH_c according to eq. (3.2), the peak areas, delimited by the points of intersection between the DSC curves and baselines, have to be determined via integration. The results are shown in Table 4.2. If the enthalpy is put into relation with that of 100% crystalline PE ($\Delta H_{m,100\%PE}$ = 293 J/g), the degree of crystallization χ is obtained.
For SG, low crystallinities between χ = 2.0 - 4.3% are obtained, where the crystallinities from the heating curves lead to slightly higher results. For cEVA, the degree of crystallization at all heating and cooling rates ranges from χ = 17.3 - 19.9% and for EVA from χ = 11.3 - 17.4%.
The relative crystallinities α in Figure 4.6 and Tables A.10 - A.12 are determined with $\alpha = \chi_T/\chi_f$ wherein χ_T is the crystallinity at the temperature T and χ_f is the final crystallinity.
It should be noted that the calculated degree of crystallization is strongly dependent on the determined integration limits (intersection of baseline and DSC curve) and thus on the curve shape. A variation in the range in which the baseline is determined can already influence the result (Mathot and Pijpers, 1983). This is especially noticeable in the heating curves, as there are fewer data points available for determining the baseline (because the melting peak occurs at higher temperatures). The results are summarized in Table 4.2.

Comparing the heating curves of SG with those of cEVA and EVA, a flat exothermic peak in the second (and higher) heating curve of SG is recognizable (Figure 4.2 (a)), which can be interpreted as cold crystallization, before the endothermic melting peak. Analogously, an increase in the relative crystallinity occurs in Figure 4.6 (a). This is further investigated in the following paragraph.

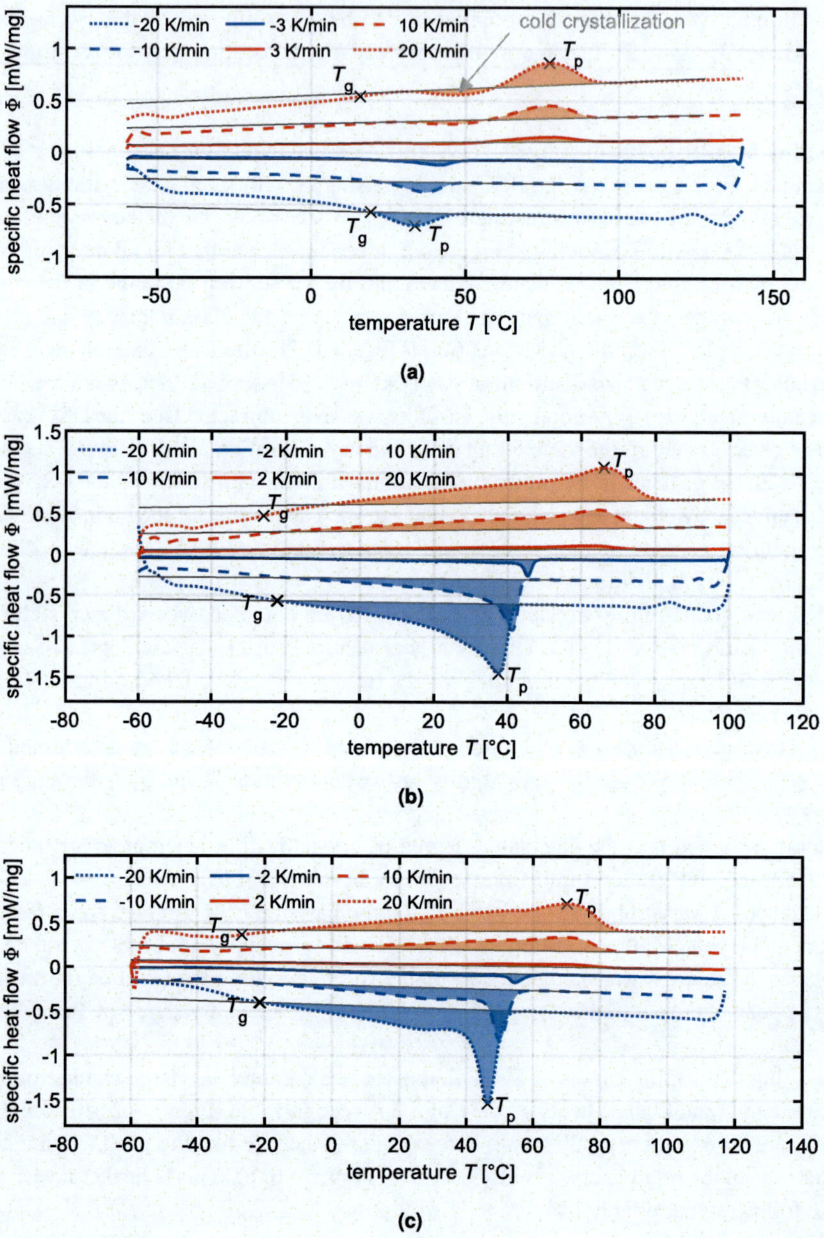

Figure 4.5 DSC baselines (a) SG, (b) cEVA, (c) EVA.

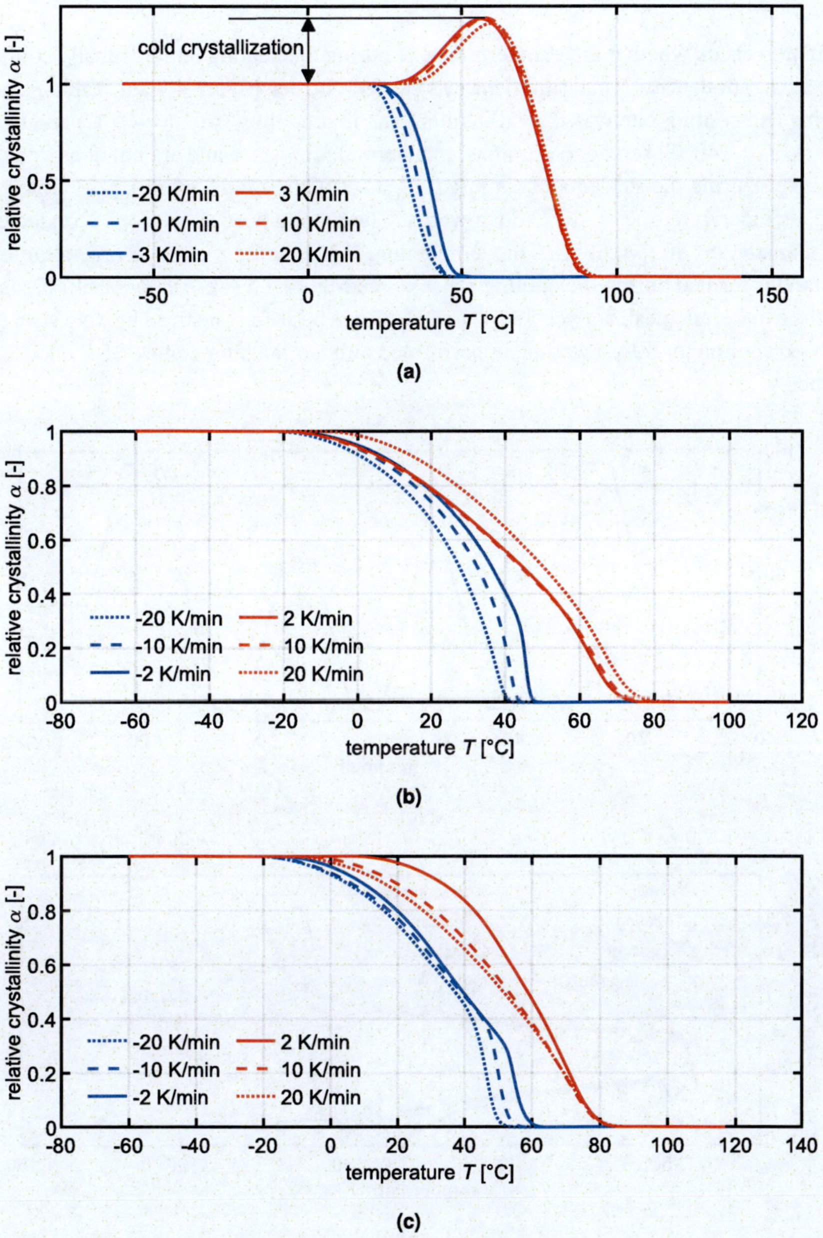

Figure 4.6 DSC relative crystallinities (a) SG, (b) cEVA, (c) EVA.

Cold crystallization of SG

In order to check whether the exothermic peak during the heating of SG is really cold crystallization, another DSC measurement was carried out, see Figure 4.7 and Table A.3. The heating and cooling rate was $C = 20$ K/min. The first cooling (to T = -60°C) and heating scan (to T = 140°C) served to eliminate the thermal prehistory and are not considered further here. During the subsequent cooling to T = -60°C (second cooling), crystallites form from a temperature of $T \approx 40$°C downwards. The second heating was only conducted up to a temperature of T = 60°C. This temperature is below the melting temperature of the crystallites formed during the cooling process, see Figure 4.5 (a). Analogous to Figure 4.5 (a), the exothermic peak between 30°C and 60°C can be seen. The third heating scan (after renewed cooling to -60°C) was again performed only until a temperature of T = 60°C was reached.

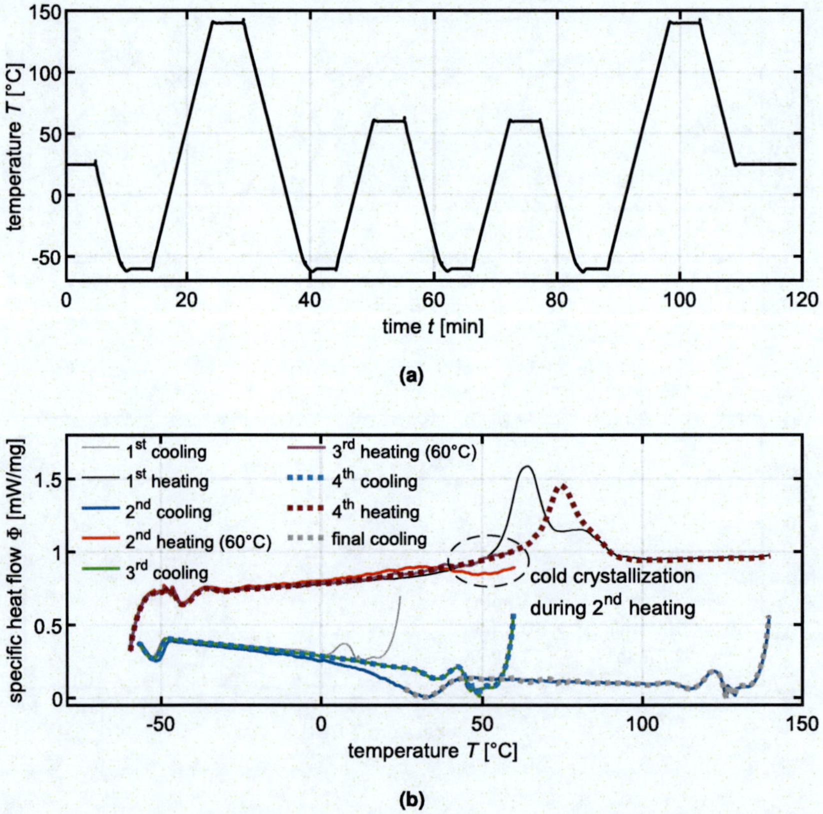

Figure 4.7 SG DSC results to check cold crystallization (a) temperature program, (b) results.

Compared to the second heating, there is no exothermic peak. In the final heating process (after renewed cooling to -60°C), a temperature of 140°C was targeted. Up to 60°C, the new heating curve is identical to the third heating curve.
From this measurement, it can be concluded that not all crystallites could be formed during the second cooling process, resulting in cold crystallization during the second heating. Since the second heating was only carried out up to 60°C, all crystallites formed up to that point are not melted. No further cold crystallization takes place in both subsequent heating processes.
If the degree of crystallization in the last heating process is considered, a melting enthalpy of ΔH_m = 26.35 J/g and a crystallinity of χ = 8.99% are obtained. Those values are only slightly below that of the first heating scan (ΔH_m = 33.34 J/g, χ = 11.37%). In the first cooling process after elimination of the thermal prehistory, however, the degree of crystallization (determined with baseline extrapolated from the melt) is only 2% (ΔH_c = 5.75 J/g). This clearly indicates the strong influence of cold crystallization.
In cases where no cold crystallization occurs during heating after cooling, the melting process is mainly dependent on the previous cooling rate. The heating rate (at the same previous cooling rate) plays only a disorderly role. However, since cold crystallization takes place in the heating process in the case of SG, the influence of the heating rate is investigated. Another DSC (see Figure 4.8 and Table A.4) was performed with the cooling rate set to $C = 10$ K/min. Subsequently, heating was performed at 0.3 K/min, 3 K/min, 10 K/min and 20 K/min. It should be noted that the results of the very low heating rate are difficult to evaluate. It can be seen at all heating rates that the melting process starts at 56 - 60°C. From a temperature of 80°C, all curves meet. As expected, the main difference is in the amount of cold crystallization. The slower the heating rate, the more time the material has to increase its crystallinity.

Comparison of results with initial crystallinities

Comparing the crystallinities with the initial crystallinities determined in the first heating scan (see Figure 4.2), it becomes obvious that for cEVA and EVA, despite the different shape of the first DSC curve compared to the following heating curves, the degree of crystallization is approximately the same. This suggests that different temperature histories result in different distributed crystal sizes without significantly changing the overall degree of crystallization. A closer look at the shape of the first heating curves reveals two peaks, indicating the presence of two different crystal structures. Compared to the second heating scans, the first heating scans show a lower endotherm in the temperature range of $T \approx 0$ - 35°C. From $T \approx 35$°C on, however, a strong increase of the endotherm follows and a peak maximum is formed at a temperature of $T_{p,m1} \approx 45$°C. This peak is also called annealing peak and occurs when the sample is stored below the melting temperature but above the glass transition temperature, is not in its thermodynamic equilibrium

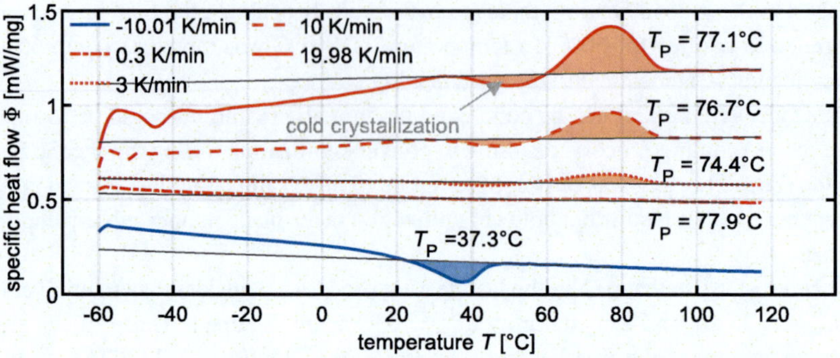

(a)

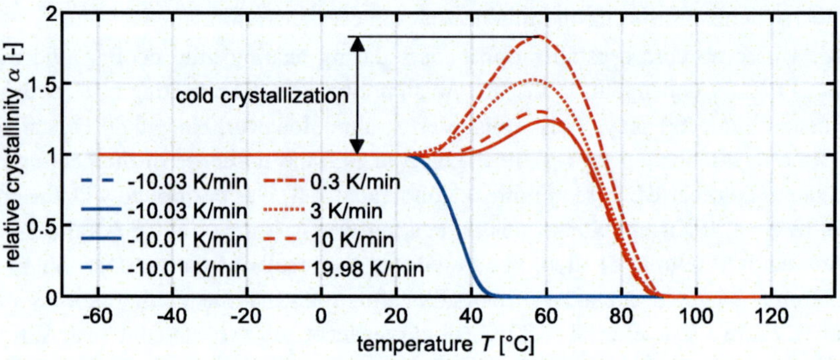

(b)

Figure 4.8 SG DSC results to check the influence of the heating rate (a) results, (b) relative crystallinities.

and tries to approach equilibrium by structural rearrangement processes. It is assumed that the structural rearrangements are crystal perfections of existing crystals. Annealing existing crystals describes the melting of smaller crystallites with low melting temperatures (with melting temperatures around the annealing temperature) and simultaneous growth or perfection of the larger crystallites, which then have a higher melting temperature. The shape of the annealing peak is strongly dependent on the storage time and the storage temperature. Basically it can be said that the peak shifts towards higher temperatures with increasing storage time. However, the peak temperature is lower than that of the main melting peak, see $T_{p,m}$ in Table 4.2, (Androsch, 1999).

Comparing the crystallinities of SG, it can be seen that the initial degree of crystallization (in the first DSC heating run, Figure 4.2) is significantly higher than that from the subsequent cooling and heating runs (see Table 4.2). This can be explained by cold crystallization. The low temperature peak in the first heating scan occurs at $T_{p,m1} \approx 62°C$.

Table 4.2 DSC results.

Sample	SG			cEVA			EVA		
C [K/min]	10.0	20.0	3.0	10.0	20.0	2.0	10.0	20.0	2.0
Glass transition									
$T_{g,heat}$ [°C]		15.7			-26.0			-26.8	
$T_{g,cool}$ [°C]		20.0			-22.4			-21.6	
Melting									
$T_{p,m}$ [°C]	77.0	77.2	74.8	64.4	66.1	62.5	69.5	70.1	69.5
ΔH_m [J/g]	10.4	9.7	12.5	58.4	49.6	53.9	42.3	48.8	33.0
χ_m [%]	3.5	3.3	4.3	19.9	16.9	18.4	14.4	16.7	11.3
Crystallization									
$T_{p,c}$ [°C]	37.1	33.5	41.5	41.3	37.6	45.6	49.7	46.4	54.5
ΔH_c [J/g]	-7.2	-5.5	-11.5	-51.2	-50.6	-52.3	-47.9	-47.3	-51.0
χ_c [%]	2.4	2.0	3.9	17.5	17.3	17.8	16.3	16.1	17.4

4.2.2 DMTA - temperature sweeps

Non-isothermal crystallization can also be investigated using DMTA temperature sweeps (DMTA-TS), in which the sample is first heated above the melting point in order to delete the thermal prehistory and then subjected to a controlled cooling and heating process. The results from one temperature sweep (cooling and heating after elimination of thermal prehistory) per sample type are presented. The tests were carried out on an *Anton Paar MCR 302* rheometer. For SG and EVA the plate-plate measuring system with a diameter of 8 mm was used, for cEVA the solid state fixture (SRF) with a torsion strip (dimensions: $\approx$ 50 mm x 10 mm) was used. Temperature ranges of [-60:+100]°C with heating and cooling rates of $C = 2$ K/min for cEVA and EVA resp. [-40:120]°C with heating and cooling rates of $C = 3$ K/min for SG were investigated. The frequency was $f = 1$ Hz and the excitation amplitude was in the linear viscoelastic range ($\gamma = 0.1\%$ for cEVA and $\gamma = 0.025\%$ for EVA and SG). The test parameters are summarized in Table A.5 in the appendix. The temperature programs for the different materials are shown in Figure 4.9 and the results in Figure 4.10.

General

It can be seen in Figure 4.10 that the heating and cooling curves do not superimpose. A hysteresis is formed, where the cooling branch results in lower moduli at a given temperature than the heating run. This is the expected result for slow crystallization (Wietor et al., 2011; Jing et al., 2019).

Determination of characteristic temperatures

The glass transition temperature is determined in the following by a peak in the loss factor in the lower temperature range. The $\tan\delta$ peak maximum coincides with a sharp drop in the storage modulus $G'(T)$. The melting or crystallization range is indicated by the hysteresis. At temperatures above the melting range, the same amorphous state prevails in both the cooling curve and the heating curve. Accordingly, the rheological behaviour is also the same and both curve sections must lie on top of each other.
In the temperature range before the hysteresis, the cooling and heating curves also lie on top of each other. This can for example mean that the degree of crystallization is constant outside the hysteresis. In that case, the beginning of the hysteresis would define the end of the crystallization process and the end point of the hysteresis would define the end of the melting process.
However, it is also possible that identical rheological behaviour occurs during cooling and heating (resulting in congruent curves) if the crystallinity changes equally in both curves. If in addition to equal crystallinities, identical crystallite structures are present in both curves, the crystals that have formed in the part of the cooling curve melt again exactly at their formation temperature; it would therefore be a case of reversing melting or crystallization. Although, the end of the melting process can be determined by the end of the hysteresis, the end of the crystallization range can no longer be determined with the help of the hysteresis. If however, different crystallite structures occur at identical degrees of crystallinities, equal curves would mean that that the rheological behaviour depends on the degree of crystallization, but not on the structure (in this case crystallite size). The results for $T_{\mathrm{g,heat}}$, $T_{\mathrm{g,cool}}$, the beginning of the hysteresis $T_{\mathrm{begin,hysteresis}}$ and the end of the hysteresis $T_{\mathrm{end,hysteresis}}$ are summarized in Table 4.3.

Table 4.3 DMTA-TS results.

Sample	SG	cEVA	EVA
$T_{\mathrm{g,heat}}$ [°C]	49.00	-24.90	-22.05
$T_{\mathrm{g,cool}}$ [°C]	41.97	-24.62	-22.05
$T_{\mathrm{begin,hysteresis}}$ [°C]	26.01	16.21	27.18
$T_{\mathrm{end,hysteresis}}$ [°C]	85.9	72.99	>80

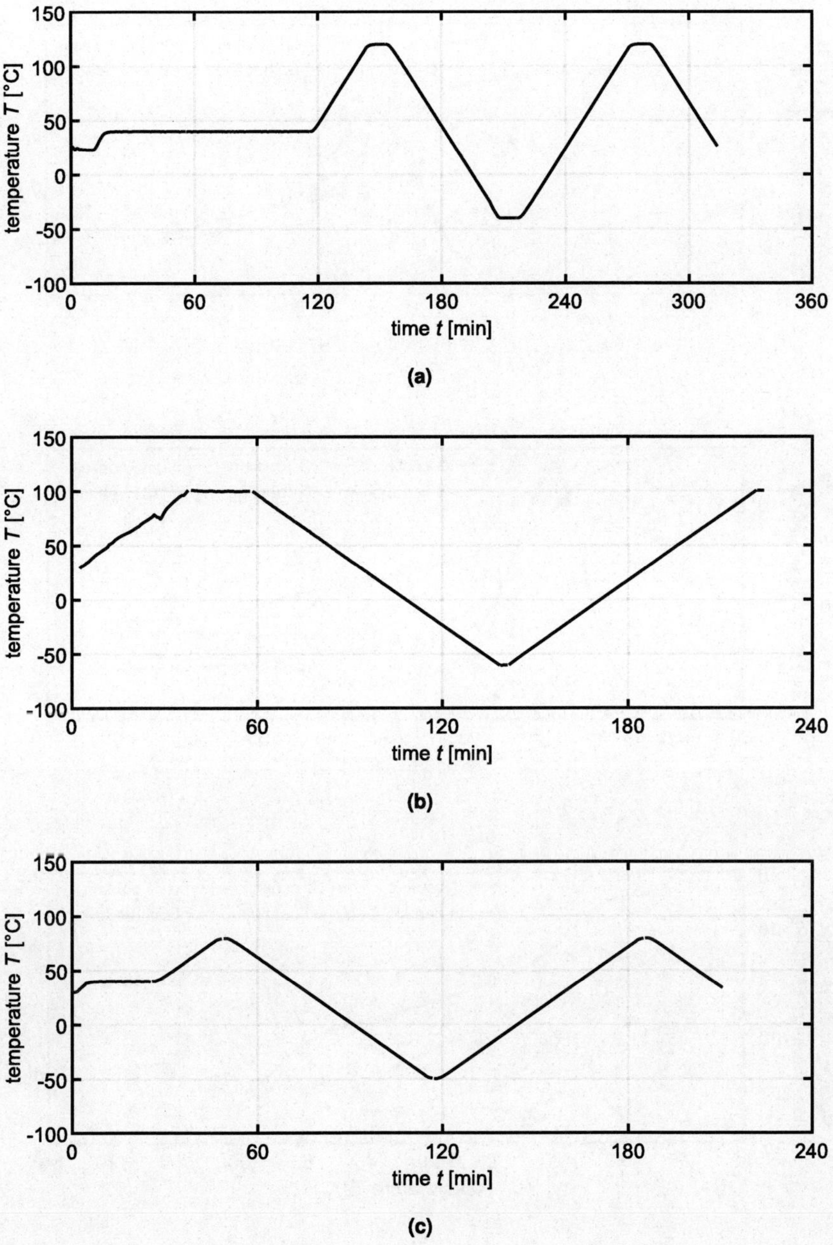

Figure 4.9 DMTA-TS temperature programs (a) SG +-3 K/min, (b) cEVA +-2 K/min, (c) EVA +-2 K/min.

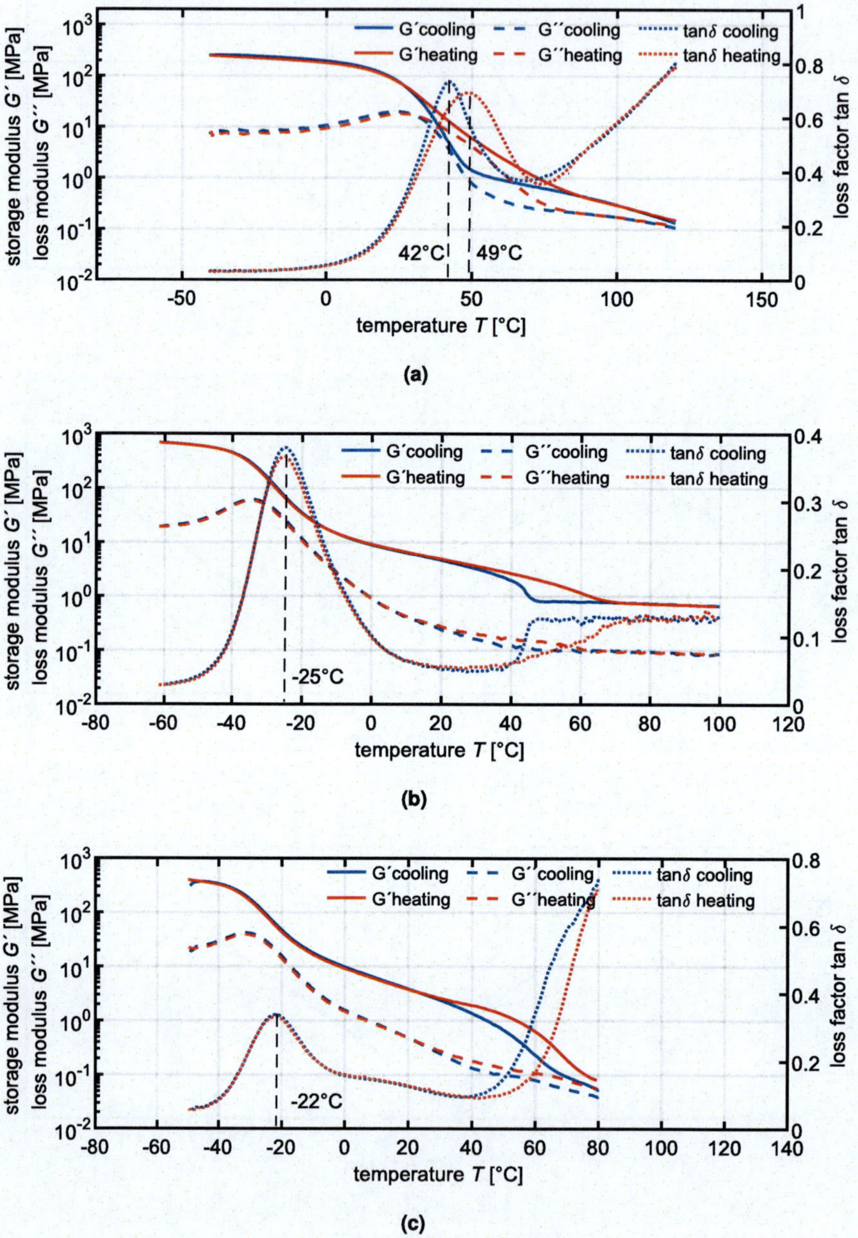

Figure 4.10 DMTA-TS results after elimination of thermal prehistory (a) SG, (b) cEVA, (c) EVA.

Comparison with DSC results

When comparing the temperature sweep curves with the DSC curves at same temperature rates (Figures 4.11 - 4.13) it can be seen that in all cases the end of the DMTA storage modulus hysteresis coincides with the end of the melting range in the melting DSC-curve, respectively with the achievement of a relative crystallinity of $\alpha = 0$. In addition, it can be seen that in all cases the DSC crystallization peak coincides with an abrupt rise in the cooling-storage modulus curve and thus the start of crystallization can be determined. For SG, the begin of the hysteresis coincides with the end of the crystallization process.

In the cEVA and EVA sample, it is apparent that the onset of the DMTA hysteresis does not coincide with the end of the crystallization peak, respectively with the achievement of a relative crystallinity of $\alpha = 1$. However, when comparing the relative degree of crystallization of the cEVA sample in Figure 4.12 (c), it can be seen that at temperatures below the onset of the DMTA hysteresis, the degree of crystallization in the heating and cooling curves is approximately the same.
Similar behaviour should be expected for EVA. However, the degree of crystallization in Figure 4.13 (c) is different in the cooling and heating curve, even though equal rheological behaviour is measured. Comparing the relative degree of crystallization at 2 K/min with those at other heating rates (Figure 4.6 (c)), the curve at a heating rate of 2 K/min deviates strongly from those at higher heating rates. It is assumed that there are inaccuracies in the baseline evaluation of the 2 K/min DSC heating curve.

Comparing the loss factor $\tan\delta$ with the DSC curves, a good agreement of the glass transition area for cEVA and EVA becomes apparent. In contrast to the evaluation of the DSC data, there is no difference between $T_{g,cool}$ and $T_{g,heat}$ in the evaluation of the DMTA data. With SG, however, the peak maximum of $\tan\delta$ depends on the temperature direction (heating or cooling). In addition, it can be seen that the $\tan\delta$ peaks are in the range of the melting and crystallization peaks of the DSC, thus strengthening the assumption of superposition and possible influence of T_g and melting or crystallization processes. This could also explain the offset of the $\tan\delta$ peaks.

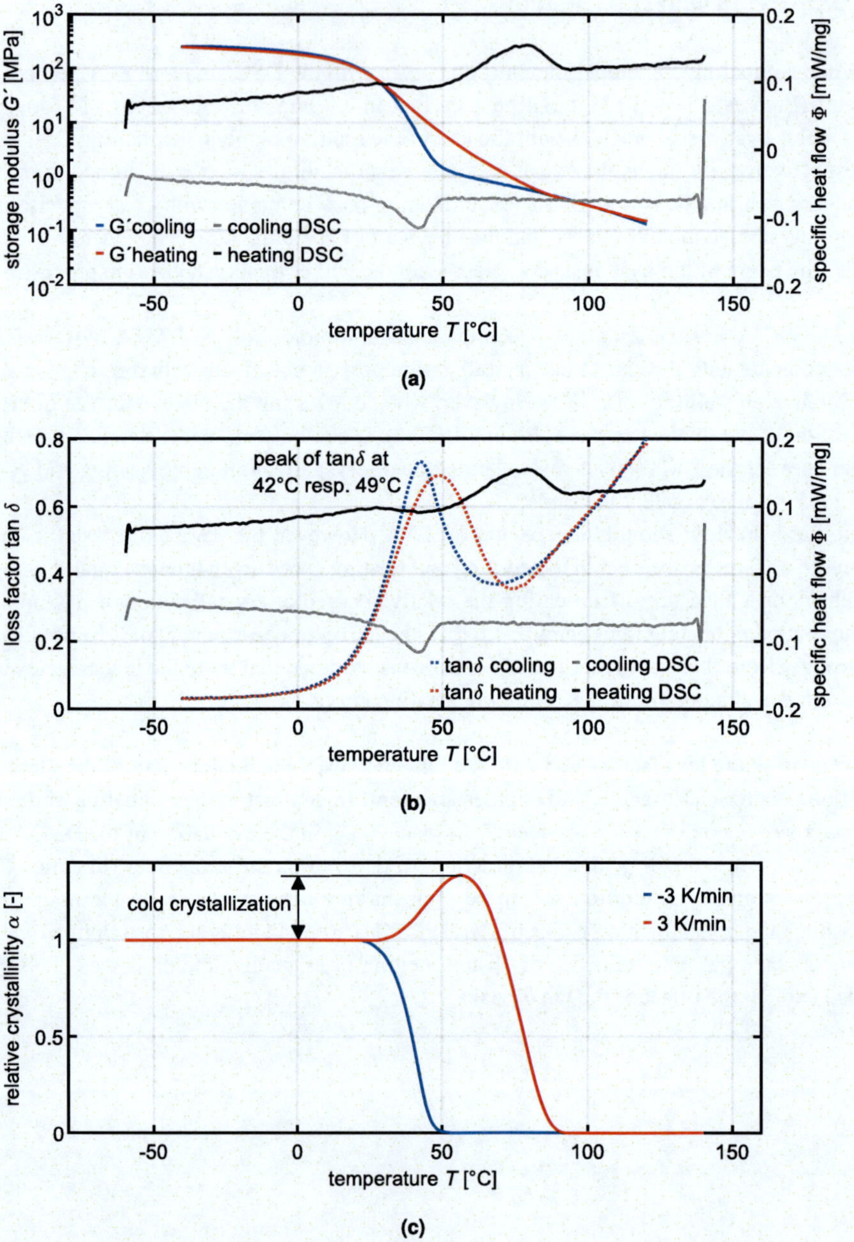

Figure 4.11 Comparison of DSC and DMTA data for SG (a) comparison DSC-DMTA storage modulus, (b) comparison DSC-DMTA loss factor, (c) relative crystallinity 3 K/min.

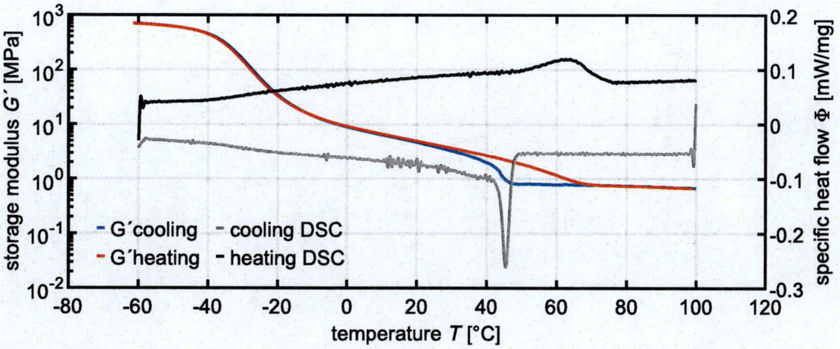

(a)

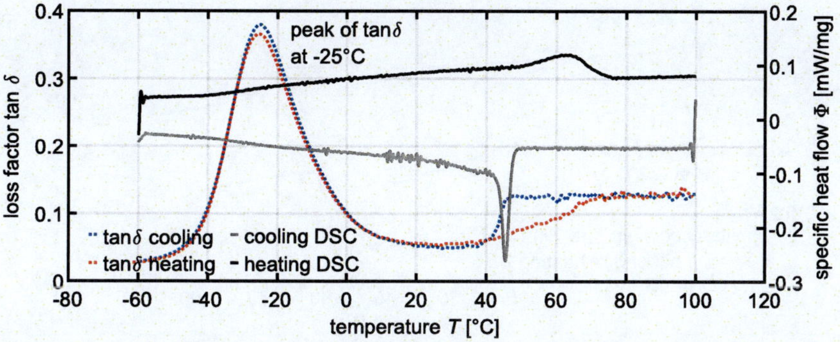

(b)

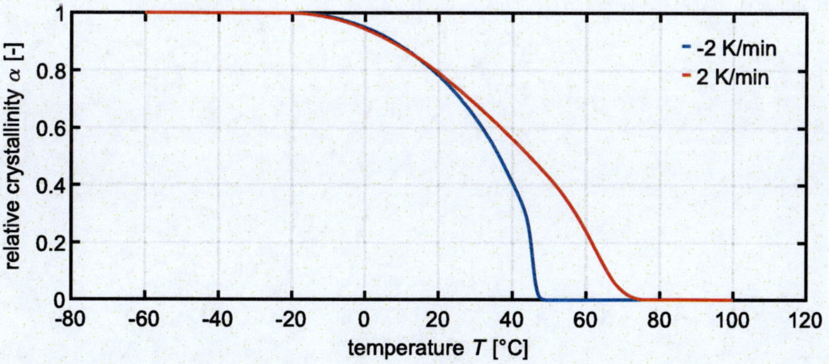

(c)

Figure 4.12 Comparison of DSC and DMTA data for cEVA (a) comparison DSC-DMTA storage modulus, (b) comparison DSC-DMTA loss factor, (c) relative crystallinity 2 K/min.

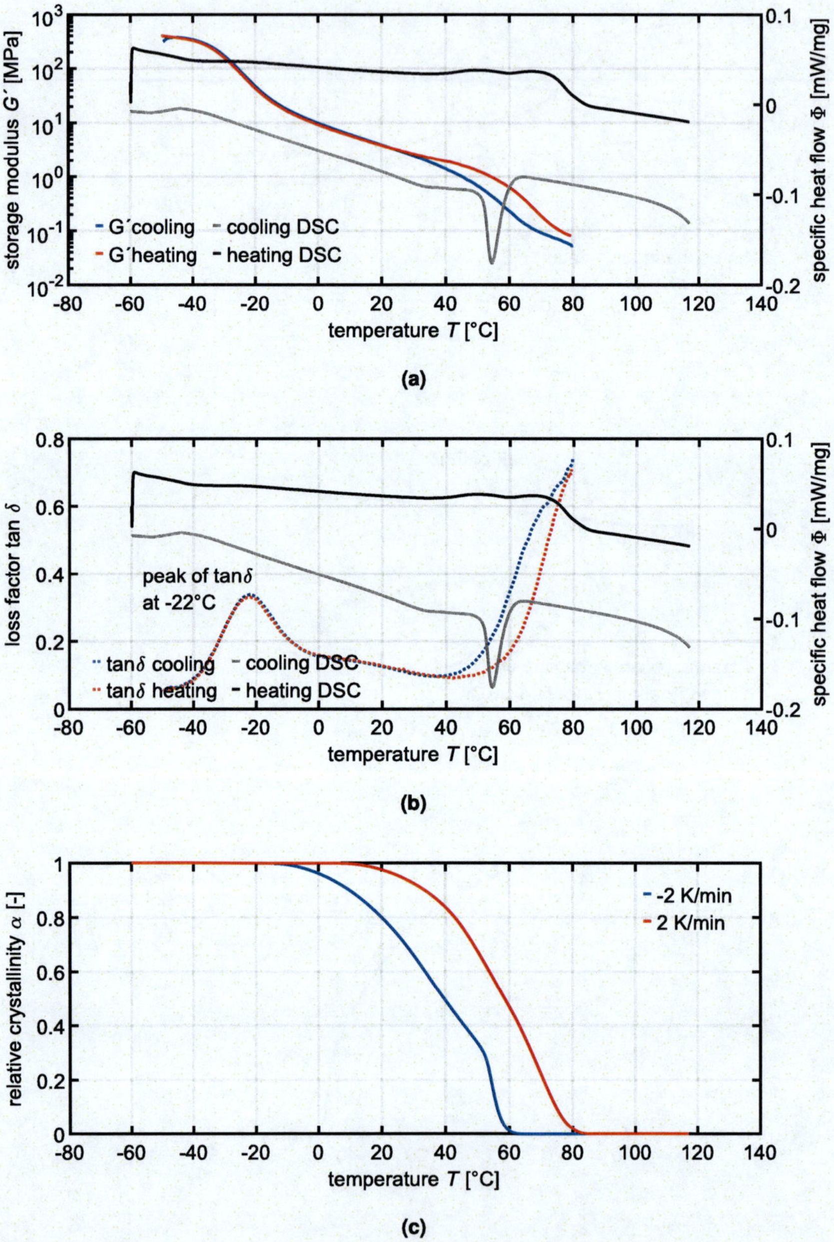

Figure 4.13 Comparison of DSC and DMTA data for EVA (a) comparison DSC-DMTA storage modulus, (b) comparison DSC-DMTA loss factor, (c) relative crystallinity 2 K/min.

4.3 Crystallization kinetics

Crystallization kinetic models are particularly useful for estimating the crystallization behaviour at new cooling rates. This is of interest for example in chapter 5, where DMTA is used to determine the viscoelastic behaviour as a function of crystallinity. Crystallization models then allow the degree of crystallization to be calculated for each data point.

The first temperature cycle (to eliminate the temperature history) is not considered in the investigations considering the non-isothermal crystallization kinetics.
For the description of the crystallization kinetics the relative degree of crystallization in dependency of temperature and temperature rate is needed for the cooling curves. With DSC data, the relative degree of crystallization at T (and temperature rate C) is calculated according to equation (3.6) by relating the partial peak area between T and T_{end} to the total peak area. Figure 4.5 shows the relative crystallinities during the crystallization processes respectively.
The following crystallization kinetic models (see also section 2.2.4.2 and 2.2.4.5) and activation energy approaches (see also section 2.2.4.4) are investigated :

- Crystallization kinetics according to Avrami;
- Crystallization kinetics according to Ozawa;
- Crystallization kinetics according to Mo-Liu;
- Investigation of z-function according to Málek.
- Activation energy according to Kissinger-Akahira-Sunose (KIS);
- Activation energy according to Ozawa-Flynn-Wall (OFW);
- Activation energy according to Friedman (FM).

Figure 4.14 - Figure 4.16 show the results of the different crystallization kinetic approaches. For all investigated materials, the modified Avrami-plots of $\ln[-\ln(1-\alpha)]$ vs. $\ln t$ exhibit a poor linear relationship. Different regions can be seen, which indicates that the modified Avrami analysis does not describe accurately the non-isothermal crystallization of the investigated materials. Similarly, especially for cEVA and EVA, the Ozawa plots do not show linear relationships between $\ln[-\ln(1-\alpha)]$ vs. $\ln C$ for the entire temperature range. The z-function proposed by Málek shows its maximum for all three materials at very low relative degrees of crystallization (less than 0.4), which also proves that the JMAK rate equation is not suitable. However, the plots of $\ln C$ against $\ln t$ at various crystallinities show a linear relationship, indicating that the Mo-Liu analysis could well describe the crystallization behaviour of the investigated materials.

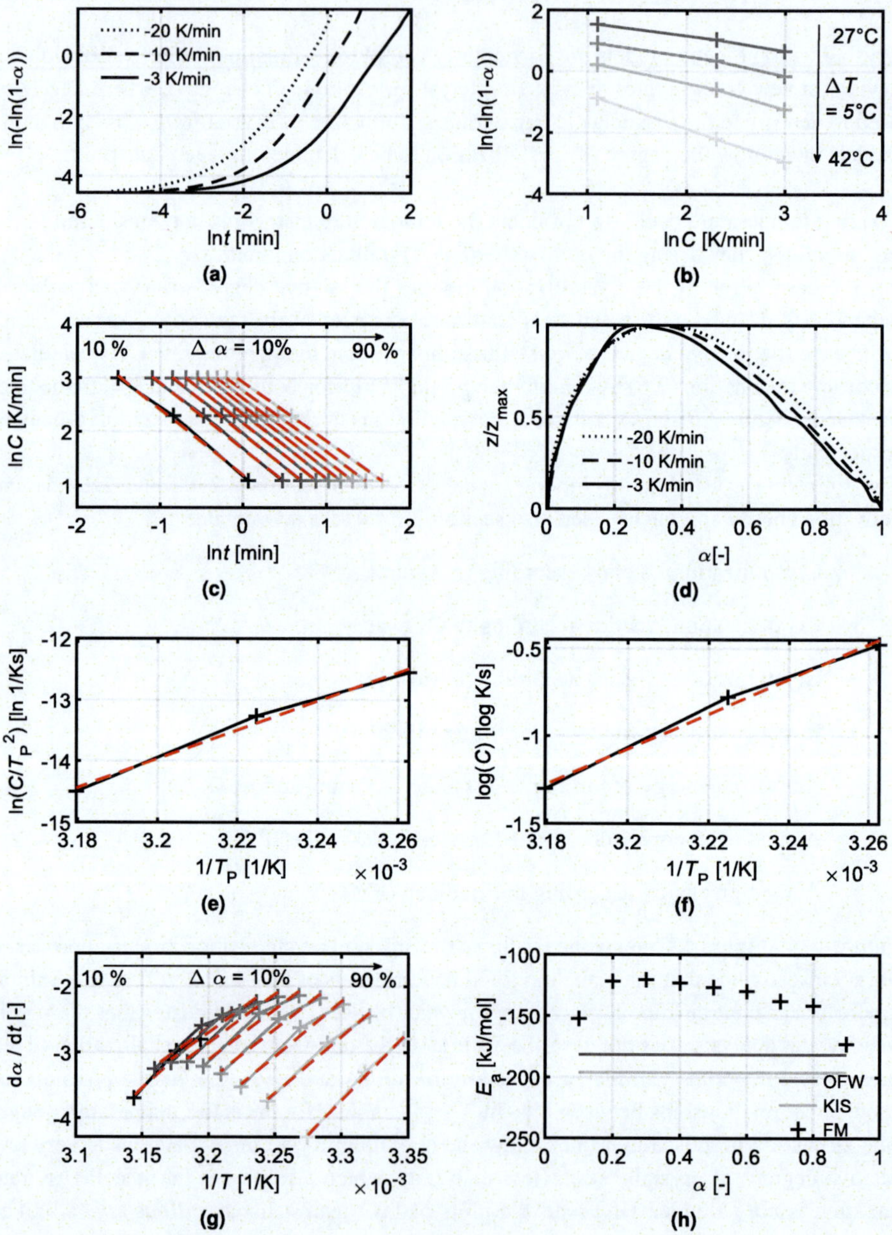

Figure 4.14 SG: Different approaches to describe the crystallization kinetics. Red lines correspond to linear fits. (a) Avrami, (b) Ozawa, (c) Mo-Liu, (d) z-function, (e) E_a KIS: E_a = -196.23 kJ/mol, (f) E_a OFW: E_a = -181.57 kJ/mol, (g) E_a FM, (h) E_a comparison.

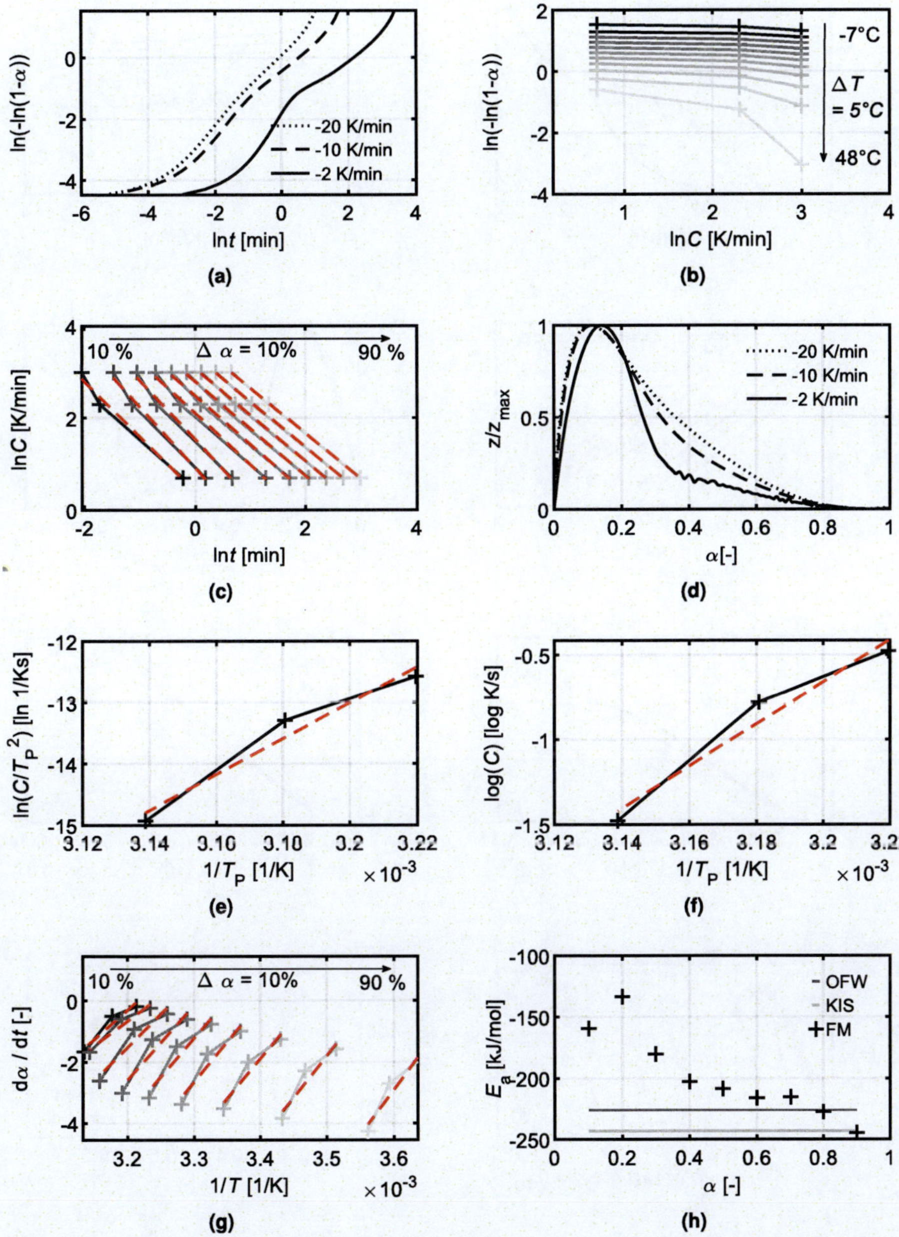

Figure 4.15 cEVA: Different approaches to describe the crystallization kinetics. Red lines correspond to linear fits. (a) Avrami, (b) Ozawa, (c) Mo-Liu, (d) z-function, (e) E_a KIS: E_a = -243.11 kJ/mol, (f) E_a OFW: E_a = -226.06 kJ/mol, (g) E_a FM, (h) E_a comparison.

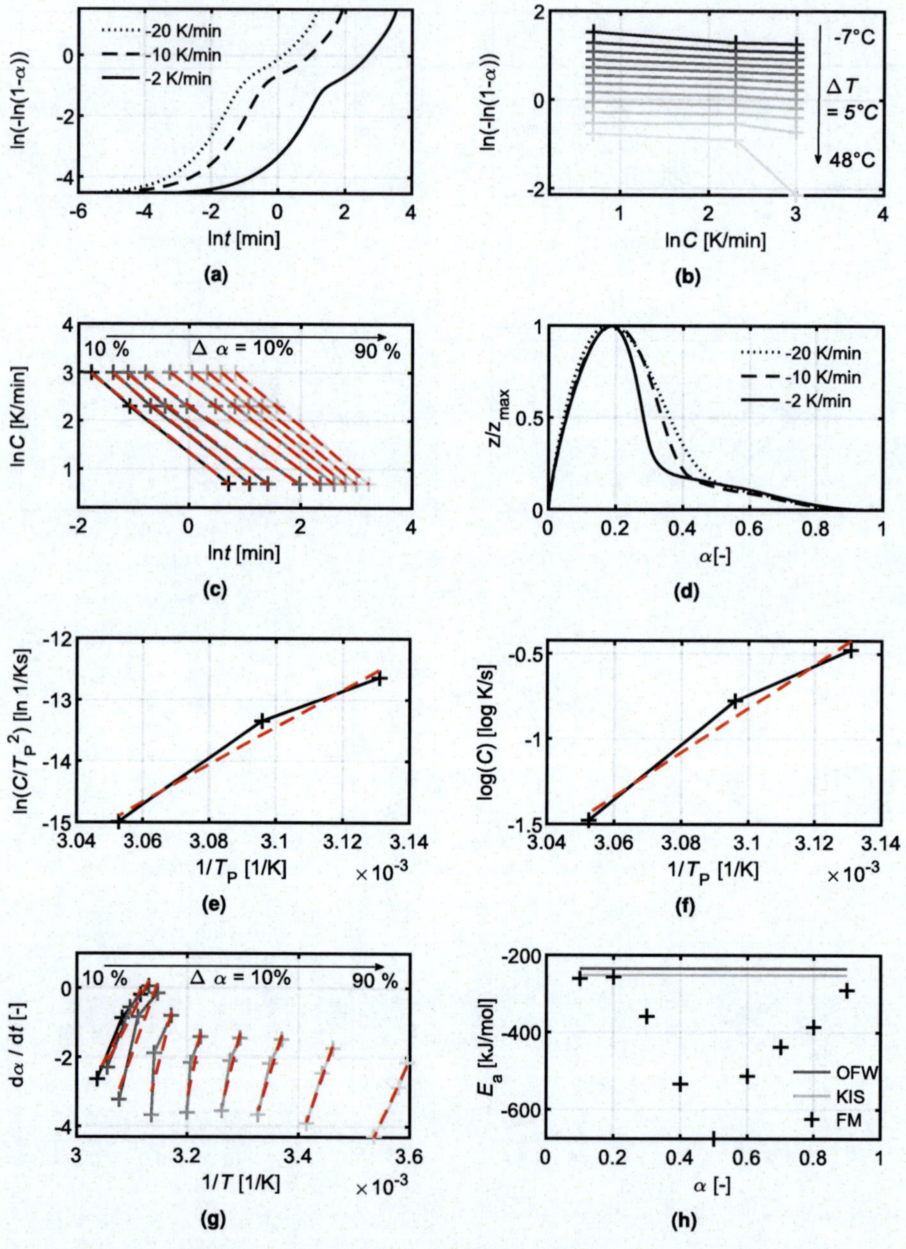

Figure 4.16 EVA: Different approaches to describe the crystallization kinetics. Red lines correspond to linear fits. (a) Avrami, (b) Ozawa, (c) Mo-Liu, (d) z-function, (e) E_a KIS: E_a = -252.24 kJ/mol, (f) E_a OFW: E_a = -234.59 kJ/mol, (g) E_a FM, (h) E_a comparison.

Also the isoconversially determined activation energy according to Friedman shows that a complex crystallization behaviour (multistep reactions) is present in all the materials investigated, since the activation energies are not independent of the relative degree of crystallization. For SG, Friedman's activation energy E_a varies between -120.29 kJ/mol and -189.77 kJ/mol, for cEVA between -133.56 kJ/mol and -227.53 kJ/mol and for EVA between -257.05 kJ/mol and -676.70 kJ/mol. The activation energies according to Kissinger and Ozawa show similar results in the case of all three materials, where the activation energy according to Kissinger delivers slightly higher absolute values. The results for the activation energies are noted in the captions of Figure 4.14 - Figure 4.16. Comparing the materials, we see that the activation energies for cEVA and EVA are in a similar range, while the activation energy for SG is somewhat lower (in absolute value).

Table 4.4 Crystallization kinetic parameters of the Mo-Liu and OFW models. The parameters of Mo-Liu are valid for C in [K/min] and t in [min]; the parameters of OFW for C in [K/s] and T_p in [K].

relative crystallinity	**SG**		**cEVA**		**EVA**	
α [-]	***a***	***b***	***a***	***b***	***a***	***b***
0.1	-0.996	1.450	-1.209	0.394	-0.919	1.344
0.2	-1.031	1.817	-1.343	0.910	-0.921	1.696
0.3	-1.059	2.041	-1.297	1.532	-0.901	1.962
0.4	-1.065	2.217	-1.123	2.107	-0.824	2.314
0.5	-1.066	2.365	-1.058	2.486	-0.857	2.706
0.6	-1.060	2.506	-1.027	2.794	-0.904	3.045
0.7	-1.049	2.636	-1.009	3.079	-0.930	3.303
0.8	-1.031	2.762	-0.995	3.355	-0.947	3.546
0.9	-1.001	2.908	-0.989	3.643	-0.965	3.816
activation energy E_a [kJ/mol]	-181.57		-226.06		-234.59	
OFW parameters	9979.9	-33.01	12425	-40.42	12894	-40.79

The crystallization behaviour at an arbitrary cooling rate is estimated in this thesis with Mo-Liu's method and Ozawa's activation energy as follows:

- Determination of the times at which certain degrees of crystallization are reached at the new cooling rate. For this purpose, the straight line equations ($y = ax + b$, with $y = \ln C$ and $x = \ln t$) according to Mo-Liu in Table 4.4 are used;

- Determination of the peak temperature with the help of the activation energy according to Ozawa: $T_p = (\log C - b_{OFW})/a_{OFW}$ [K]. a_{OFW} and b_{OFW} can be found in Table 4.4;

- Conversion of the times determined with Mo-Liu into temperatures. For this purpose, the cooling rate $C = \Delta T / \Delta t$ is used ($\Delta T = C \cdot \Delta t$). To obtain the absolute temperatures for certain degrees of crystallization, ΔT is added to the previously determined peak temperature T_p.

In addition to the approaches already presented, DMTA-TS data is used to determine the relative degree of crystallization (see chapter 3). The following approaches will be investigated and compared to the DSC results (with the same temperature rate):

- Approach according to Khanna (1993);
- Approach according to Pogodina et al. (1999);
- Suspension approach according to Boutahar et al. (1998);
- Colloid approach according to Boutahar et al. (1998);
- Approach according to Basu and Banerjee (1994).

According to Minale et al. (2019) the boundary curves for the completely molten material G'_m and for the completely crystallized material G'_s can be determined by approximating the storage modulus temperature curves in the corresponding regions with Arrhenius equations, see eq. (3.15). For G'_m, the range to be fitted is easy to determine, since the onset of crystallization is accompanied by a sharp increase in G' in the cooling data. G'_s, however, is more difficult. Minale et al. (2019) use the data points before (and the heating data at the beginning of) the hysteresis for this purpose. It should be noted, however, that as shown in Figure 4.12, congruent heating and cooling curves do not necessarily correspond to a relative crystallinity of 1. The results are shown in Figure 4.17. Boutahar's approach for colloids did not produce any reasonable results, which is why they are not shown.
For SG, it can be seen that crystallization can be well approximated with the Khanna, Pogodina and Boutahar (suspension) approaches, with the Pogodina approach giving the results that are most similar to the DSC measurements. The relative crystallinity during the heating process, however, cannot be reproduced.
The course of the relative crystallinity of cEVA can be well mapped for values between 0 and 0.3 in both the cooling and heating paths. For higher relative degrees of crystallization, however, the Khanna and Pogodina approaches are too steep compared to the DSC measurements. Boutahar's suspension approach is flatter and closest to the DSC results.
For EVA, Boutahar's suspension approach led to implausible results, which is why it is not shown here. For relative crystallinities between 0 and 0.4 in both the cooling and heating paths, Khanna's approach gives the results closest to the DSC measurements. At higher crystallization levels there is no good agreement.

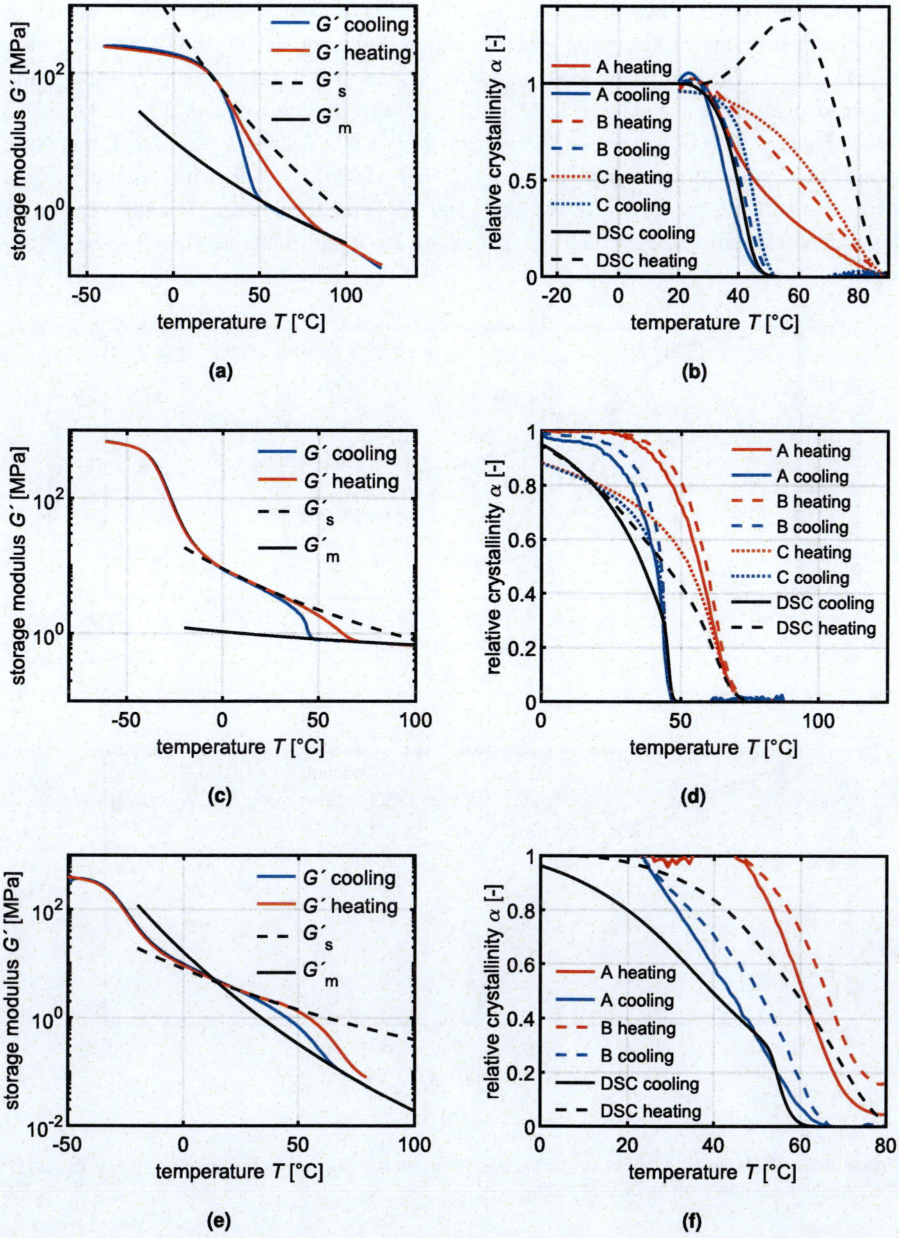

Figure 4.17 Crystallization kinetics with DMTA-TS. Left column shows the determination of G'_s and G'_m (parameters in Table A.9). The right column shows the relative crystallinity according to different approaches: A represents the approach of Khanna (1993), B of Pogodina et al. (1999) and C of Boutahar et al. (1998) for suspensions, (a) and (b): SG, (c) and (d): cEVA, (e) and (f): EVA.

Due to the difficulties in identifying the start of the melting process with DMTA, Basu's approach will only be investigated using cEVA as an example (Figure 4.18). The storage modulus just before reaching the glass transition range equals $G'_{T_g} = 443.29$ MPa (value at T = -39°C). The end of the glass transition region is assumed to be at T = 0°C (since the $\tan\delta$ peak has decayed here). Here, the storage modulus equals G'_{T_m} = 9.04 MPa. Hence $\log G_{T_g} - \log G'_{T_m} = \log 443.29 - \log 9.04 = 1.69$. Applying eq. (3.20), this corresponds to a crystallinity of $\chi = [1-(1.69/2.46)] \cdot 100\% = 31.3\%$, which is much higher than the results determined with the DSC and XRD measurements. For this reason, the denominator in eq. (3.20), which is assumed by Basu and Banerjee (1994) to be 2.46, is revised.

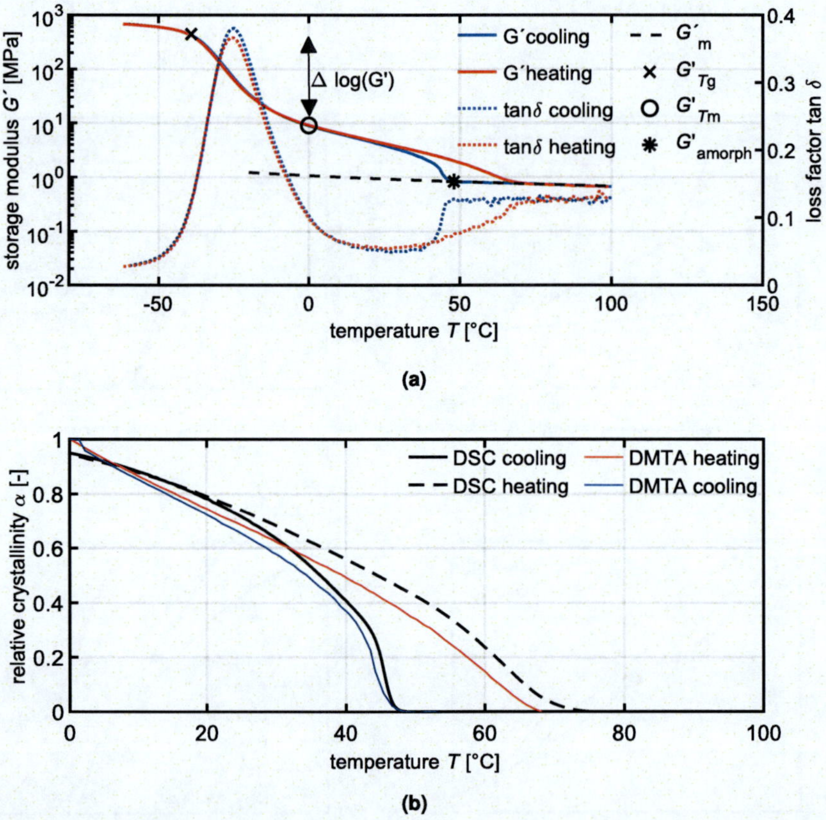

Figure 4.18 Relative crystallinity of cEVA based on eq. (3.20), (a) definition of $\Delta \log G'$, (b) relative crystallinity in comparison to DSC results.

For this purpose, G'_{amorph} was determined to $G'_{\text{amorph}} = 0.83$ MPa, which corresponds to the value from the cooling curve at the onset of crystallization. Hence, the drop of the storage modulus from the beginning of the glass transition region until the amorphous state is reached corresponds to $\Delta \log G' = \log 443.29 - \log 1.06 = 2.62$ and $\chi = [1 - (1.69/2.72)] \cdot 100\% = 37.9\ \%$, which is even higher and seems to be incorrect. Nevertheless, this value was used to calculate the relative degree of crystallization, see Figure 4.18 (b). Despite the high overestimation of the absolute degree of crystallization, there is a surprisingly good agreement in the relative degree of crystallization with the DSC values. It should be noted, that in this approach, G'_{amorph} is assumed to be constant. However, it is expected that this value is temperature dependent and therefore the use of G'_{m} from Figure 4.17 (c) would be useful. In this case, for the calculation of the degree of crystallization at the temperature T, G'_{amorph} would also have to be determined at this temperature, see Figure 4.18 (a). Since G'_{m} is very flat in the present sample cEVA, only small differences are to be expected. The advantage of this method compared to the one explained by Minale et al. (2019) is that only G'_{m} is needed, which is much easier to determine than G'_{s}.

Validity check of crystallization parameters for SG

Mo-Liu seems to be a suitable model to predict the crystallization behaviour at a specific cooling rate C. To validate this, a prediction for the crystallization behaviour at -5 K/min and -0.3 K/min is made for SG as an example based on the parameters given in Table 4.4. This prediction is checked by another DSC (test parameters in Table A.6). The temperature program included temperature rates of +-20 K/min, +-5 K/min and +-0.3 K/min. The results of the 20 K/min curves have been performed only to check comparability with the previous DSC measurements and are not shown here. The results of the +-5 K/min and +-0.3 K/min cycles are shown in Figure 4.19 (a).

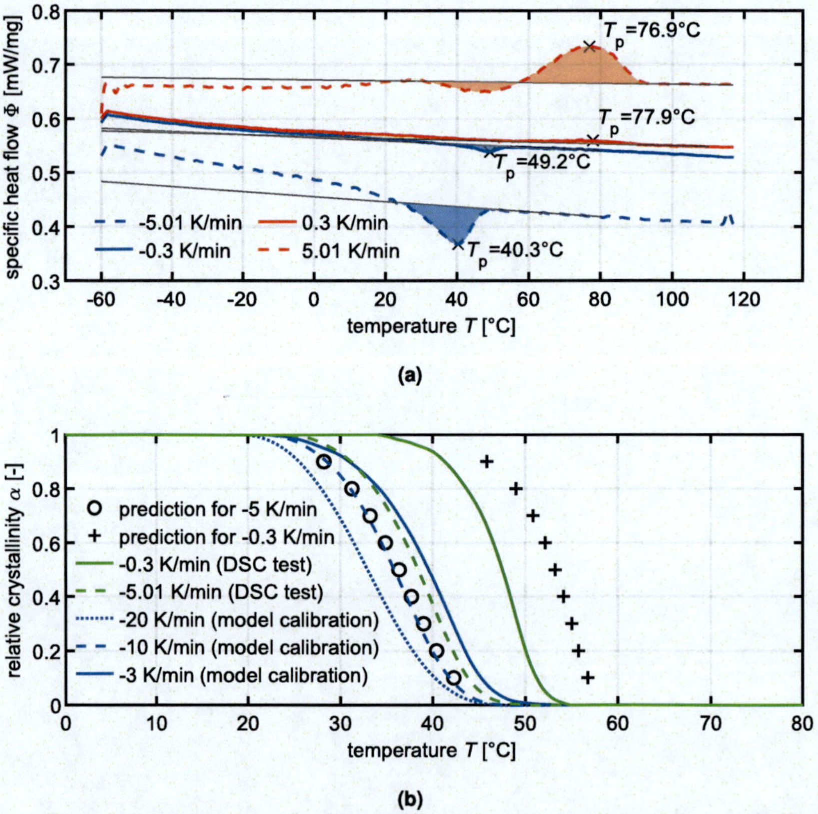

Figure 4.19 SG: Validation of the kinetic model (Mo-Liu and Ozawa's activation energy), (a) DSC results, (b) comparison of prediction with DSC results. The blue lines correspond to the DSC measurements that were used for model calibration.

Figure 4.19 (b) compares the predicted crystallization behaviour (markers) with the measured crystallinities (green lines). At a cooling rate of $C =$ -0.3 K/min, a deviation of $\approx$ 5°C and for a cooling rate of $C =$ -5 K/min a deviation of $\approx$ 2.5°C occurs. Due to the deviations, measured data is used in the following if possible.

4.4 Influence of physical aging

At the beginning of this chapter it was shown that the melting behaviour during the first heating cycle differs from the subsequent heating cycles. The influence of the physical aging was further investigated for the SG samples in DSC, DMTA-TS and DMTA-tS. As shown in Figure 4.1 (a) and Table A.1, the SG sample was heated with 3 K/min from 23 to 50°C after being conditioned for 24 h at 23°C in the DSC device. In addition, another DSC test (test parameters in Table A.7) was carried out, in which the first and second temperature cycles were each carried out at 3 K/min. This results in three DSC heating curves with different physical age, namely "0h", "24h" and "as received". Moreover, DMTA-TS from 23 to 100°C were performed with the same heating rate of 3 K/min after different residence times (0h, 24h, 4d) at 23°C. In addition, a TS from cold to warm was performed without zeroing the physical age of the sample in advance (physical age "as received"). The SRF measurement system had to be used for this. With the PP measuring system, high temperatures are required to establish adhesion between the sample and the measuring system, whereas with the SRF measuring system, the sample is clamped mechanically. The sample was mounted at -40°C and the TS was started with 3 K/min. The test parameters of the DMTA-TS are summarized in Table A.8. Finally, a DMTA-tS (test parameters in Table A.13) was performed for 5 days, during which the stiffness increase (increase in storage modulus) at 23°C was measured once every hour.

4.4.1 Differential Scanning Calorimetry and DMTA - temperature sweeps

Figure 4.20 (a) shows that the DSC-thermograms of the physically aged samples have in contrast to the unaged sample two endothermic peaks. An increase of the aging time shifts the lower melting peak to a higher temperature, increases the peak area and makes its appearance sharper. The peak position of the high temperature peak does not seem to be affected by the aging time. Looking at the DMTA-TS results in Figure 4.20 (b) resp. (c), it becomes clear that with increasing physical age, the shape of the shear modulus curve changes significantly. With increasing physical age, the high shear modulus range extends over a wider temperature range and the subsequent stiffness drop becomes significantly steeper. From a temperature of approx. 55°C the individual heating curves meet again. Comparing the DSC thermograms with the DMTA temperature sweeps, it can be seen that the significant storage modulus decreases collapse with the lower peak temperature.

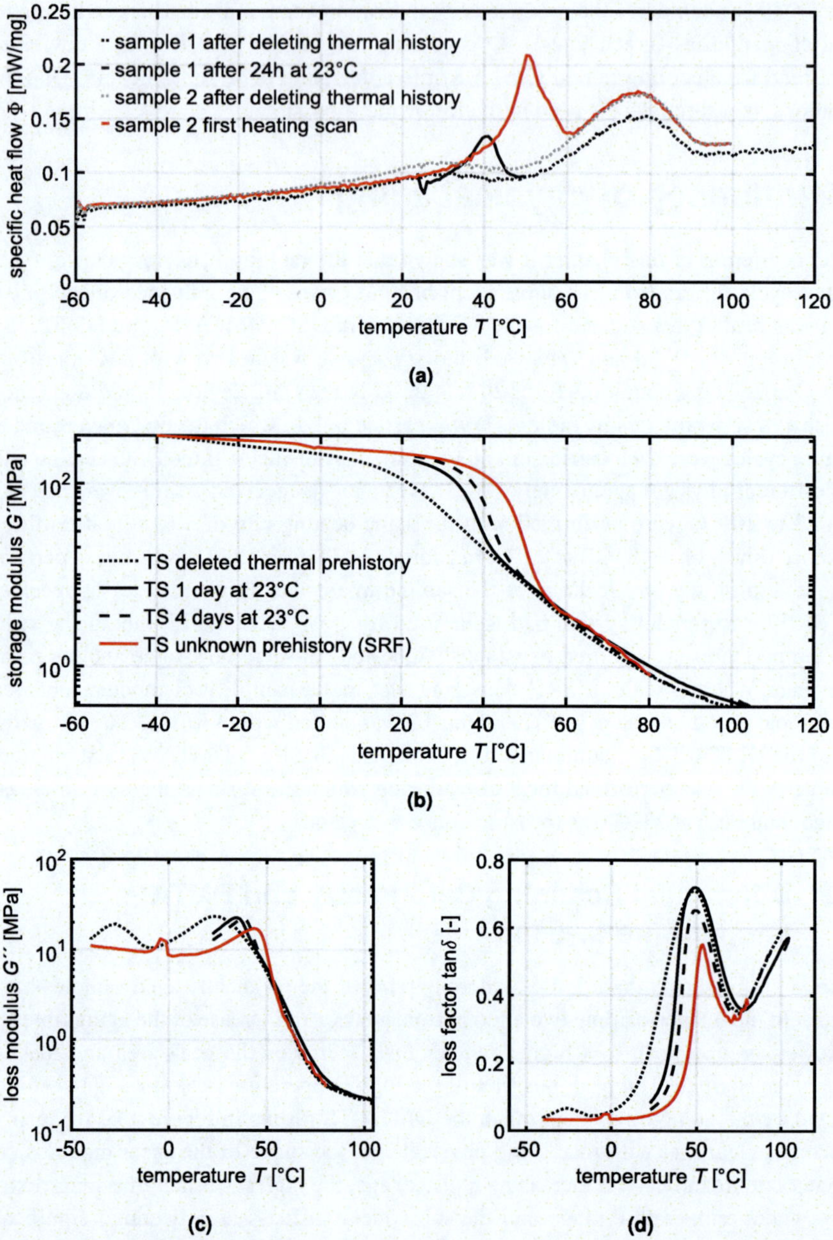

Figure 4.20 SG: Impact of physical aging at room temperature - investigation with a heating rate of 3 K/min (a) DSC, (b) DMTA storage modulus, (c) DMTA loss modulus, (d) DMTA loss factor. All DMTA results have the same legend.

This has already been reported in literature before, e.g. (Kohzaki et al., 1987; Hirasawa et al., 1989; Loo et al., 2005) and (Wakabayashi and Register, 2006). However, there is some disagreement in the literature about the origin of the low temperature peak. Marx et al. (1973); Wakabayashi and Register (2006) and Loo et al. (2005) explain the phenomenon by the melting of small secondary PE crystallites, which slowly form after the primary crystallization. Gómez et al. (2019) describes it as a structural reordering / perfection of existing “quasi-crystals”. Quasi crystals are amorphous polyethylene chain segments, which are packed together. They undergo a morphological transformation, which increases the structural order of the PE chain segments in the quasi-crystals structures. Kohzaki et al. (1987), however, have attributed the low-temperature endotherm to an order–disorder transition within the ionic aggregates.

4.4.2 DMTA - time sweeps

In addition to the already shown temperature sweeps, one time sweep of 5 days was conducted in order to observe the storage modulus increase at room temperature. Figure 4.21 (a) shows that the shear modulus increases during the temperature sweep at 23°C. The increase is very strong especially in the first 40 h and slowly decreases thereafter. However, it can also be observed that the shear modulus still increases slightly even after 5 days, but the increase within the last 20 h was less than 1%. The results of the time sweep fit qualitatively well to the starting points of the temperature sweeps, see Figure 4.21 (b).

4.5 Summary and comparison with literature

In this section, the results are summarized and compared with literature references. In addition, the need for further research is presented.

Crystallinities

It was experimentally proven that SG, cEVA and EVA have semicrystalline domains. The initial crystallinity of all three materials in the "as received" state was determined using DSC and XRD. The two types of experiments provide good agreement. SG has the lowest crystallinity with $\approx$ 9.5 - 9.86 %. For cEVA a value of 17.8 - 18.23% and for EVA a value of 16.9 - 16.92% was determined.
There is almost no literature data on crystallinity for SG, which is why other ionoplastics, which are similar to SG, are also used for comparison. Wakabayashi and Register (2006) have investigated random E/MAA copolymers from DuPont with different MAA contents and different neutralisation levels by DSC and determined degrees of crystallization between 6 - 26% (with $\Delta H_{\mathrm{m,100\%PE}}$ = 278 J/g). The samples were first stored at room temperature for 6 days and then heated from [0:120]°C with a heating rate of 5 K/min.

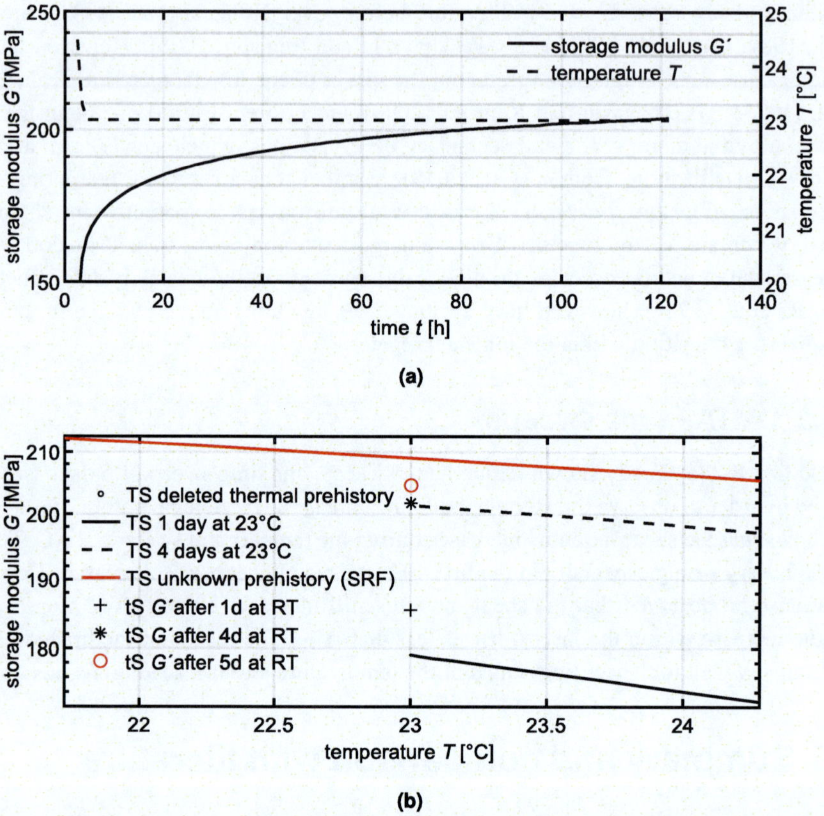

Figure 4.21 SG: Impact of physical aging at room temperature - investigation in a time sweep (a) time sweep results, (b) storage modulus values of the time sweep in comparison to the temperature sweeps.

Gómez et al. (2019) have investigated the crystallinity of the ionoplast Surlyn PC2000 from the company DuPont. This ionoplast has 5.4% MAA and was neutralised to 67%. The material was first stored at room temperature for 2 weeks and then investigated in DSC at a heating rate of 10 K/min in a temperature range of [20:100]°C. A crystallinity of 17% was determined (with $\Delta H_{m,100\%PE}$ = 278 J/g). In his work, Trösch (2015) dealt, among other things, with the laminated glass interlayer SG (at that time still from DuPont) and shows a thermogram of a DSC, which extends over a temperature range of [10:140]°C. No information is available on the heating rate or crystallinity. However, the melting enthalpies of the first heating run are given. Using $\Delta H_{m,100\%PE}$ = 293 J/g, a crystallinity of 5.24% is obtained.

The literature data is summarized in Table 4.5. The crystallinity of approximately 9.5 - 9.86% determined in this thesis is plausible.

Table 4.5 SG: Comparison of crystallinities with literature.

Author and material	Test set-up and temp. prog.	χ [%]
Gómez et al. (2019), Surlyn PC2000, DuPont: 5.4 %MAA, 67% neutralized	DSC, [20:100]°C 10 K/min Sample stored for 2 w at RT	17
Wakabayashi and Register (2006), E/MAA statistical co-polymers, DuPont, different %MAA and neutralization levels	DSC, [0:120]°C 5 K/min Sample stored 6 d at RT	6-26
Trösch (2015)[1], SentryGlas, DuPont	DSC, [10:140]°C, heating rate not specified	5.24
Schuster, SG6000, Kuraray	DSC, [-60:140]°C 20 K/min	9.5
	XRD, RT	9.86

[1] The initial crystallinity was calculated from the ΔH values depicted in a graph with $\Delta H_{m,100\%PE} = 293$ J/g.

Although only a few publications specifically investigate the crystallinity of EVA-based laminated glass interlayers, there is quite a lot of literature on EVA and cEVA in general. However, it is not always apparent whether the samples are crosslinked or non-crosslinked material. Moreover, in some publications the crystallinity was determined with the first heating runs of the DSC (without elimination of the thermal history) and some with the second heating runs (after elimination of the thermal history). Since in the present work there were no major differences in crystallinities between the crosslinked and non-crosslinked material, nor between the first and second heating runs, the comparisons with the literature data are made bundled for cEVA and EVA and independent of the heating cycle. Since only few data are available on laminated glass interlayers made of EVA, care is taken in the literature comparison to ensure that a similar wt% VAC content is present.

Johnsen and Nachtrab (1969) investigated EVA samples with different wt% VAC in DSC measurements in a temperature range of [-80:140]°C with a heating rate of 16 K/min. The thermal history of the samples was eliminated beforehand. In addition, XRD measurements were performed at temperatures below the glass transition temperature. For 30.7 wt% VAC, he determined a crystallinity of 17.4% by DSC and 17.1% by XRD.

Kamath and Wakefield (1965) investigated EVA samples with different wt% VAC by XRD and established a functional relationship between the wt% VAC and the crystallinity. For EVA with 28 wt% VAC, this results in a degree of crystallization of 21%.

Bistac et al. (1998) investigated EVA with 28 wt% VAC and different melt indices by

means of DSC measurements. The first heating run was evaluated, which was conducted with 10 K/min in a temperature range of [-150:150]°C. Crystallinities between 9.9% (at MFI 800 g/10min) and 17.8% (at MFI 3 g/10min) were measured. $\Delta H_{m,100\%PE} = 280$ J/g was used. After consultation with the interlayer manufacturer, the EVA investigated in the present work has an MFI of ≈ 5 - 8 g/10min.
Wang and Deng (2019) investigated crosslinked and non-crosslinked Elvax3175 LG from DuPont with 28 wt% VAC (used as shape memory polymer for biomedical applications) in DSC and by wide-angle XRD (special form of XRD). The crosslinked samples had a gel rate of 96%. The crosslinking was carried out by Wang and Deng (2019) using diculym peroxide. DSC measurements were performed from [-100:200]°C with a heating rate of 20 K/min. It is not specified whether the crystallinity was determined from the first or the second heating scan. A degree of crystallization of 14.3% for the non-crosslinked EVA and 16.4% for the crosslinked cEVA (with $\Delta H_{m,100\%PE} = 287.3$ J/g) was determined. The XRD measurements took place at room temperature and resulted in a value of 27.6% for EVA and 15.2% for cEVA. The rather large deviation between the XRD result and the DSC result for uncrosslinked EVA is explained by Wang and Deng (2019) by the fact that the integrated melting enthalpy in the DSC measurement is overlapped by the glass transition.
Sessini et al. (2019) are investigating bionanocomposites based on ethylene vinyl acetate, which is why studies were also carried out on pure EVA. An EVA with 19 wt% VAC from Exxon Mobil Chemical Company was used. The degree of crystallization was determined to 18% in the second heating cycle of a DSC measurement of [-80:150]°C with 10 K/min (with $\Delta H_{m,100\%PE} = 293$ J/g).
Adothu et al. (2019) investigated the crystallinity of non-crosslinked and crosslinked EVA, which is used as an encapsulant for PV modules. The investigated material consists of 29 wt% VAC and 89.35% gel rate. The degree of crystallization was determined between [-70:280]°C at a heating rate of 10 K/min. It is assumed that the first heating cycle was used. The crystallinity for EVA was determined to 15.06 % and for cEVA to 10.04% (with $\Delta H_{m,100\%PE} = 288$ J/g).
Agroui et al. (2012) investigated the crosslinked EVA A9918P from Specialized Technology Resources with 33 wt% VAC and a gel content of 85%. This EVA is often used for photovoltaic applications. The degree of crystallization was determined by DSC in a temperature range of [-80:85]°C and a heating rate of 7 K/min from the second heating run to 5%. The value is considerably lower than the previously mentioned degrees of crystallization. $\Delta H_{m,100\%PE} = 288$ J/g was used.
The literature data is summarized in Table 4.6. Overall, it can be stated that for comparable EVA, crystallization values between 10 - 18% are represented in literature, which fits well with the results of this chapter.

Table 4.6 EVA and cEVA: Comparison of crystallinities with literature.

Author and material	Test set-up and temp. prog.	χ [%]
Johnsen and Nachtrab (1969), EVA 30.7 wt% VAC	DSC, [-90:100]°C 16 K/min XRD, < T_g	17.4 17.1
Kamath and Wakefield (1965), EVA 28 wt% VAC	XRD	21
Bistac et al. (1998), EVA 28 wt% VAC with various MFI	DSC, [-150:150]°C 10 K/min	9.9-17.8
Wang and Deng (2019), EVA and cEVA Elvax3175LG, DuPont 28 wt% VAC, gel content 96%	DSC, [-100:200]°C 20 K/min XRD, RT	14.3[1] 16.4[2] 27.6[1] 15.2[2]
Sessini et al. (2019), EVA Exxon, Mobil Chemical Company 19 wt% VAC	DSC, [-80:+150]°C 10 K/min	18
Agroui et al. (2012), cEVA A9918P, Specialized Technology Resources 33 wt% VAC, gel content 85%	DSC, [-50:85]°C 7 K/min	5
Adothu et al. (2019), cEVA and EVA 29 wt% VAC, gel content 89.35%	DSC, [-70:280]°C 10 K/min	15.03[1] 10.04[2]
Schuster, cEVA and EVA EV200, Glaast 28 wt% VAC, EVA and cEVA	DSC, [-60:120]°C 20 K/min	11.3-17.4[1] 16.7-18.4[2]
	XRD, RT	16.92[1] 18.23[2]

[1] non crosslinked / uncured EVA.
[2] crosslinked / cured cEVA.

Characteristic temperatures

The characteristic temperatures T_g, T_m and T_c were determined with DSC and DMTA-TS. The results are compared with literature data on comparable materials.

For SG it could be shown by means of DSC as well as by means of DMTA that the glass transition is superimposed by the melting or crystallization of the crystallites. Therefore, the glass transition temperature could only be estimated by DSC. The evaluation of the DMTA-TS with +-3 K/min allowed a detailed determination of the glass transition. In the heating process the glass transition temperature was determined to $T_{g,heat}$ = 49°C and

in the cooling process to $T_{g,cool}$ = 41°C (values at 1 Hz). The melting temperature after elimination of the thermal prehistory is only very slightly dependent on the heating rate and achieves values between $T_{p,m}$ = 74.8°C and $T_{p,m}$ = 77.2°C in the DSC heating process. The crystallization temperature after cooling from the melt is highly dependent on the cooling rate and lies between $T_{p,c}$ = 33.5°C and $T_{p,c}$ = 41.5°C depending on the cooling rate (3 - 20 K/min). The highest cooling rate led to the lowest $T_{p,c}$. The low temperature melting peak during the first heating process (sample "as received") was measured at $T_{p,m1} \approx$ 62°C. However, in the section on the influence of physical aging it could be shown that the position, shape and size of this peak depend on the physical age and the annealing temperature.
The glass transition of cEVA was determined in DSC at +-20 K/min to $T_{g,heat}$ = -26°C and $T_{g,cool}$ = -22.4°C. Similar values ($T_{g,heat}$ = -26.8°C, $T_{g,cool}$ = -21.6°C) were obtained for EVA. The glass transition temperature determined by DMTA-TS at +-2 K/min and 1 Hz for cEVA is $T_g \approx$ -25°C and for EVA $T_g \approx -22$°C. In contrast to SG, there is no dependency on the temperature direction (heating or cooling). The peak melting temperature is between $T_{p,m}$ = 63 - 66°C for cEVA and $T_{p,m} \approx$ 70°C for EVA. The crystallization peak temperatures again show a strong dependence on the cooling rate. For cEVA, peak crystallization temperatures between $T_{p,c}$ = 38°C and $T_{p,c}$ = 46°C and for EVA between $T_{p,c}$ = 46°C and $T_{p,c}$ = 55°C were measured. The low temperature melting peak in the first heating process (sample "as received") occurred at $T_{p,m1} \approx$ 45°C for both, cEVA and EVA.

Trösch (2015) used the DMTA-TS data (1 Hz, three-point-bending mode) depicted in (Puller, 2013) to determine the glass transition temperature of SG to T_g = 59°C. Melting and crystallization were observed in DSC measurements, conducted in a temperature range of [10:140]°C with unknown temperature rate. $T_{p,m1}$ = 53.47°C (low temperature melting peak in first heating cycle), $T_{p,m}$ = 81.81°C and T_c = 4.21°C were obtained.
In his dissertation, Kothe (2013) investigated the aging behaviour of polymer interlayer materials in the construction industry. For SG (at that time still from DuPont) he determined T_g = 59.7°C with a DMTA-TS in tensile mode with a frequency of 1 Hz in a temperature range of [-50:100]°C and a heating rate of 3 K/min.
Kuntsche (2015) also investigated the material behaviour of different interlayers in the construction industry. For SG (SG5000 from Kuraray) he determined T_g = 55.1°C in a DMTA-TS with a frequency of 1 Hz in a torsional strip mode. The heating rate was 1 K/min, and the temperature interval of [-60:80]°C was investigated from cold to warm.
The results of the characteristic temperatures in literature are summarized in Table 4.7. It can be noted that the characteristic temperatures determined in this chapter are slightly lower than those from literature. However, it should be noted that the material is not exactly the same. In (Schneider, 2019) it was shown that SG6000 has a lower heat deflection temperature than SG5000, so the results obtained in this chapter are considered realistic.

Table 4.7 SG: Comparison of characteristic temperatures with literature.

Author and material	T_g [°C]	$T_{p,m1}$ [°C]	$T_{p,m}$ [°C]	T_c [°C]
Trösch (2015), SentryGlas, DuPont	59	53	82	41
Kothe (2013), SentryGlas, DuPont	60	/	/	/
Kuntsche (2015), SG5000, Kuraray	55	/	/	/
Schuster, SG6000, Kuraray	15-49	62	77	33-42

Arsac et al. (1999) studied a series of EVA copolymers with different wt% VAC with DSC and DMTA. The DSC were carried out at a temperature rate of 10 K/min in [-60:140]°C. For a 28 wt% VAC EVA, the low temperature peak was determined to $T_{p,m1}$ = 43.5°C and the high temperature melting peak to $T_{p,m}$ = 73.5°C. The glass transition temperature T_g = -25°C was determined with DMTA in a torsional strip mode, at an angular frequency of 1 rad/s in a temperature range of [-140:melting point]°C.
Johnsen and Nachtrab (1969) defined the glass transition and melting as regions with start and end temperatures in DSC measurements with 16 K/min between [-90:100]°C after a fast cooling from the melt. The glass transition region of a 30.7 wt% VAC EVA starts at a temperature of -40°C and ends at -25°C. The melting region begins directly after the glass transition, at a temperature of -25°C and extends up to a temperature of 83°C.
Wang and Deng (2019) investigated crosslinked and non-crosslinked Elvax3175 LG from DuPont. For 28 wt% VAC EVA with a gel content of 95%, a glass transition temperature of T_g = -30°C was determined in DSC measurmenet with 20 K/min in a temperature range of [-100:200]°C. The melting temperature after elimination of the thermal prehistory was determined to T_m = 73°C for EVA and to T_m = 65°C for cEVA. The crystallization temperature, determined with -20 K/min, was determined to T_c = 45°C for EVA and T_c = 37°C for cEVA.
Sung et al. (2005) studied the effects of crystallinity and crosslinking on the thermal and rheological properties of EVA copolymers with different wt% VAC. The glass transition temperature was determined with DMTA measurements in oscillating torsional mode in a temperature range of [-100:100]°C at a frequency of 1 Hz and at a heating rate of 5 K/min. For non-crosslinked EVA with 26 %wt VAC, a glass transition temperature of T_g = -16°C was determined. This result remained unchanged for different levels of crosslinking. It should be noted, that the crosslinking was done by the author with dicumyl peroxide.
Sessini et al. (2019) performed DSC tests in a heat-cool-heat mode in a temperature range of [-80:150]°C and a temperature rate of 10 K/min. They determined the glass transition temperature of an EVA with 19 wt% VAC from Exxon Mobil Chemical Company to T_g = -26°C. The melting peak temperature was determined to T_m = 86°C and the crystal-

lization peak temperature to $T_c = 70°C$.
Kothe (2013) examined Etimex 486 and STR Photocap 15420 EVA (both presumably crosslinked) with DMTA-TS in tensile mode with a frequency of 1 Hz in a temperature range of [-50:100]°C and a heating rate of 3 K/min. For both materials he obtained $T_g \approx -28°C$.
Agroui et al. (2012) determined the glass transition temperature of cEVA A9918P from Spezialized Technology Resources (33 wt% VAC and gel content 85%) to $T_g = -25°C$, the low melting temperature to $T_{p,m1} = 47.5°C$, the high melting temperature to $T_{p,m} = 55°C$ and the crystallization temperature to $T_c = 36°C$. The melting and crystallization temperature values were determined in a DSC heat-cool-heat cycle in a temperature range of [-50:85]°C with a heating rate of 7 K/min and a cooling rate of 40 K/min. The glass transition temperature was determined with DMTA-TS in tensile mode at a heating rate of 7 K/min and a frequency of 1 Hz.
Adothu et al. (2019) investigated the characteristic temperatures of EVA and cEVA with 29 wt% VAC and 89.35% gel content with DSC in a temperature range of [-70:280]°C and with a heating rate of 10 K/min. The glass transition temperature for EVA was determined to $T_g \approx -39°C$ and for cEVA to $T_g \approx -37°C$. The low temperature melting peak was determined to $T_{p,m1} = 45°C$ for EVA and to $T_{p,m1} = 46.5°C$ for cEVA. The high temperature melting temperature for EVA equals $T_{p,m} = 69°C$ and for cEVA $T_{p,m} = 62°C$.
Pander et al. (2011) investigated cEVA for PV application (without further specification). The glass transition range was determined in DMTA-TS to [-20:+20]°C and the melting range to [40:60]°C.
Eitner et al. (2010) also investigated a cEVA for PV application (without further specification). The glass transition temperature $T_g = -35°C$ was determined with DMTA in torsional strip mode. A temperature range of [-60:100]°C was investigated with an unknown heating rate at a frequency of 1 Hz.
Ledergerber et al. (2016) examined the melting peak temperature of Evatane 28-05 (27 - 29 wt% VAC, MFI 5 - 8 g/10min) from Arkema to $T_m = 72°C$ with DSC in a temperature range of [-50:150]°C and a heating rate of 10 K/min during the second heating cycle.
Kuntsche (2015) determined the glass transition temperature of EVASafe G77 from Bridgestone with different DMTA modes. In the DMTA torsional strip mode, $T_g = -21°C$ was obtained at a frequency of 1 Hz and a heating rate of 1 K/min in [-80:80]°C.

Schuster et al. (2018) investigated evguard from Folienwerk Wolfen (crosslinked EVA, with 28 wt% VAC) and found a glass transition temperature of $T_g \approx = -23°C$ with DMTA-TS (torsional strip and tensile mode). The tests were performed from temperatures above the melting temperature down to at least -60°C with a cooling rate of 1 K/min (in the torsional mode) resp. 2 K/min (in the tensile mode). The melting temperature after elimination of the thermal prehistories and the crystallization temperature were determined in DSC tests with a temperature rate of +-20 K/min in a temperature interval of [-60:100]°C to $T_m = 63°C$ and $T_c = 38°C$. In the same paper, EVASafe G77 from Bridgestone was

Table 4.8 EVA and cEVA: Comparison of characteristic temperatures with literature.

Author and Material	T_g [°C]	$T_{p,m1}$ [°C]	$T_{p,m}$ [°C]	T_c [°C]
Arsac et al. (1999), EVA 28 wt% VAC, M = 58.3 g/mol	-25	46	74	/
Johnsen and Nachtrab (1969), EVA 30.7 wt% VAC	[-40:-25]	/	[-25:83]	/
Wang and Deng (2019), Elvax3175LG, DuPont 28 wt% VAC, gel content 96%	-30	/	73[1] 65[2]	45[1] 37[2]
Sung et al. (2005), EVA26	-16	/	/	/
Sessini et al. (2019), Exxon Mobil Chemical Company 19 wt% VAC	-26	/	86	70
Kothe (2013), Etimex 486 STR Photocap 15420	-28	/	/	/
Agroui et al. (2012), A9918P Specialized Technology Resources 33 wt% VAC, gel content 85%	-25	47.5	55	36
Adothu et al. (2019), cEVA and EVA 29 wt% VAC, gel content 89.35%	-39[1] -37[2]	45[1] 47[2]	69[1] 62[2]	/
Pander et al. (2011), PV cEVA	[-20:20]	[40:60]	/	/
Eitner et al. (2010), PV cEVA	-35	/	/	/
Schaible et al. (2016), Evatane 28-05 Arkema	/	/	72	/
Kuntsche (2015), EVASafe G77 Bridgestone	-21	/	/	/
Schuster et al. (2018), evguard Folienwerk Wolfen, 28%VAC, crosslinked	-23	/	63	38
Schuster et al. (2018), EVASafe G77 Bridgestone	-16	/	71	43
Schuster, EV200, Glaast 28 wt% VAC	-(22-27)[1] -(23-26)[2]	45[1,2]	70[1] 63-66[2]	47-55[1] 38-46[2]

[1] non-crosslinked / uncured EVA.
[2] crosslinked / cured cEVA.

investigated in a similar manner. T_m = 71°C and T_c = 43°C were obtained from DSC measurements with +-20 K/min and T_g = -16°C was obtained from DMTA-TS (double shear mode) with 2 K/min in [-65:80]°C in heating direction.
The results from literature are summarized in Table 4.8. Considering that the characteristic temperatures are influenced by many factors (wt% VAC, melt flow index, temperature program, evaluation methods, etc.) and that not all parameters are mentioned in most of the literature, there is a good agreement with the characteristic temperatures determined in this chapter.

Crystallization kinetics

Crystallization kinetics are based on the relative degree of crystallization, which can be determined either from DSC or DMTA. Both methods have been investigated. The determination of relative crystallinities from DMTA data proved to be very difficult and was not pursued further. Different kinetic models were examined using the DSC data, whereby Mo-Liu's model has been found to be the most suitable for the investigated materials. Nevertheless, a comparative DSC measurement on SG showed that deviations from the model occur and that the model needs to be approved. The investigation of different kinetic models proved for all three materials, that their crystallization processes consist of several processes, which can not be described accurately with simple models.

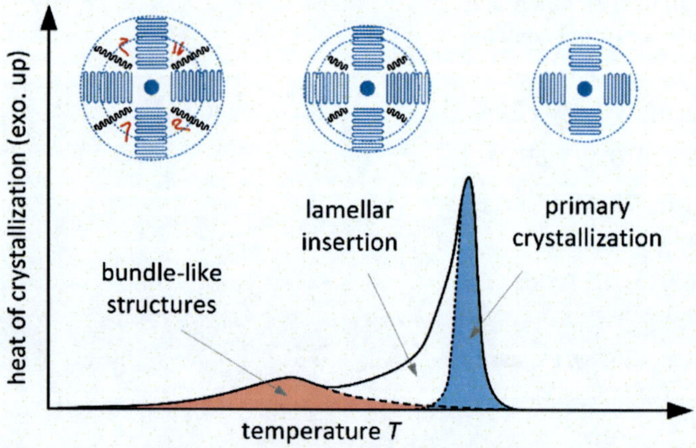

Figure 4.22 Decomposition of the crystallization process according to van Drongelen et al. (2015).

van Drongelen et al. (2015) gave a suggestion how to investigate the quite broad crystallization peaks on the example of a metallocene LLDPE (ExxonMobil), which is a polyolefin. They divided the crystallization process into three steps, see Figure 4.22. The primary crystallization process can be detected by the symmetric fraction of the sharp peak at high temperatures and can be fitted with a Gaussian function (eq. 3.23). Primary crystallization is assigned to long chains, which fold into stacked lamellae that span the spherulites from the melt. The second step, secondary crystallization, is assigned to shorter chain sections that form lamellar stacks in the amorphous regions between the primary lamellae inside the spherulites. This process follows rapidly after primary crystallization, which generates the asymmetric shape of the primary peak. The third part in the crystallization process, also called second stage of secondary crystallization can be explained by the fact, that the already formed lamellae restrict the movements of the remaining sequences, which results in a granular morphology consisting of small bundled structures or loosely packed sequences. In (van Drongelen et al., 2015) this third process was detected via a second, much wider, peak at much lower temperatures and was fitted to a Lorentzian (eq. 3.24) function. Additionally it was assumed that the enthalpy not included in both peak functions can be assigned to the first step of secondary crystallization and can be described by n Gaussian functions, representing multiple lamellar insertion processes with different thickness.

Figure 4.23 exemplary shows a crystallization peak deconvolution of the DSC (-10 K/min) data for all three materials. The peak deconvolution was performed with MATLAB®. Two to three Gaussian functions were used to fit the data. The results of the peak deconvolution are shown in Table 4.9.

Table 4.9 XRD results.

Sample	SG	cEVA	EVA
Peak 1			
x_c [°C]	32.34	19.25	17.71
w [°C]	10.86	32.16	39.57
A [°C mW/mg]	0.59	4.14	3.21
H [mW/mg]	0.0511	0.1208	0.0765
Peak 2			
x_c [°C]	38.52	35.46	38.77
w [°C]	8.52	10.46	23.80
A [°C mW/mg]	0.62	2.45	2.28
H [mW/mg]	0.0674	0.2202	0.0901
Peak 3			
x_c [°C]	/	41.03	50.08
w [°C]	/	3.72	4.81
A [°C mW/mg]	/	1.97	2.57
H [mW/mg]	/	0.4960	0.5018

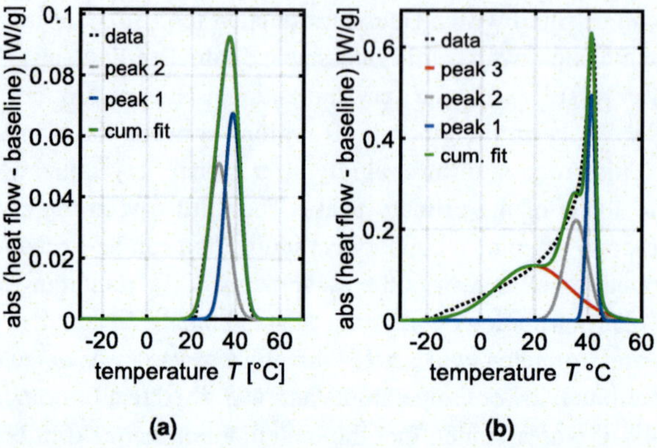

Figure 4.23 Decomposition of the crystallization process. The decomposition was performed with *peakfit.m version 9.0* in MATLAB® (a) SG, (b) cEVA, (c) EVA.

Analogous to the crystallization during the cooling process, crystallization kinetics could be established for cold crystallization in the heating cycle, which would be interesting for SG, and for secondary crystallization, which occurs with increasing physical age. However, due to insufficient data, this is not investigated within the scope of the present work. For the latter, it is proposed to perform the following DSC investigations: Heat samples above the melting temperature, and cool at a defined cooling rate. The samples should then be stored at RT for different durations. After defined storage times, the sample is to be cooled down and a DSC heating run with a defined heating rate is to be performed. Secondary peaks of different sizes are to be expected, which increase in area with increasing storage time and whose peak temperature moves towards higher temperatures. In addition to storage at room temperature, different storage durations at elevated temperatures (but below the melting temperature of the primary crystallites) are to be investigated.

5 Identification of time-superposition principles for EVA and ionoplastic interlayers

This chapter refers to the materials SG and cEVA. In contrast to the previous chapter, non-crosslinked EVA is not investigated.
The aim of the following investigations is to determine the linear viscoelastic material behaviour in dependency of temperature and crystallinity. DMTA-TFS are performed and evaluated based on chapter 2.3.1, chapter 2.4.1 and chapter 3. In the following, first the DMTA-TFS data are used to check the characteristic temperatures found in the previous chapter. Subsequently, mastercurves are created, Prony parameters and time-superposition principles are identified.

5.1 Temperature-frequency and frequency-temperature sweeps

The temperature-frequency-sweeps (TFS) have been carried out from warm to cold if not otherwise specified. This eliminates the unknown thermal prehistory. In addition, since the crystallization temperature is lower than the melting temperature, the material is in the molten, purely amorphous state for a wider temperature range. Since the amorphous state is the "worst case" for the stiffness, it was assumed that this avoids an overestimation of the shear modulus at high temperatures. In addition, a comparative cold-to-warm TFS was performed for SG prior the warm-to-cold TFS (but after removal of thermal prehistory). Finally, to investigate the influence of physical aging, the SG sample, which was investigated in the 5 day time sweep was used (directly following the tS, without sample removal) to perform a final TFS from cold to warm. The test parameters for the performed DMTA-TFS on SG are summarized in Table A.13.
For cEVA, only a warm-to-cold TFS has been performed. Additionally, in the higher temperature range, a frequency-temperature sweep (FTS) with a constant temperature rate of 2 K/min was conducted in both directions (warm to cold and cold to warm). The test parameters are summarized in Table A.14 and Table A.15.

M. Schuster, *Characterization of Laminated Safety Glass Interlayers*,
Mechanik, Werkstoffe und Konstruktion im Bauwesen 66,
https://doi.org/10.1007/978-3-658-39821-7_5

5.1.1 Frequency- and temperature-dependent material behaviour

If the measured quantities for different frequencies are plotted as a function of temperature, the frequency-independent thermodynamic transformations can be distinguished from the frequency-dependent kinetic processes. Figure 5.1 (a), (c) and (e) show the results for SG in the cooling mode. The steep rise in the storage modulus starting from the high temperatures occurs at a temperature of about 55°C for all frequencies. As already shown in the previous section, this steep rise in the storage modulus corresponds to the beginning of the crystallization process. The glass transition temperature T_g is investigated by means of the maximum loss factor $\tan\delta$. A strong dependency of the frequency can be seen. At 0.1 Hz: $T_g = 45°C$, at 1 Hz: $T_g = 50°C$ and at 10 Hz: $T_g = 55°C$.
Figure 5.1 (b), (d) and (f) show the results for SG in the heating mode (after deleting the thermal prehistory). This time, two maxima can be seen in the loss factor $\tan\delta$; one at about $T_1 \approx$ -30°C and one at about $T_2 \approx$ 50°C. Both are frequency dependent (for 0.1 Hz: T_1 = / T_2 = 45°C, 1 Hz: T_1 = -30°C T_2 = 55°C, 10 Hz: T_1 = -25°C T_2 = 65°C) and thus represent relaxation processes. These were also mentioned in (Wakabayashi and Register, 2006) and explained as the glass transition of the ion-poor and ion-rich regions. The melting process cannot be identified in the illustrations.
Figure 5.2 shows the results for SG in the heating mode after a 5 d storage at T = 23°C. Again, two frequency-dependent maxima can be seen in the loss factor curve at $T_1 \approx$ -30°C and $T_2 \approx$ 50°C (for 0.1 Hz: T_1 = -25 T_2 = 45°C, 1 Hz: T_1 = -30°C T_2 = 55°C, 10 Hz: T_1 = -30°C T_2 = 60 - 65°C). An initial drop in stiffness is observed in the storage modulus curve G' for all frequencies in the temperature range of 25 - 35°C. This is attributed to the melting of the secondary crystallites, which formed during the 5 d storage at room temperature. However, it should be noted that the measurement points of a curve were recorded in a 5°C grid and the specified values are therefore not particularly accurate.

Figure 5.3 shows the results of the cEVA TFS in the cooling mode performed with an approx. 8 mm thin sample ((a), (c), (e)) and an approx. 1.6 mm thick sample ((b), (d), (f)). In both cases, the measurement results are not stable for all measurement frequencies, which is why only reduced data sets were used for evaluation. The reason for the unstable measurements is still unclear; one possibility would be the appearance of resonance effects. The thicker sample shows lower storage modulus values at high temperatures, which fits better with the results from the temperature sweep in the previous chapter. The onset of crystallization is in both cases recognizable by a frequency-independent increase in the storage modulus. This occurs at temperatures around $T \approx$ 50 - 55°C.

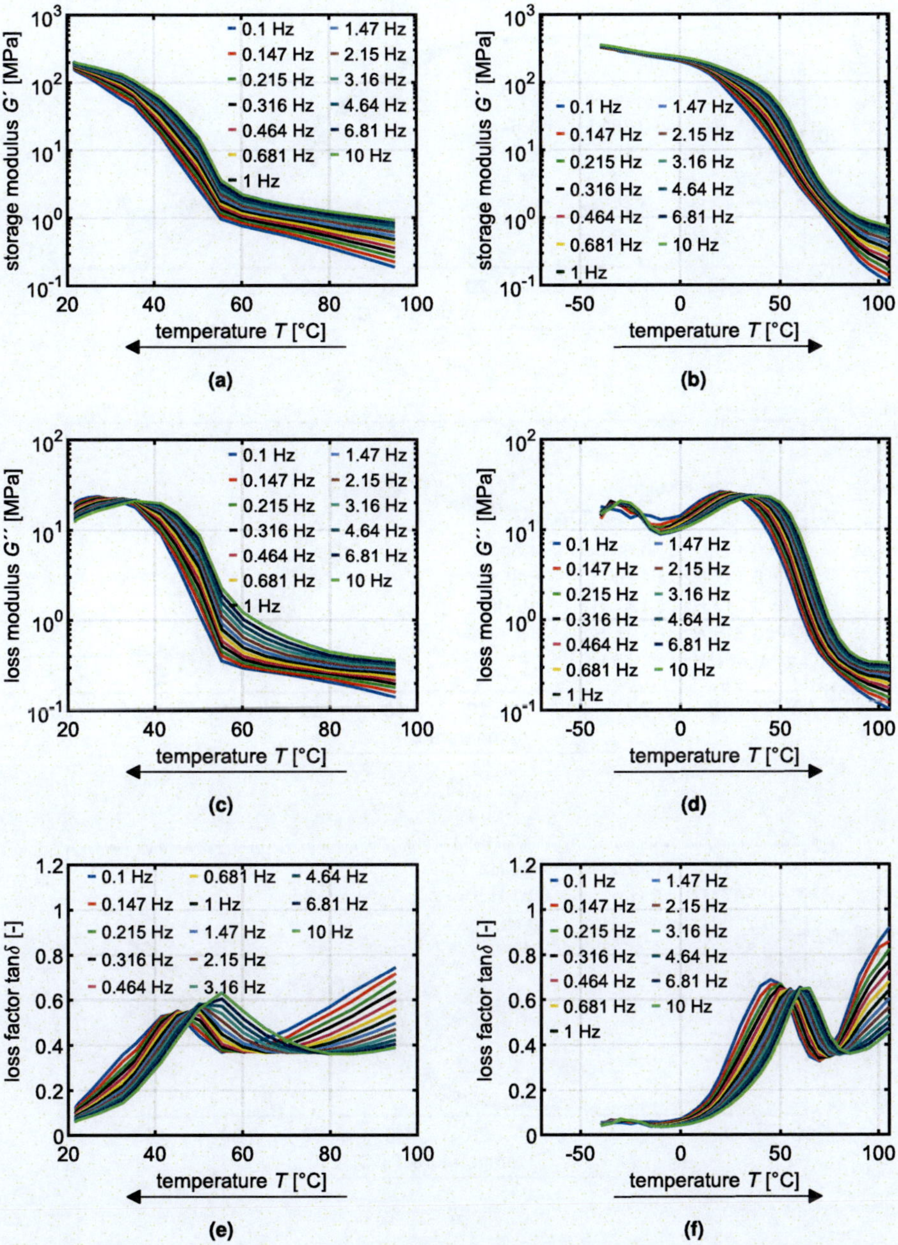

Figure 5.1 SG: Temperature dependent material behaviour for different frequencies (a) storage modulus in TFS cooling, (b) storage modulus in TFS heating, (c) loss modulus in TFS cooling, (d) loss modulus in TFS heating, (e) loss factor in TFS cooling, (d) loss factor in TFS heating.

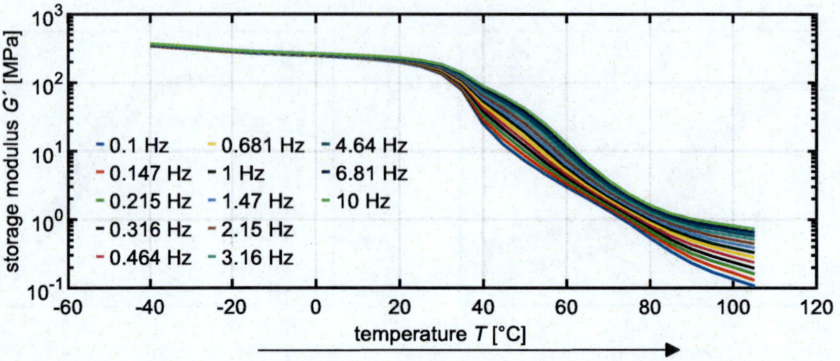

(a)

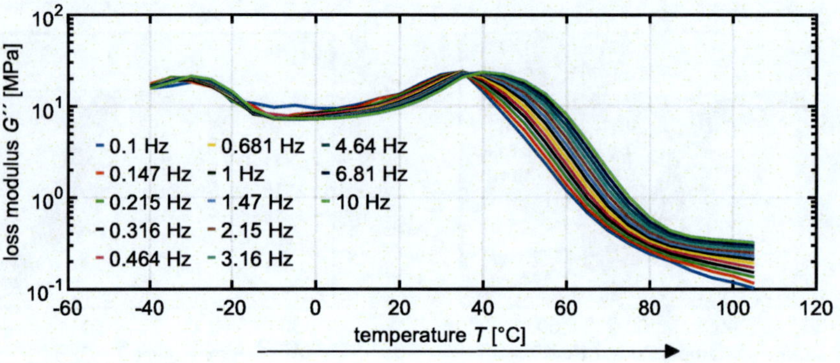

(b)

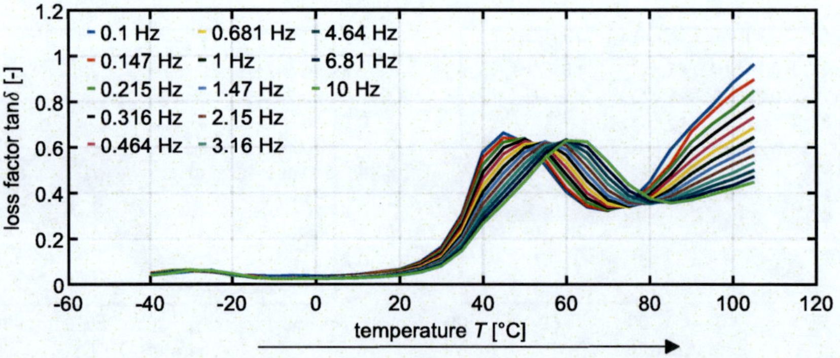

(c)

Figure 5.2 SG: Temperature dependent material behaviour for different frequencies in TFS heating after a 5 d storage at 23 °C (a) storage modulus, (b) loss modulus in TFS heating, (c) loss factor.

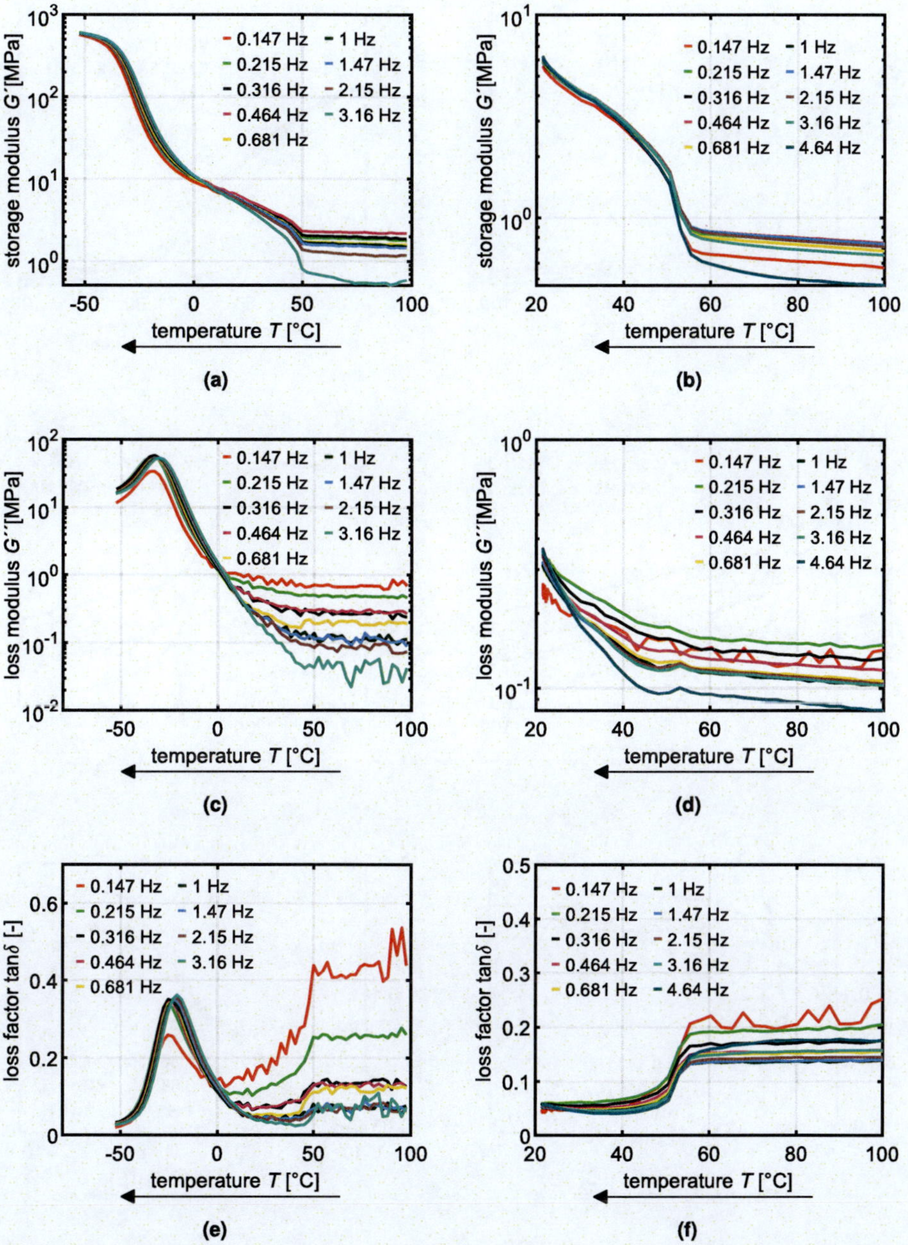

Figure 5.3 cEVA: Temperature dependent material behaviour for different frequencies measured in TFS (a) storage modulus, (b) storage modulus thicker sample, (c) loss modulus, (d) loss modulus thicker sample, (e) loss factor, (d) loss factor thicker sample.

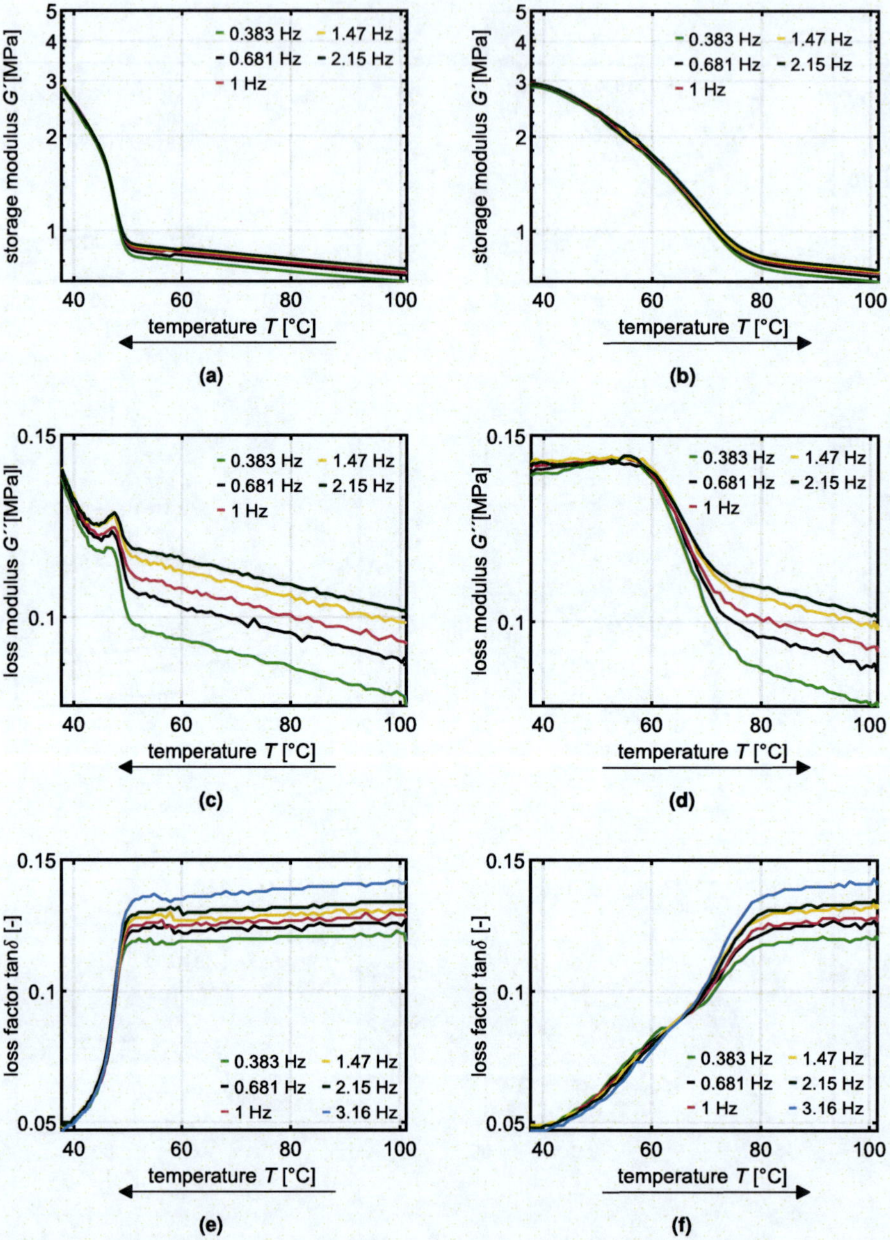

Figure 5.4 cEVA: Temperature dependent material behaviour for different frequencies (a) storage modulus in FTS cooling, (b) storage modulus in FTS heating, (c) loss modulus in FTS cooling, (d) loss modulus in FTS heating, (e) loss factor in FTS cooling, (d) loss factor in FTS heating.

Again, the glass transition temperature T_g is shown by a frequency-dependent maximum in the loss factor tan δ at approx. -20°C (0.215 Hz: T_g = -25°C, 1 Hz: T_g = -22.5°C, 3.16 Hz: T_g = -20°C). The measurement points of a curve were recorded in 2.5°C steps.
The results of the FTS are shown in Figure 5.4. Here too, not all measurement points could be used for further evaluation. The onset of crystallization is clearly seen at a temperature of 50°C in the cooling storage modulus curves.

5.1.2 Comparison with temperature-sweep data

In order to compare the TFS data with the TS data, the step-shaped temperature programs of the TFS are linearized. The TFS temperature steps had a height of 5°C. The dwell time at one step lasted approx. 7.7 min and the duration between two steps for temperature regulation lasted approx. 10 - 12 min, which results in a linearized heating or cooling rate of $\approx$ 0.3 K/min, see Figure 5.5 (a) and (b).

Comparing the storage modulus of the linearized temperature curve with that from the TS in Figure 5.5 (c), deviations are observed in the range of 20 - 50°C in both the heating and cooling data. At the same temperature, higher storage modulus values are present in the linearized temperature curve. This is plausible, since the temperature rate in the linearized temperature program is significantly lower, which leads to a higher cold crystallization during heating (see Figure 4.8), and to a faster crystallization during cooling (see Figure 4.19).

Comparing the storage modulus of the linearized temperature programs without thermal age and with a thermal age of 5 days (see Figure 5.5 (e)), higher storage moduli can be seen in the heating curve of the aged test specimen in a temperature range of approximately 0 - 40°C. Above 40°C, the heating curves are congruent. This was predictable based on the investigation into physical age carried out in the previous chapter (Figure 4.20). It is interesting that the storage modulus of the cooling curve below 45°C lies between the two heating curves. This could possibly be a measurement inaccuracy, or it could mean that the degree of crystallization in the cooling curve is between those of the two heating curves.

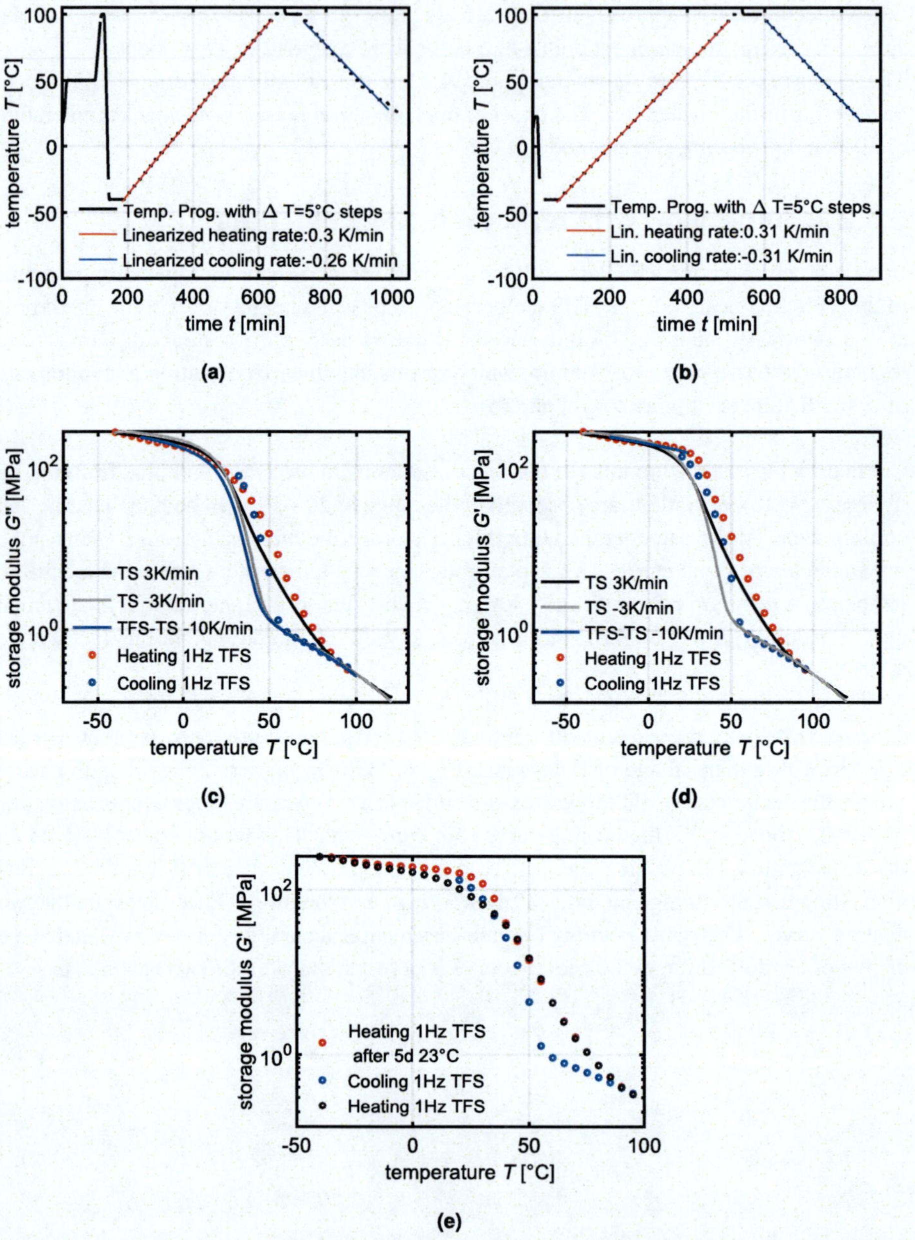

Figure 5.5 Comparison of 1 Hz TFS with TS data, (a) TFS temperature program 1, (b) TFS temperature program 2: after 5 d at 23 °C, (c) comparison of 1Hz storage modulus data, (d) comparison of 1 Hz storage modulus data after 5 d at 23 °C, (e) comparison of 1 Hz TFS data with and without 5 d at 23 °C.

For cEVA, the TFS temperature steps corresponded to 2.5°C. For the thin sample, the dwell time at one step was approx. 1.7 min and the duration between two steps for temperature regulation was approx. 4 min. As shown in Figure 5.6 (a), the linearized cooling rate corresponded to -0.42 K/min.

For the thicker sample, the dwell time at one step lasted again 1.7 min. However, since in this experiment, no nitrogen was used, the duration of the temperature regulation between two steps unfortunately took longer (approx. 7 min). This reduces the linearized cooling rate down to approx. -0.22 K/min, see Figure 5.6 (b).

Since the high temperature data of the thin cEVA sample seem to be too high (compared to the temperate sweep data in the previous chapter), the thin sample data was only used in the low temperature range ([25:-50]°C). In the high temperature range ([100:25]°C) the thick sample data was used. As shown in the previous chapter, the prevailing degree of crystallization is dependent on the cooling rate. Accordingly, caution is required when combining the two data sets. The data set switches at a temperature of 25°C. Remembering Figure 4.6 (b), a relative crystallinity of 73% is reached at 25°C with a cooling rate of -2 K/min. Since here, the cooling rates are considerably smaller (-0.42 K/min resp. -0.22 K/min), it is assumed that the relative crystallinity at 25°C is in both cases close to 1 and that the data sets hence can be combined. Figure 5.6 (c) confirms that the storage modulus of both tests coincide for temperatures lower than 50°C.

Figure 5.6 (e) compares the combined TFS data at a frequency of 1 Hz, with the data of the FTS at 1 Hz (heating rate 2.1 K/min, cooling rate: -1.57 K/min as shown in Figure 4.6 (d)) and the DMTA-TS data with a temperature rate of 2 K/min. The small difference in cooling rate (-2 K/min for the TS and -1.57 K/min in the FTS) already leads to a shift in the onset of the crystallization process. As expected, the process starts at higher temperatures for lower cooling rates.

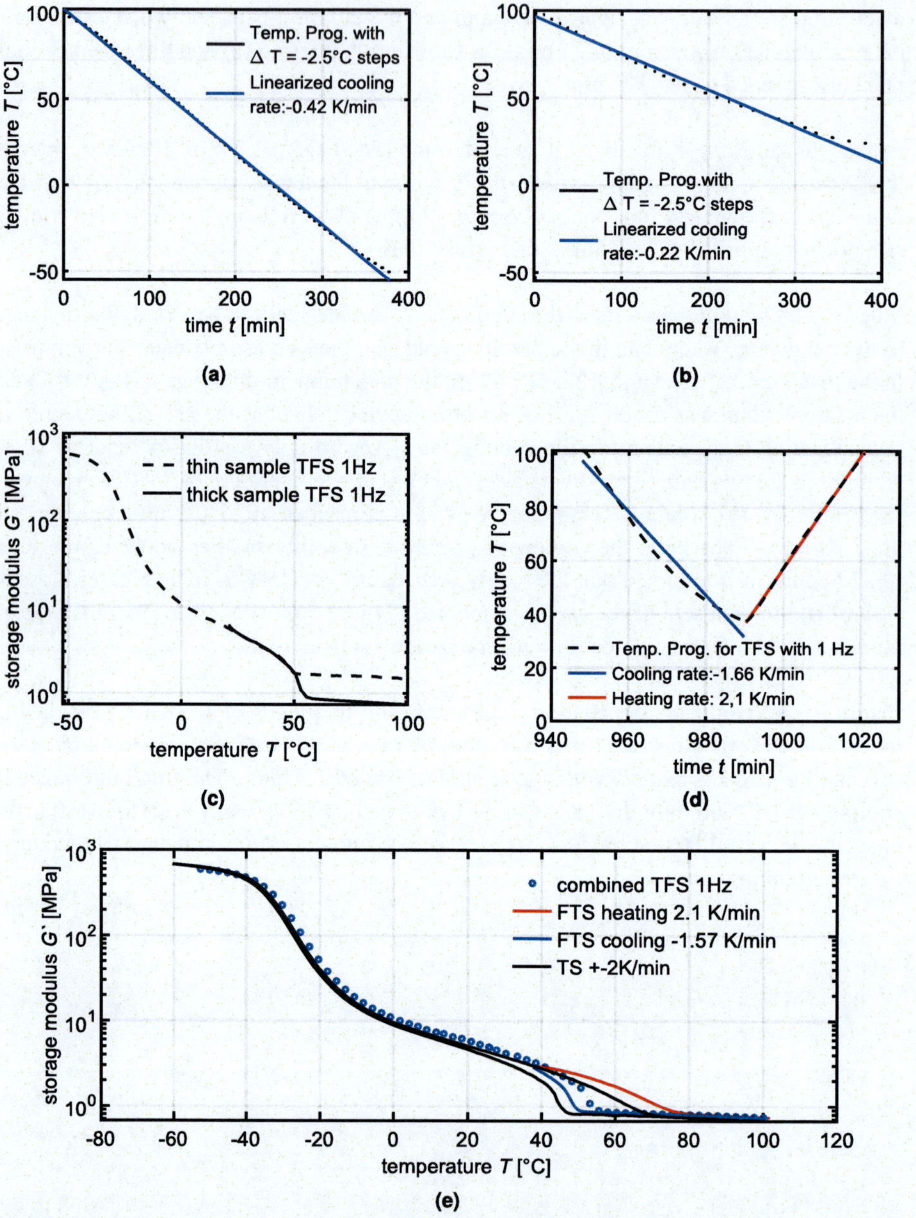

Figure 5.6 Comparison of 1 Hz TFS and FTS with TS data, (a) TFS temperature program thin sample, (b) TFS temperature program thick sample, (c) comparison of TFS thin and thick storage modulus at 1 Hz, (d) FTS temperature program, (e) comparison of 1 Hz storage modulus.

5.2 Creation of mastercurves

To determine the Prony parameters and time-superposition principles, mastercurves for a certain reference temperature T_{ref} have to be created from temperature-frequency sweep (TFS) or frequency-temperature-sweep (FTS) curves. If the Cole-Cole and van-Gurp-Palmen plots each show a closed course, the mastercurve can be created by a purely horizontal shift of the individual curves. A MATLAB® script, that incrementally shifts the individual complex modulus $|G^*|$ frequency curves until a continuous curve is created is used (see also chapter 3). The horizontal shift factors are then examined in more detail and mathematically approximated by means of time-superposition principle functions (chapter 2.4.1). The mastercurves of the storage G' and loss G'' modulus at its reference temperatures are mathematically approximated by a Prony series (chapter 2.3.1).

5.2.1 Mastercurves for SG

The reference temperature has been chosen to T_{ref} = 95°C for SG. This temperatures is above the melting temperature of the crystallites, so that at this temperature the samples are in the amorphous state. Figures 5.7 and A.4 show the results for SG in the TFS cooling mode, Figures 5.8 and A.5 for the TFS heating mode (after deleting the thermal prehistory) and Figures 5.9 and A.6 for the TFS heating mode after a 5 days storage at T = 23°C. In all three cases, the creation of a continuous complex modulus $|G^*|$ curve was possible by purely horizontal shifting the individual curves. Looking at the Cole-Cole and van-Gurp-Palmen plots in the appendix, there are only slight deviations from a clear course and the mastercurves of the storage G' and loss modulus G'' as well as the loss factor $\tan\delta$ are also plausible. Figure 5.10 shows the mastercurves of the complex modulus $|G^*|$ from the heating and cooling processes. In the range between 10^3 and 10^8 Hz (corresponds to the cooling curves in the temperature range [35:50]°C), the mastercurve from the cooling curves is above that from the heating curves. Figure 5.11 shows that mastercurves of the complex modulus $|G^*|$ from both heating processes with different thermal histories are congruent. If, for example, T = 20°C had been selected as the reference temperature, different (horizontally shifted) mastercurves would result, see Figure 5.12. Overall, the three mastercurves at T_{ref} = 95°C are very similar, so that in the following it is assumed that one and the same mastercurve (independent of the temperature history) is produced at reference temperatures in the amorphous range by purely horizontal shifting of the measurement curves. However, the shift factors of the individual curves differ considerably: E.g., the shift factor at T = 20°C corresponds to $\log a$ = 9.63 for the heating data after elimination of the thermal history and to $\log a$ = 13.38 for the heating curve after 5 days at T = 23°C. This must be caused by the different temperature programs, which lead to different crystallinities. Hence, it is assumed, that the horizontal shift factor is not only dependent on temperature but also on the prevailing degree of crystallization. The shift factors are listed in Table A.16 - Table A.18.

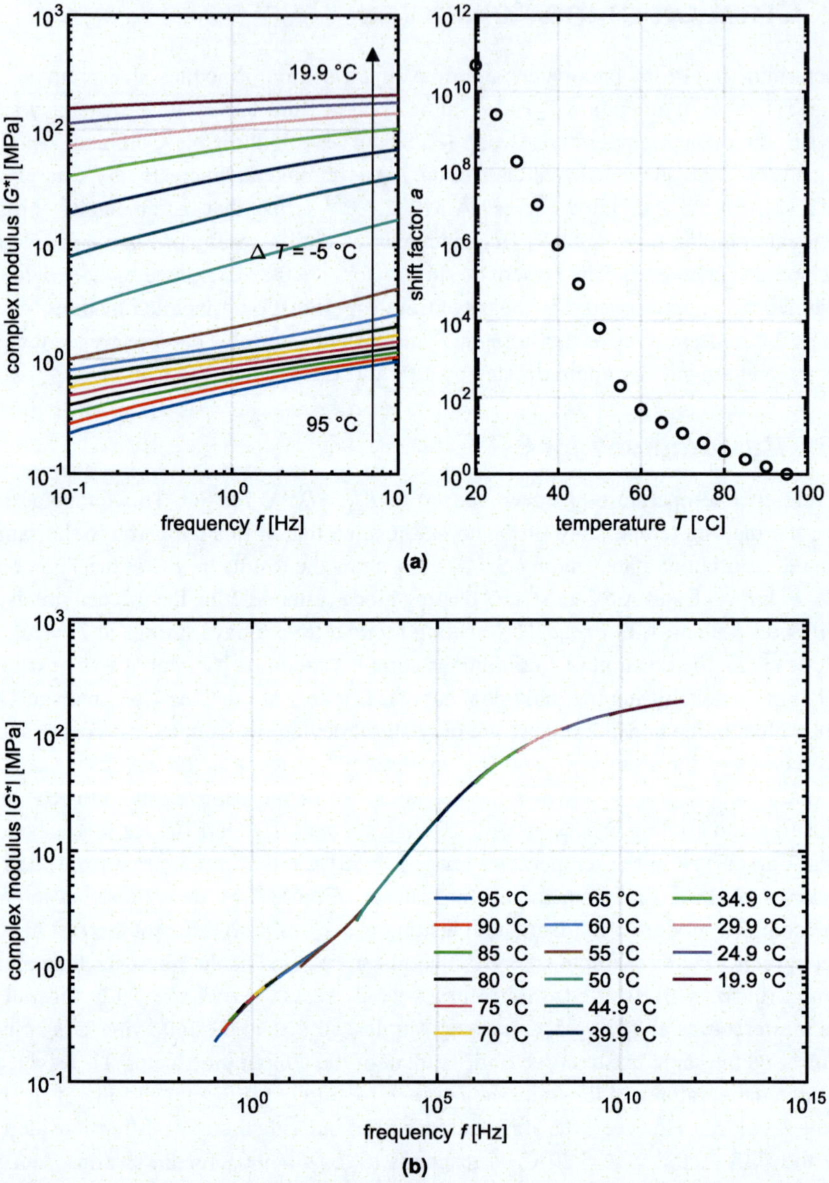

Figure 5.7 SG: mastercurving process with TFS cooling data, (a) raw data and shift factors, (b) complex modulus mastercurve at T_{ref} = 95 °C.

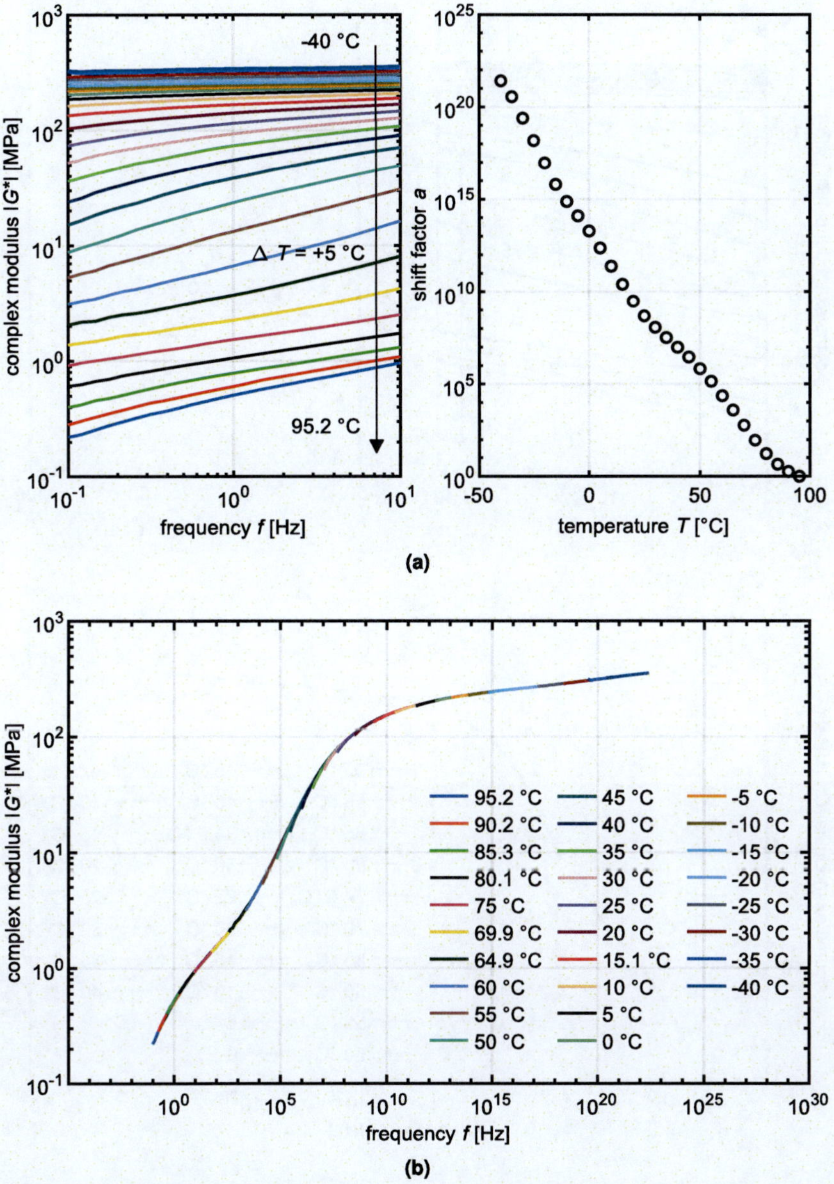

Figure 5.8 SG: mastercurving process with TFS heating data, (a) raw data and shift factors, (b) complex modulus mastercurve at $T_{\mathrm{ref}} = 95\,°\mathrm{C}$.

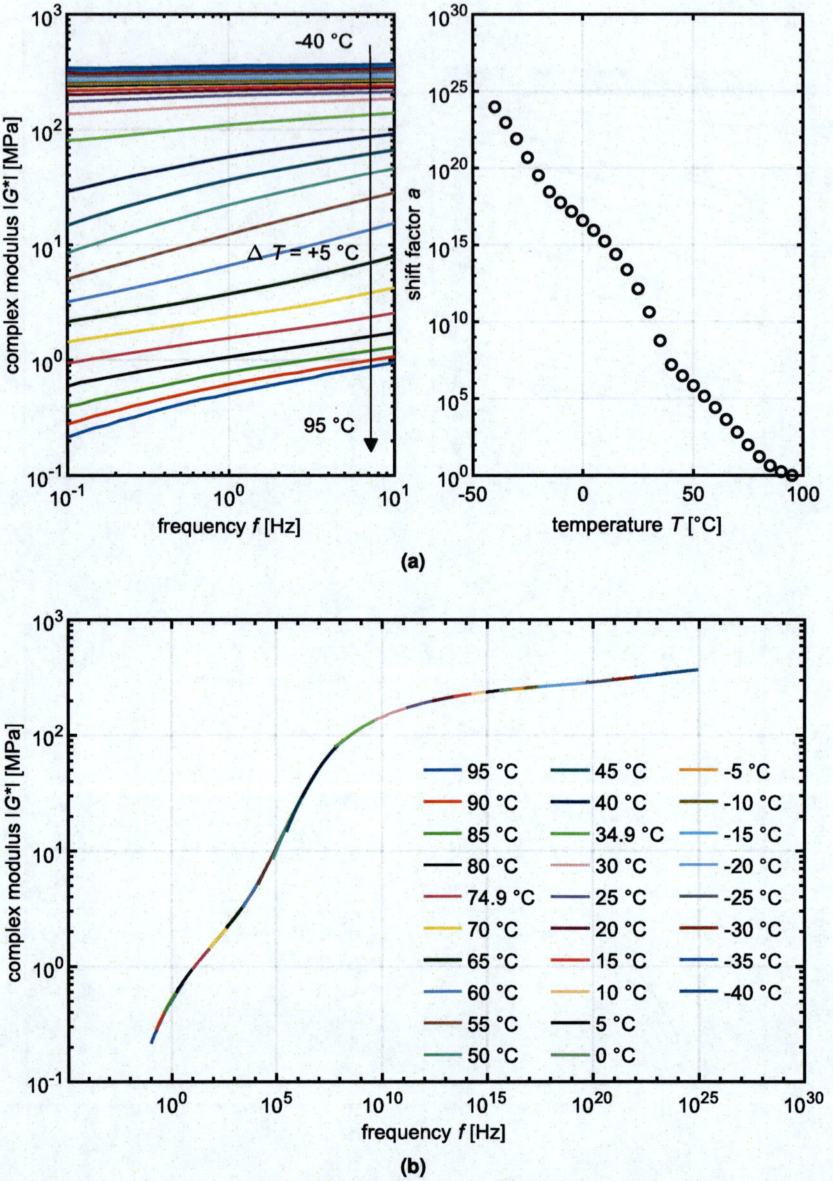

Figure 5.9 SG: mastercurving process with TFS heating after 5 d at 23 °C data, (a) raw data and shift factors, (b) complex modulus mastercurve at T_{ref} = 95 °C.

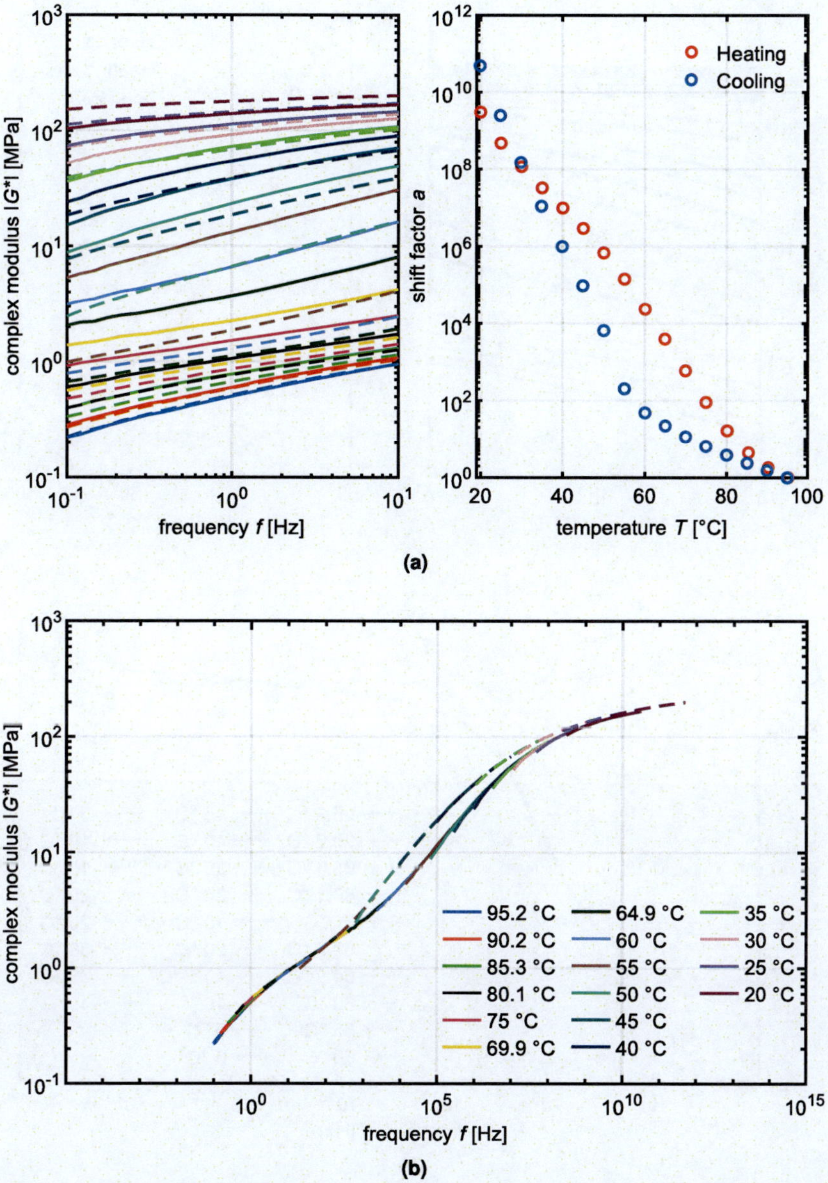

Figure 5.10 SG: comparison of heating and cooling mastercurves at T_{ref} = 95 °C, dashed lines: cool, solid lines: heat, (a) raw data and shift factors, (b) mastercurves.

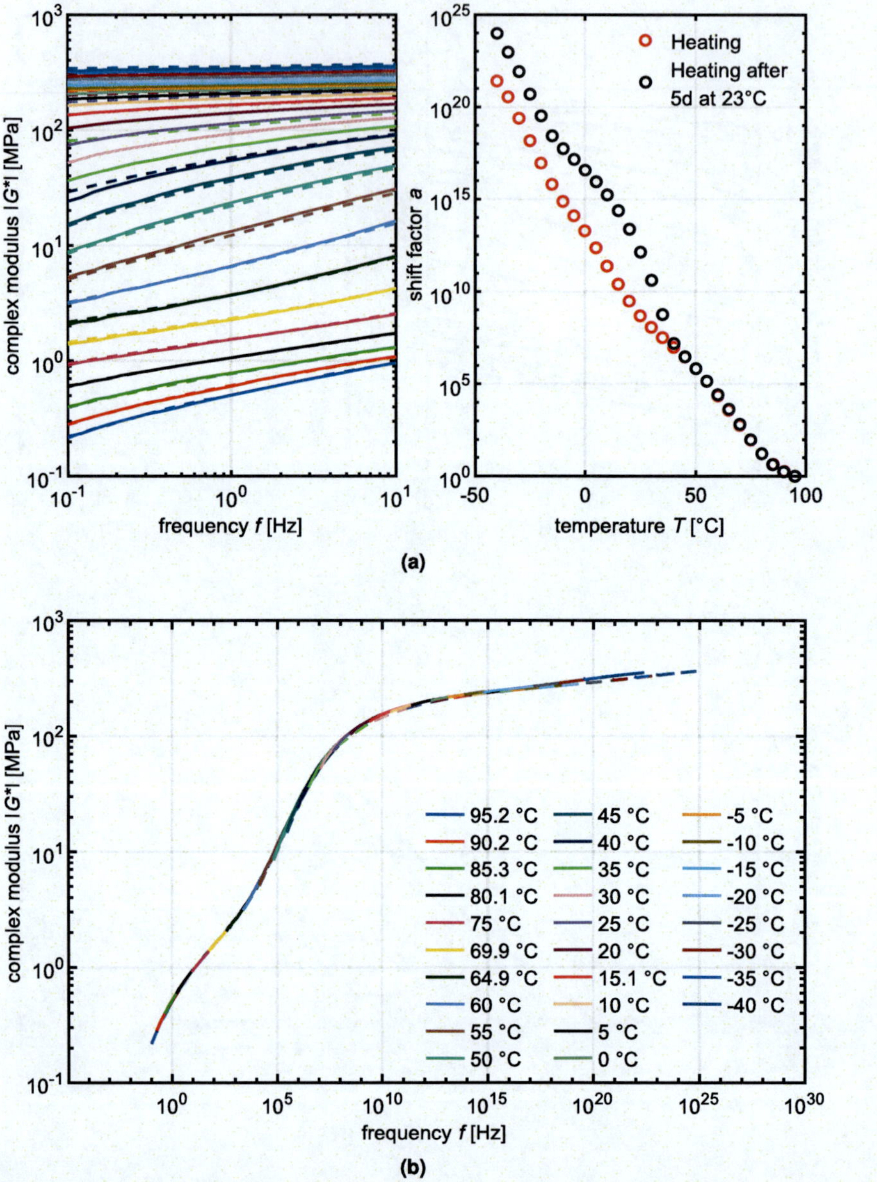

Figure 5.11 SG: comparison of heating and heating after 5 d at 23 °C mastercurves at T_{ref} = 95 °C, dashed lines: heat after 5 d, solid lines: heat, (a) raw data and shift factors, (b) mastercurves.

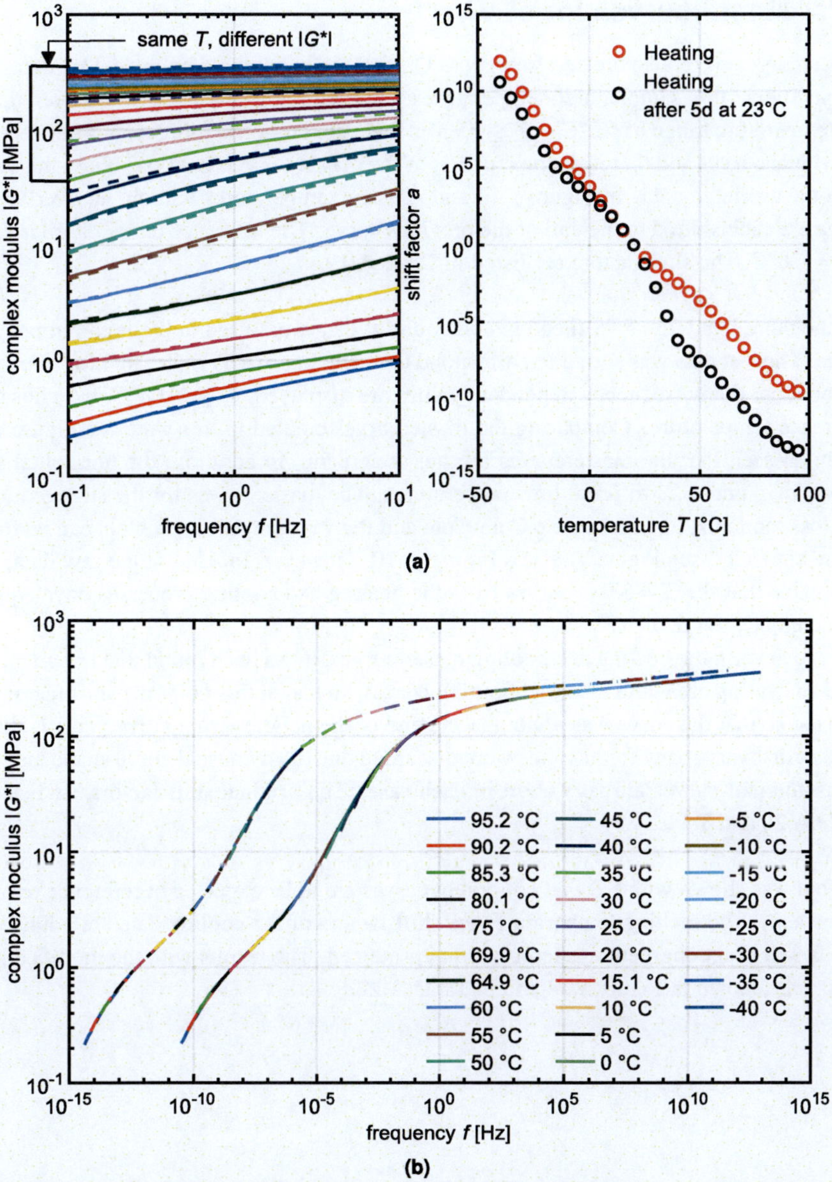

Figure 5.12 SG: comparison of heating and heating after 5 d at 23°C mastercurves at T_{ref} = 20°C, dashed lines: heat after 5 d, solid lines: heat, (a) raw data and shift factors, (b) mastercurves.

5.2.2 Mastercurves for cEVA

Mastercurves were also created for cEVA. On the one hand, the combined TFS data (low temperatures: thin sample, high temperatures: thick sample) were used. It turns out that in the temperature range of $6.5°C \leq T \leq 55.5°C$ there are no overlapping ranges between the individual curves and therefore the creation of the mastercurve is not possible in this range with the available data. In Figure 5.13 and A.7, the temperatures without an overlapping range are each shifted to the end of the previous curve. The reference temperature is set to T_{ref} = -20°C. The shift factors are listed in Table A.19.

On the other hand, the FTS (heating and cooling) data were used in the high temperature range. The test data was sorted to correspond to the output of a TFS, i.e. that for a constant temperature many frequency dependent values are displayed. T_{ref} = 100°C was chosen as reference temperature. Comparing the mastercurves created in this way, see Figure A.8, it can be seen that the mastercurves are not congruent. In addition, the horizontal shift factors a_{heat} and a_{cool} differ at low temperatures. The mastercurves for the storage modulus, loss modulus, loss factor, the Cole-Cole and the van-Gurp-Palmen plots can be found in the appendix, see Figure A.9 and Figure A.10. However, looking at the raw data, it is noticeable that the T = 37°C curves from the heating and cooling processes have equally high complex modulus $|G^*|$ values.
After reviewing the heating and cooling mastercurves, it has been found that not all curves have an overlap area with the neighbouring curves, so that at this point the shift factor was selected so that the curve was shifted to the end of the neighbouring curve. This is due to the fact that very many frequencies were discarded due to an unusual curve shape and thus the isothermal curves are very short in each case. The original shift factors are listed in Table A.20 and Table A.21.

Since it was shown with SG that congruent curves are to be expected at reference temperatures above the melting temperature, the shift factors of the cooling data are adjusted in such a way that congruent mastercurves are produced. The results with the modified shift factors are shown in Figure 5.14 and in Table A.22.

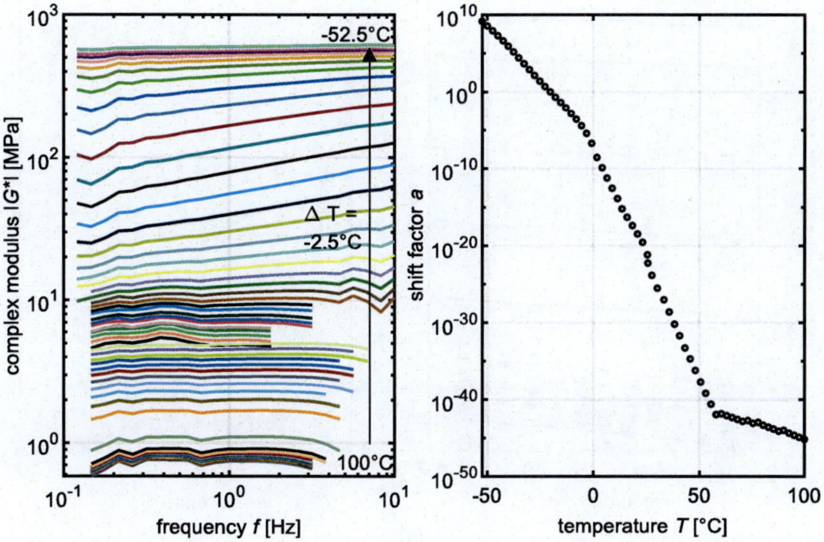

(a)

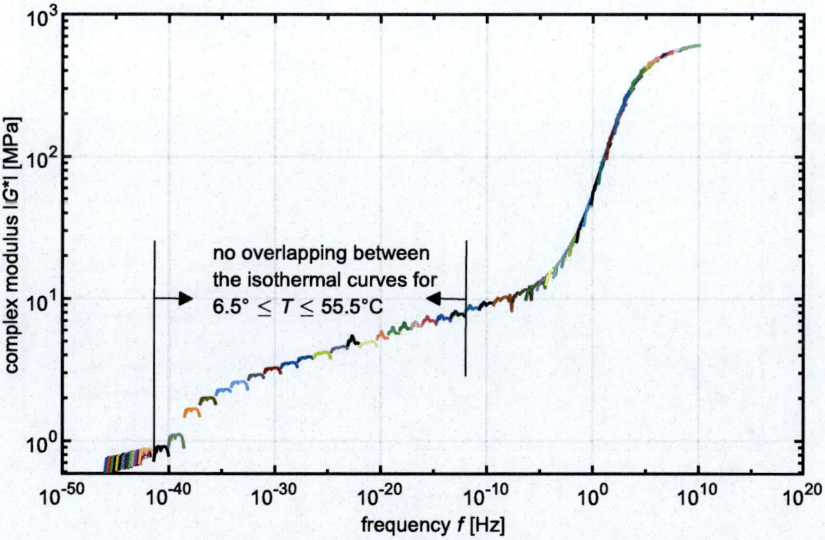

(b)

Figure 5.13 cEVA: mastercurve at T_{ref} = -20 °C, (a) raw data and shift factors, (b) mastercurve.

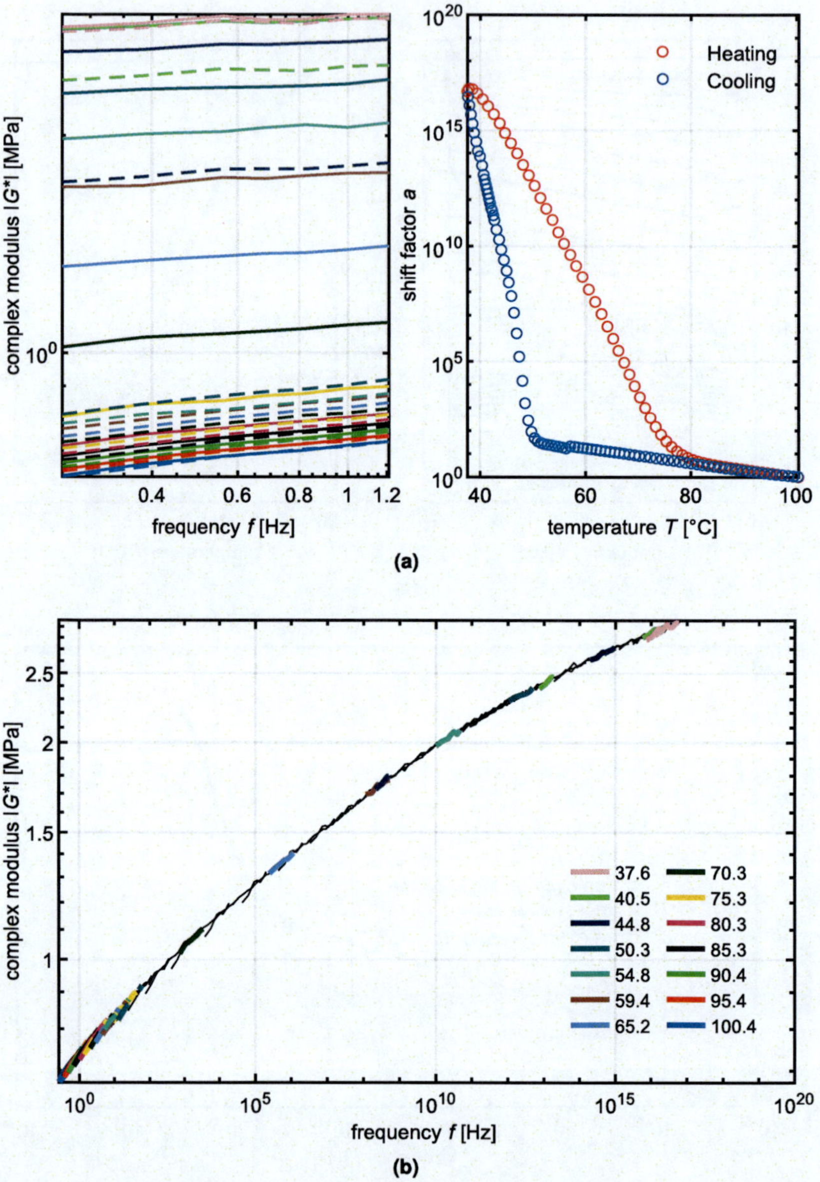

Figure 5.14 cEVA: comparison of the heating and cooling mastercurves from FTS, (a) raw data (solid lines: heating, dashed lines: cooling) and shift factors, (b) mastercurves at $T_{\text{ref}} = 100\,°\text{C}$. For the sake of clarity, only a selection of temperatures is shown in the raw data.

5.3 Decomposition of the horizontal shift factor

The fact that identical mastercurves but different shift factors are obtained suggests that the shift factor is composed of two components: a temperature-dependent (a_T) and a crystallinity-dependent (a_χ) component. A horizontal displacement approach considering different degrees of crystallization has already been addressed in the literature, e.g. (Lamberti et al., 2007; Carotenuto et al., 2019a,b; Minale et al., 2019) and will be further investigated in the following subsections.

5.3.1 Time-superposition principles for SG

As described in chapter 2.4, it is assumed that the total horizontal shift factor a can be divided into a solely temperature dependent a_T and a solely crystallinity dependent a_χ part: $a = a_\mathrm{T} \cdot a_\chi$ resp. $\log a = \log a_\mathrm{T} + \log a_\chi$.
For SG, crystallization occurs at temperatures below T = 60°C, hence the crystallinity equals $\chi = 0$ for all data points in the cooling TFS that were measured at $T >$ 60°C and the shift factor is only affected by temperature changes. Melting in the heat TFS takes place at temperatures above T = 60°C. However, the degree of crystallization is constant only for temperatures below T = 30°C (relative crystallinity = 1), since cold crystallization occurs in the temperature range 30°C $< T <$ 60°C (see also chapter 4).
If no cold crystallization occurred, a constant degree of crystallization would prevail up to T = 60°C, so that in this temperature region, the shift factors in the heating TFS would depend only on temperature. In that case, the temperature dependent shift factor (which is independent of the degree of crystallization) could simply be determined by combining the incremental shift factors of the heating TFS in the temperature range $T <$ 60°C and those of the cooling TFS in the temperature range $T >$ 60°C. Subsequently, the individual incremental shift factors are combined to the absolute shift factors under consideration of the reference temperature (multiplication of the a values, or summation of the $\log a$ values). The rest of the shift factor could then be attributed to the degree of crystallization. However, this method is not possible here, since in a temperature range of 30°C$< T <$60°C the degree of crystallization changes in both the heating TFS and the cooling TFS. In order to divide the total shift factor a into a purely temperature-dependent component a_T and a purely crystallinity-dependent component a_χ, the quotient of both total shift factors ($a_\mathrm{heat}/a_\mathrm{cool}$, or $\log a_\mathrm{heat} - \log a_\mathrm{cool}$) is calculated at the same temperature T, see Figure 5.15 (a) and (b). The result represents the displacement component caused by the difference between the two degrees of crystallization ($\chi_\mathrm{heat} - \chi_\mathrm{cool}$), see Figure 5.15 (c) and (d). For each temperature step, a constant degree of crystallization is assumed. The crystallinities were obtained from DSC measurement with the corresponding temperature rate (here: temperature rates of the linearized temperature program in Figure 5.5). If no DSC measurements are available at the temperature rate present in the TFS, crystallization kinetics (e.g. Mo-Liu see chapter 4) can be used as an alternative to deter-

mine the degrees of crystallization. This procedure was repeated for all temperatures so that the purely crystallinity-dependent shift factor a_χ could be approximated by a function (time-crystallinity-superposition TCSP), see Figure 5.16 (a). Accordingly, the purely temperature-dependent shift factor a_T can be calculated by dividing the total shift factor by the respective crystallinity shift factor ($a_T = a/a_\chi$), or by subtracting the logarithmic values from each other ($\log a_T = \log a - \log a_\chi$). a_T is then approximated (time-temperature-superposition TTSP), see Figure 5.16 (b).

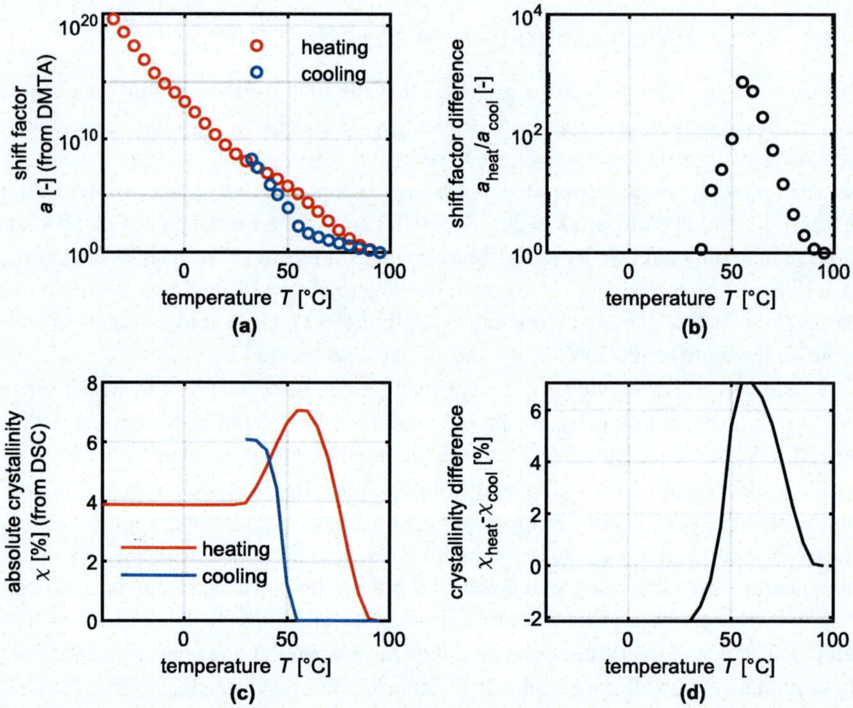

Figure 5.15 SG: decomposition of the horizontal shift factor (a) shift factors determined with DMTA, (b) shift factor difference, (c) absolute crystallinities determined with DSC, (d) crystallinity difference.

Since no DSC data are available for a thermal prehistory of 5 d at $T = 23$°C, the procedure explained above will be used to back-calculate the degree of crystallization from the TFS data. For this purpose, a temperature-dependent shift factor a_T is first determined for each temperature step using the TTSP function shown in Figure 5.16 (b). The degree of crystallization-dependent shift factor is then calculated back ($a_\chi = a/a_T$) and finally the degree of crystallization is determined using the TCSP function in Figure 5.16 (a). The result is shown in Figure 5.17.

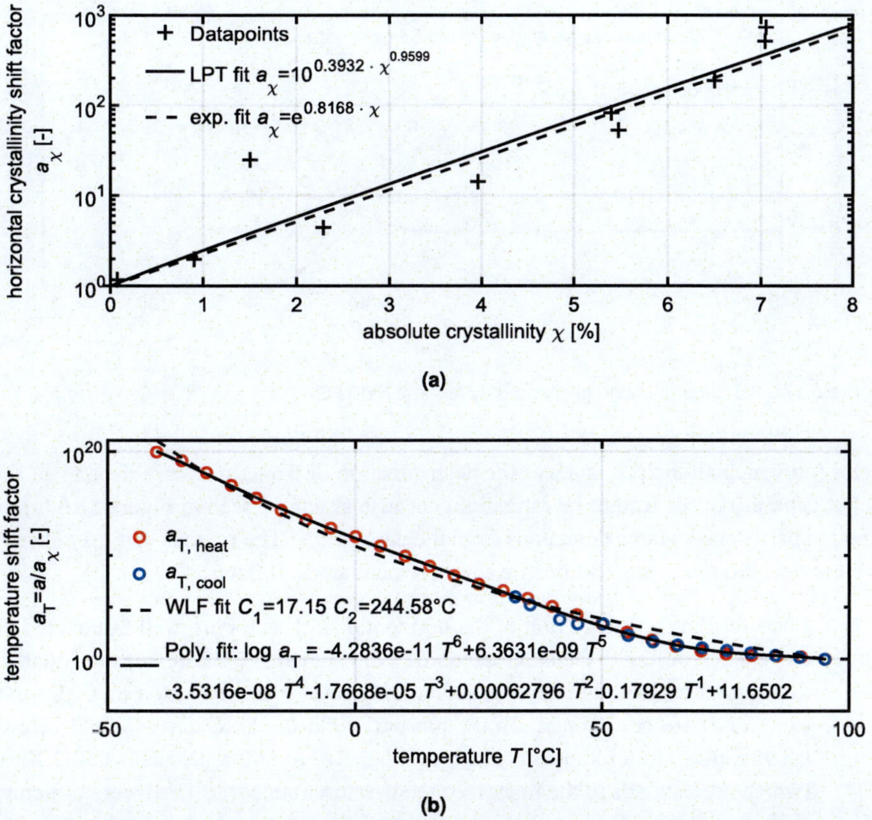

Figure 5.16 SG: shift factors, (a) in dependency of crystallinity, (b) in dependency of temperature.

It can be seen that the absolute degree of crystallinity up to a temperature of -10°C is close to 10%. Slight fluctuations are attributed to the a_T function. Between -10°C and 15°C, cold crystallization takes place, reaching a maximum crystallinity of about 15%. Between 15°C and 40°C, a melting process takes place, with crystallinity dropping to about 7%. This is immediately followed by the second melting process, until finally all crystallites are melted at 90°C. This second melting process is identical to that of the TFS in heating mode (without 5 d storage at 23°C). However, the maximum crystallinity value of 15% determined here is higher than the crystallinity determined in the first DSC heating run (9.6%). At this point, it is unclear whether such a high degree of crystallinity can actually occur due to the temperature history and cold crystallization, or whether it is an overestimation of the crystallinity. The latter could be due to inaccuracies in the determined shift functions. Especially the function for the determination of the crystallinity-dependent shift

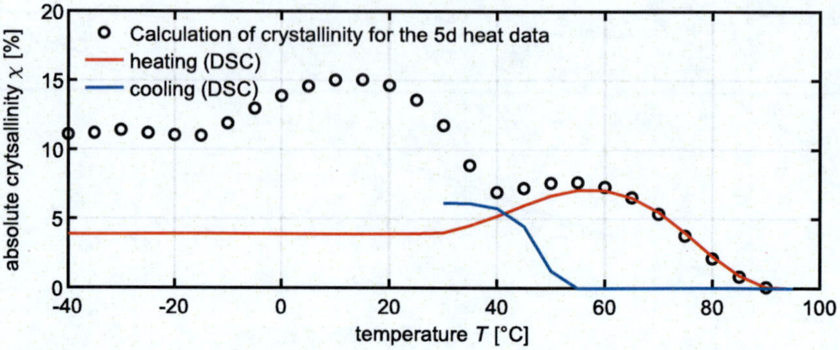

Figure 5.17 SG: determination of crystallinity for heat 5 d TFS.

factor (on which all the rest, such as the determination of the temperature-dependent shift factor, is based) needs further investigation, because as can be seen in Figure 5.16 (a), the function partly shows large deviations from the data points. The reasons that may have led to these deviations and suggestions for improvement are as follows:

- Basis for the decomposition of the horizontal shift factor are well evaluable DSC curves. According to the DSC standard EN ISO 11357-3, the degree of crystallization is best evaluated at heating and cooling rates that are not too low. However, exact rates are not defined. In the present work, the DSC curves could be evaluated well up to a temperature rate of 2 - 3 K/min. Measurements at 0.3 K/min (which corresponds to the linearized temperature rate of the TFS) were difficult to evaluate, which may introduce errors into the crystallization curve determined in this way. Therefore, at such low temperature rates, it is recommended to use an instrument calibrated specifically for small strain rates.
- Since the DMTA-TFS was performed in 5°C increments, there are only 7 data points that can be evaluated in the temperature range of 30 - 60°C. This could be improved by examining smaller temperature steps. However, the linearized temperature rate will then be even smaller.
- It was assumed that there is a constant degree of crystallization at each temperature level. This is a simplification and can lead to an inaccurate evaluation if crystallization proceeds very quickly. It is assumed that this is also the reason why the mastercurves from the heating and cooling TFS (Figure 5.10) are not completely congruent. Especially the initial crystallization in the cooling curve is quite fast (see Figure 4.19, steep curves), so that the degree of crystallization of a temperature step is not necessarily constant and the curve would then theoretically have to be shifted not with a constant but a variable shift factor.

- The DMTA-TFS can be replaced by a DMTA-FTS in the crystallization range. In this case, many temperature sweeps are performed for different (but constant) frequencies. The advantage here is that the temperature rate can be specified. On the one hand, this avoids the staircase-like temperature profile of the DMTA-TFS being compared with the linear temperature profile of the DSC, and on the other hand, temperature rates can be set that are well feasible with the DSC. Another advantage is that the temperature sampling rate can be adjusted so that measurements can be made in 1°C steps, for example.

5.3.2 Time-superposition principles for cEVA

The same procedure was used for the decomposition of the total horizontal shift factor of the cEVA sample.

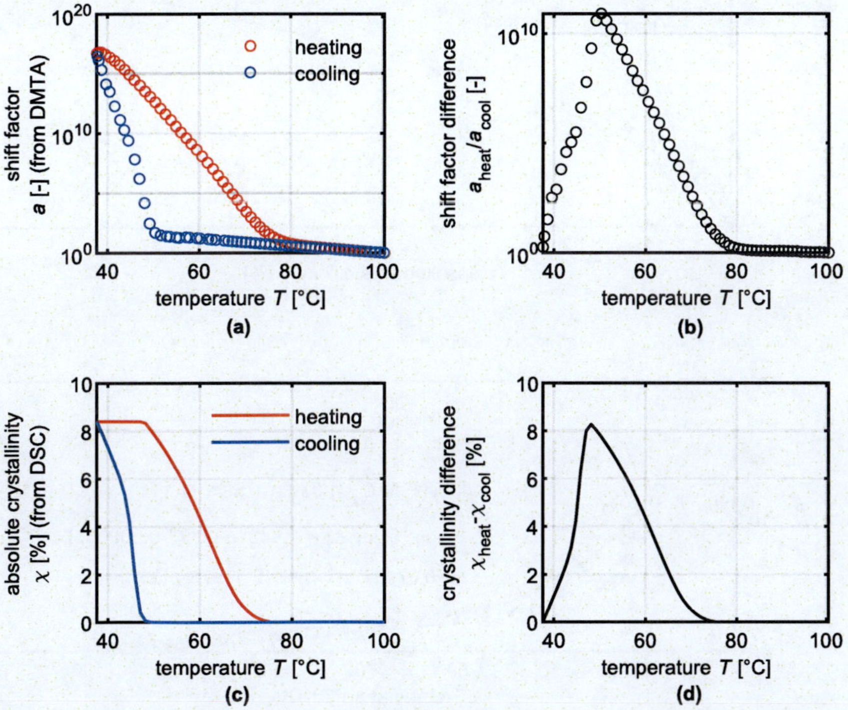

Figure 5.18 cEVA: decomposition of the horizontal shift factor, (a) shift factors determined with DMTA, (b) shift factor difference, (c) absolute crystallinities determined with DSC, (d) crystallinity difference.

Figure 5.18 (a) shows the total horizontal shift factors that were used to construct the mastercurves from FTS. In Figure 5.18 (b), the quotient of both total shift factors ($a_{\text{heat}}/a_{\text{cool}}$, resp. $\log a_{\text{heat}} - \log a_{\text{cool}}$) is represented. The absolute crystallinity in Figure 5.18 (c) was determined with DSC data recorded at an equal temperature rate as in the DMTA-FTS (+2K/min). Comparing Figure 5.18 (c) with the original Figure 4.12 (c), it can be seen that the degree of crystallization of the heating curve was assumed to be constant in the temperature range of [37.5: 48]°C. This was decided because the TFS measurements were only carried out in the very small temperature interval [37.5:100]°C, and therefore the crystallization process cannot take place completely when cooling down from the melt. The degree of crystallization in the heating curve was therefore assumed to be constant at the maximum crystallinity reached during cooling. Figure 5.18 (d) shows the difference in absolute crystallinities in the heating and cooling runs.

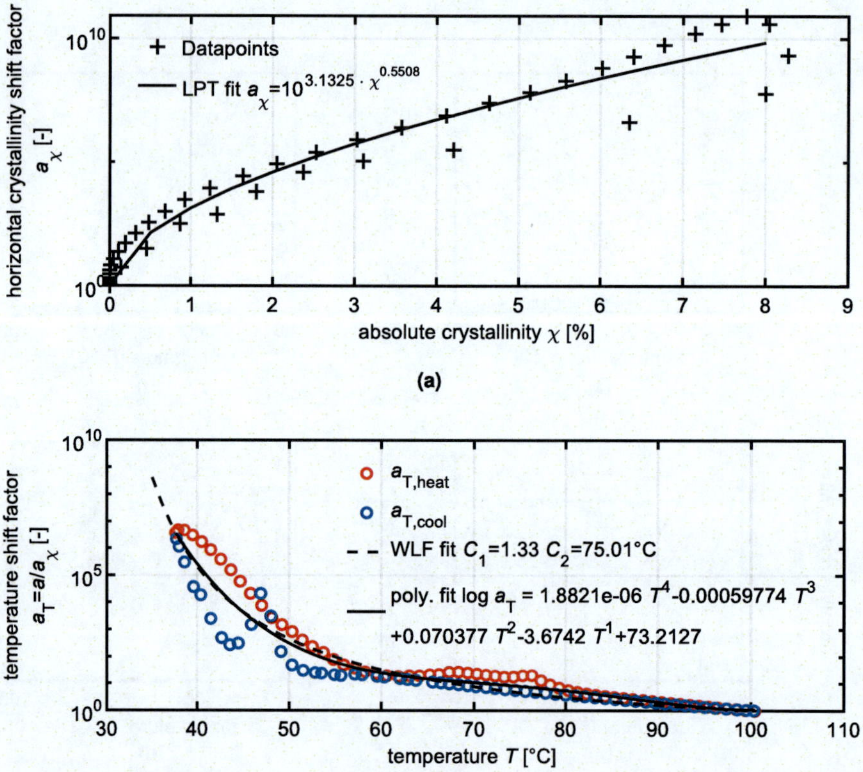

Figure 5.19 cEVA: shift factors, (a) in dependency of crystallinity, (b) in dependency of temperature.

Comparing the difference of the shift factor in Figure 5.18 (b) with that of the degree of crystallization in Figure 5.18 (d), the TCSP function for the crystallinity-dependent shift factor a_{χ} shown in Figure 5.19 (a) can be determined. Finally, the purely temperature-dependent shift factor a_{T} shown in Figure 5.19 (b) can be derived and approximated with a TSSP.
It can be seen that there is a much greater deviation here than with SG. This can be explained by the fact that several complications or assumptions had to be made for cEVA. Firstly, it has already been shown that the DMTA results for cEVA are not particularly stable and many data had to be discarded due to a non-physical course. This led to the fact that not all isothermal curves have an overlap area with the adjacent curve, making it difficult to create the mastercurve. Accordingly, the shift factors from the cooling data were manually adjusted so that the resulting mastercurve is congruent with that from the heating data. A further assumption is that the course of the degree of crystallization is constant in the temperature range [37.5:48]°C. It was not investigated whether this could lead to cold crystallization, which changes the degree of crystallization.

5.4 Determination of Prony series

In the previous section it was shown that the mastercurves of different temperature histories (from warm to cold, cold to warm after elimination of thermal history, cold to warm with previous storage at room temperature) are equal when the reference temperature is chosen above the melting point. For this reason, the Prony series for SG is determined for a reference temperature of T_{ref} = 95°C.
To determine the Prony series, Kuntsche's two-stage optimization problem (Kuntsche, 2015) is used, see also chapter 3.

The Prony parameters for SG are shown in Table 5.1. In addition, the previously determined temperature-time-superposition principle and crystallization-time-superposition principle are given. A shorter Prony series, with 10 Maxwell elements is specified in the appendix (Table A.23). This can be used for engineering applications. A comparison of the long and short Prony series with the mastercurves are shown in Figure A.11.

Since it was not possible for cEVA to create a continuous mastercurve over the whole temperature range, two different Prony series are determined. The first one (Table 5.2) is generated from the combined TFS data (Figure 5.13) for the low temperature range ($T \leq$ 6.5°C) with T_{ref} = -20°C. It is assumed that the relative crystallinity is close to 1. In addition to WLF, a polynomial function is chosen to approximate the temperature dependent shift factors (TTSP).
The second Prony series (Table 5.3) is created from the high temperature FTS data with T_{ref} = 100°C and a relative crystallinity of 0 (Figure 5.14). Functions for a purely tempera-

Table 5.1 SG: TTSP, TCSP and Prony series. Note that the indicated relaxation times are very small. This is due to the high reference temperature.

Time-Temperature-Superposition for -40°C $\leq T \leq$ 95°C

$$\log a_{\mathrm{T}} = -4.2836 \cdot 10^{-11} \cdot T^6 + 6.3631 \cdot 10^{-9} \cdot T^5 - 3.5316 \cdot 10^{-8} \cdot T^4 - 1.7668 \cdot 10^{-5} \cdot T^3 + 6.2796 \cdot 10^{-4} \cdot T^2 - 0.17929 \cdot T + 11.6502$$

Time-Crystallinity-Superposition for 0% $\leq \chi \leq$ 8%

$$\log a_{\chi} = 0.3932 \cdot \chi^{0.9599}$$

Prony series for T_{ref} = 95°C, χ_{ref} = 0%

G_0	312.98	MPa
i	τ_i [s]	g_i
1	1.00E+01	1.5565E-04
2	1.00E+00	5.5886E-04
3	1.00E-01	1.1826E-03
4	1.00E-02	1.5691E-03
5	1.00E-03	2.5716E-03
6	1.00E-04	3.4810E-03
7	1.00E-05	7.0988E-03
8	1.00E-06	2.0829E-02
9	1.00E-07	7.8781E-02
10	1.00E-08	8.8094E-02
11	1.00E-09	1.1157E-01
12	1.00E-10	9.1198E-02
13	1.00E-11	7.3971E-02
14	1.00E-12	6.7564E-02
15	1.00E-13	9.0724E-02
16	1.00E-14	3.8361E-02
17	1.00E-15	3.6017E-02
18	1.00E-16	9.0148E-02
19	1.00E-17	1.2595E-02
20	1.00E-18	3.1613E-02
21	1.00E-19	2.6000E-02
22	1.00E-20	4.1657E-02
23	1.00E-21	8.0555E-02
24	1.00E-22	2.0333E-03
25	1.00E-23	1.5588E-03

Table 5.2 cEVA: TTSP and Prony-Series at T_{ref} = -20 °C.

Time-Temperature-Superposition: WLF for $T \leq 6.5$°C		
C_1	236504.15	
C_2	603728.43	°C

Time-Temperature-Superposition: Polynomial for $T \leq 6.5$°C

$$\begin{aligned}\log a_{\mathrm{T}} = & -3.5608 \cdot 10^{-5} \cdot (T - T_{\mathrm{ref}})^3 \\ & -2.3216 \cdot 10^{-3} \cdot (T - T_{\mathrm{ref}})^2 \\ & -3.1984 \cdot 10^{-1} \cdot (T - T_{\mathrm{ref}}) + 5.4987 \cdot 10^{-2}\end{aligned}$$

Prony series for T_{ref} = -20°C

G_0	565.55	MPa
i	τ_i [s]	g_i
1	1.E+14	7.0328E-03
2	1.E+13	5.3336E-04
3	1.E+12	1.6892E-03
4	1.E+11	1.3240E-03
5	1.E+10	1.5628E-03
6	1.E+09	1.6452E-03
7	1.E+08	2.1699E-03
8	1.E+07	2.1888E-03
9	1.E+06	3.1271E-03
10	1.E+05	3.0644E-03
11	1.E+04	6.2294E-03
12	1.E+03	7.9062E-03
13	1.E+02	1.0549E-02
14	1.E+01	1.4040E-02
15	1.E+00	2.7722E-02
16	1.E-01	5.0818E-02
17	1.E-02	7.0708E-02
18	1.E-03	1.0715E-01
19	1.E-04	1.4342E-01
20	1.E-05	1.4060E-01
21	1.E-06	1.2235E-01
22	1.E-07	9.6088E-02
23	1.E-08	5.7747E-02
24	1.E-09	5.0099E-02
25	1.E-10	6.9457E-02

Table 5.3 cEVA: TTSP, TCSP and Prony series. Note that the indicated relaxation times are very small. This is due to the high reference temperature.

Time-Temperature-Superposition for 37.5°C $\leq T \leq$ 100°C

$$\begin{aligned}\log a_{\mathrm{T}} =& 1.8821 \cdot 10^{-6} \cdot T^4 - 5.9774 \cdot 10^{-4} \cdot T^3 \\ &+ 7.0377 \cdot 10^{-2} \cdot T^2 - 3.6742 \cdot T \\ &+ 73.2127\end{aligned}$$

Time-Crystallinity-Superposition for 0% $\leq \chi \leq$ 8%

$$\log a_{\chi} = 3.1325 \cdot \chi^{0.5508}$$

Prony series for T_{ref} = 100°C, χ_{ref} = 0%

G_0	3.69	MPa
i	τ_i [s]	g_i
1	1.00E+01	7.2715E-02
2	1.00E+00	1.9890E-02
3	1.00E-01	3.9379E-02
4	1.00E-02	3.7276E-02
5	1.00E-03	3.8781E-02
6	1.00E-04	4.0510E-02
7	1.00E-05	4.2607E-02
8	1.00E-06	4.5755E-02
9	1.00E-07	4.8175E-02
10	1.00E-08	5.0587E-02
11	1.00E-09	5.2810E-02
12	1.00E-10	5.3775E-02
13	1.00E-11	5.3738E-02
14	1.00E-12	5.3989E-02
15	1.00E-13	5.4571E-02
16	1.00E-14	5.4732E-02
17	1.00E-15	5.5009E-02
18	1.00E-16	5.5009E-02
19	1.00E-17	5.9556E-02

ture -dependent shift factor a_T (TTSP) and a purely crystallinity-dependent shift factor a_χ (TCSP) are given. Comparisons of the Prony series with the respective mastercurves are shown in Figure A.12.

5.5 Summary and comparison with literature

In this section, the results for the mastercurves and the time-superposition principles are summarized. In addition, the cEVA results are compared with results from literature. For SG, only very few literature data is available, so that no comparison is made.

Summary

In this chapter, the results of DMTA-TFS and DMTA-FTS were presented first. It was shown how frequency-independent transformations can be distinguished from frequency-dependent relaxations with the help of these data. The frequency-dependent maximum in the loss factor of SG confirmed that the glass transition temperature is indeed at $T_g \approx$50°C, as suspected in the previous chapter, and is thus superimposed on the melting or crystallization process. In addition, a second glass transition was observed at $T_g \approx$-30°C. This observation was described previously in (Wakabayashi and Register, 2006). The higher glass transition temperature was attributed to the relaxation of the ion-rich region and the lower one to the relaxation of the ion-poor region.
The glass transition temperature of cEVA was also confirmed by the frequency-dependent maximum in the loss factor to $T_g \approx$ -25°C.
Subsequently, mastercurves were generated for SG from TFS with three different temperature histories (from warm to cold, from cold to warm after eliminating the thermal history, and from cold to warm after 5 days at 23°C). It could be shown that almost congruent mastercurves result when the reference temperature is chosen to be above the melting temperature. However, the horizontal shift factors required to generate the mastercurves differ depending on the temperature history. It was concluded that the horizontal shift factor depends not only on the prevailing temperature but also on the degree of crystallization and can be decomposed into two parts: a purely temperature-dependent part and a purely crystallinity-dependent part.
The crystallinity-dependent fraction was determined by comparing the difference of the shift factors (from DMTA) of two different temperature histories with the difference of the degrees of crystallization (from DSC) of the same two temperature histories. The factors determined were successfully approximated mathematically using an LPT approach. Subsequently, the purely temperature-dependent shift factor was determined and approximated with a polynomial function. Finally, a Prony series was determined for the mastercurve.
The same procedure was followed for cEVA. However, it was not possible to obtain reliable DMTA results over the entire temperature range of [100:-40]°C. Many data points

had to be discarded, as a result of which some curves had no overlap area to the neighboring curve and the mastercurve generation failed at temperatures above $\approx$ 6.5°C. For this reason, the Prony series was created only for temperatures up to a maximum of 20°C and at a reference temperature of T_{ref} = -20°C. In this temperature range, the relative degree of crystallization is 1. In order to investigate the influence of the degree of crystallization, additional FTS (from warm to cold and from cold to warm) were performed in the higher temperature range [37:100]°C. However, here again so many data points had to be sorted out because of non-physical course that the subsequent decomposition into a purely temperature-dependent shift factor and a purely crystallinity-dependent shift factor contains many uncertainties. Nevertheless, both a LPT and a WLF function and a second Prony series for the high temperature range were derived.

Comparison with literature

A large number of studies trying to characterize cEVA interlayers over a wide temperature range can be found in literature. A selection of literature is shown in Table 5.4. In all mentioned publications, mastercurves were determined for a specific reference temperature. The type of mastercurve and the reference temperature are given in Table 5.4.
In addition, in almost all literature (all except Eitner et al. (2010)) the incremental shift factors are shown. Many publications tried to approximate the incremental shift factors over the whole temperature range with a WLF function, but all authors consider the quality of the WLF fit over the whole temperature range as poor. The resulting WLF parameters can also be found in Table 5.4.
Figure 5.20 (a) shows the incremental shift factors and WLF equations of all publications listed in Table 5.4 at a reference temperature of T_{ref} = -20°C. The data of Hána et al. (2020) could not be mapped, since -20°C is outside the investigated range. The data of Kuntsche (2015) refers to a reference temperature of T_{ref} = 20°C; in order to be able to compare this with the others, the reference temperature was set to -20°C by dividing all incremental shift factors determined at 20°C by the value at -20°C.
It can be seen that in the range of low temperatures (up to approx. 20°C) the incremental shift factors are relatively close to each other. Above a temperature of about 20°C, however, there are large deviations in the shape of the curve and the size of the shift factors. Most of the curves show an almost horizontal course in the shift factors at high temperatures. However, the temperature at which this plateau occurs varies. Furthermore, it is confirmed that the mentioned WLF parameters describe the shift factors only poorly.
Figure 5.20 (b) shows a qualitative comparison of the mastercurves at a reference temperature of T_{ref} = -20°C. For this purpose, the data of Kuntsche (2015) was shifted to -20°C (multiplication with shift factor). Since different mastercurves were shown in the different publications (e.g. $G'(f), G'(\omega), G(t)$), the data from the publication had to be post-processed to approximate the time course of the shear modulus $G(t)$. For this purpose, the

simplified relation ($G(t) \approx G'(1/\omega)$ or, for the data of Kuntsche (2015) $G(t) \approx G^*(1/\omega)$) was used. It is aware that this is only a very rough conversion, but it is considered sufficient for qualitative comparison. The original mastercurves can be found in the corresponding literature. Again, large deviations are visible, especially for long times.

Table 5.4 Comparison of TTSP data and mastercurves.

Author and material	Test set-up and temp. prog.	Master-curve	C_1 [-]	C_1 [°C]	T_{ref} [°C]
Kraus (2018)[1], evguard Folienwerk Wolfen, 28 wt% VAC	DMTA tensile mode, [100:-90]°C, ΔT = -2°C	$G'(f)$, $G''(f)$, Prony series	256	951	-20
Schuster et al. (2018), evguard Folienwerk Wolfen, 28 wt% VAC	DMTA torsion strip, [80:-70]°C, ΔT = -5°C	$G^*(f)$	71	295	-20
Pander et al. (2011), cEVA for PV modules	Creep, [-20:115]°C, ΔT = +5°C	$G(t)$	69	170	-20
Eitner et al. (2010), cEVA for PV modules	Tensile creep and relaxation, [-35, -28, -18, 1, 20, 41, 49, 60, 80, 100, 119, 139]°C	$G(t)$	48	173	-20
Hána et al. (2020), EVASafe, Bridgestone	DMTA single slap laminates, [20:50]°C, ΔT = +5°C	$G'(f)$, Prony series	230	1150	20
Kuntsche (2015), EVASafe G77. Bridgestone	DMTA torsion strip, [-80:80]°C, ΔT = +10°C resp. ΔT = +5°C	$G^*(f)$	/	/	20
Schuster, EV200, Glaast 28 wt% VAC	DMTA torsion strip, [100:-40]°C (TFS)[2], ΔT = -2.5°C	$G^*(f)$	236504	603728	-20

[1] Mean WLF values are used.
[2] No overlapping between 6.5°C $\leq T \leq$ 55.5°C.

The reason for the deviations at high temperature is probably the influence of the degree of crystallization of the investigated interlayer. The degree of crystallization depends, among other things, on the temperature history, e.g. the test direction (from warm to cold, or cold to warm, with or without elimination of the thermal prehistory), and the time (how long does the measurement take at a certain temperature). In (Kraus, 2018; Schuster et al.,

2018) and in the present document, the tests were performed from warm to cold, in contrast to all the other publications listed. The horizontal plateau in the shift factor occurs at lower temperatures than in the experiments with increasing temperature. This fits with the fact that the crystallization temperature (peak temperature in DSC) is lower than the melting temperature (peak temperature in DSC).

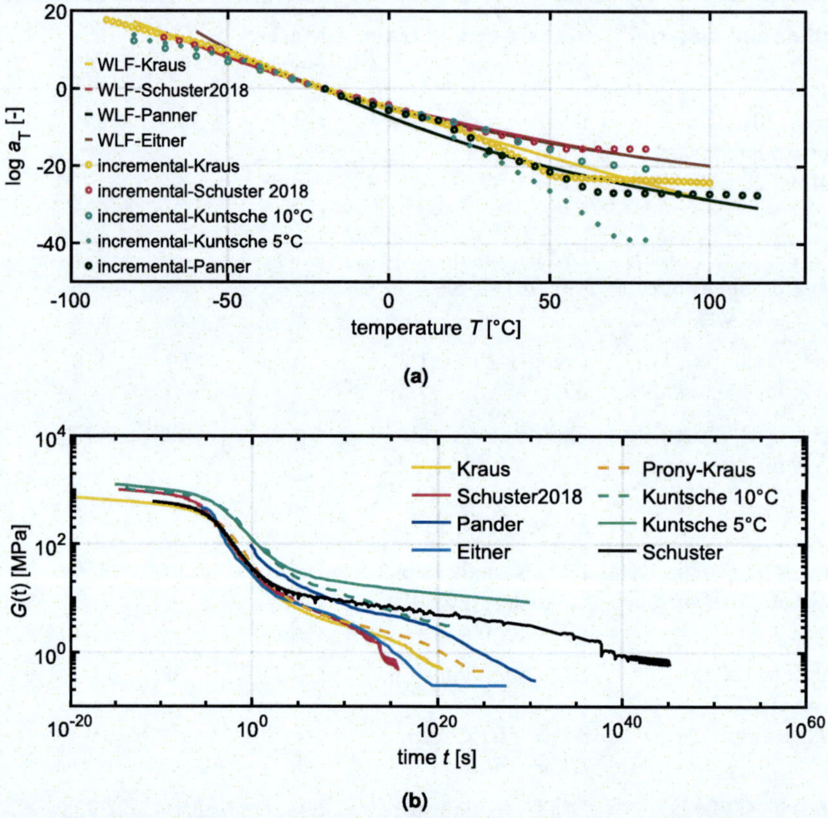

Figure 5.20 cEVA: Comparison with literature, T_{ref} = -20℃ (a) shift factors, (b) mastercurves.

Since large deviations in the mastercurves and shift factors occurred in the range of high temperatures or long loading durations, the range of high temperatures was investigated in more detail in this chapter, taking crystallinity into account, and a purely temperature-dependent and a purely crystallinity-dependent shift factor were derived. Only a few publications with a comparable material can be found on this topic. Arsac et al. (1999) determined a purely temperature-dependent shift factor on EVA with different VAC contents (it is supposed that the EVAs are not crosslinked) by investigating only temperatures above the melting point. They found that the shift factors can be formed into a curve independent of the VAC content, which can be fitted very well using the WLF equation.

Carotenuto et al. (2019a) and Minale et al. (2019) also determined a purely temperature-dependent and a crystallinity-dependent shift factor for two different EVAs (it is supposed that the EVAs are not crosslinked). In both publications, the temperature-dependent shift factor could be well fitted by a WLF equation. The WLF parameters and the reference temperatures are shown in Table 5.5.

Table 5.5 Comparison of TTSP data and mastercurves - Part 2.

Author and material	Test set-up and temp. prog.	Master-curve	C_1 [-]	C_1 [°C]	T_{ref} [°C]
Williams et al. (1955), universal	$[T_g{:}T_g{+}100]$°C	/	17	51.6	T_g
	$[T_{ref}\text{-}50{:}\ T_{ref}{+}50]$°C		8.9	102	T_g+50
Arsac et al. (1999), EVA [5:40] wt% VAC	DMTA disc-plate, [50:200]°C	$G'(\omega)$ $\mu^*(\omega)$	3.5	161	110
Carotenuto et al. (2019a), P28800, Repsol, 28 wt% VAC	DMTA plate-plate, [100:40]°C ΔT = -10°C	$G'(\omega)$ $G''(\omega)$ $\mu^*(\omega)$	5	48	40
Carotenuto et al. (2019a)[1], P28150, Repsol 28 wt% VAC	DMTA plate-plate, [100:40]°C ΔT = -10°C	$G'(\omega)$ $G''(\omega)$ $\mu^*(\omega)$	4.4	30	40
Schuster, EV200, Glaast 28 wt% VAC	DMTA torsion strip, [100:+37]°C (FTS)	$G^*(f)$	1.33	75.01	100

[1] P2880 and P2815 differ in melt flow index values and densities.

Comparing the WLF equations at a reference temperature of T_{ref} = 100°C (the values of Carotenuto et al. (2019a); Minale et al. (2019) were obtained by dividing all shift factors at the reference temperature of 40°C by the value at 100°C), it can be seen that the shift functions are close. Figure 5.21 and Table 5.5 also contain the universal WLF constants for polymer solutions, organic glass forming liquids, inorganic glasses and amorphous polymers from Williams et al. (1955). A glass transition temperature of -20°C is supposed. There is even less literature available on the crystallization degree dependent shift factor for EVA, so in addition to (Carotenuto et al., 2019a; Minale et al., 2019), (Roozemond et al., 2012) was also considered. Roozemond et al. studied the effect in HDPE. It is presumed that the results are comparable, since it concerns both with HDPE and with EVA the formation of PE crystallites. Contrary to the investigations performed in this thesis, in literature, the LPT parameters are given in dependency of the relative degree of

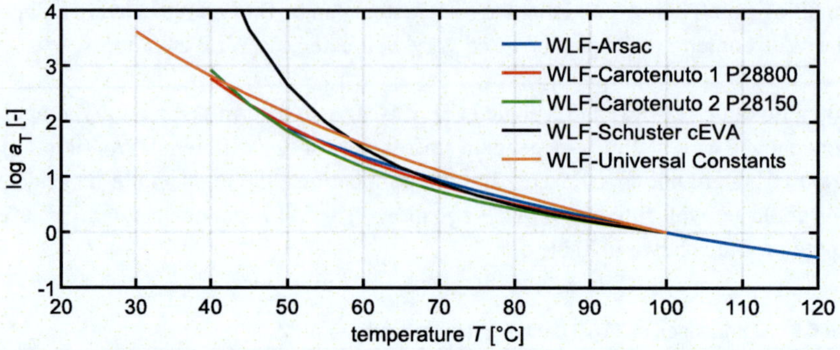

Figure 5.21 Comparison TTSP with literature, T_{ref} = 100 °C.

crystallization α. Hence, the here determined LPT parameters have to be transformed to be comparable with the literature values.
It applies:

$$\log a_\chi = a_1' \cdot \chi^{a_2'} = a_1' \cdot (\alpha \cdot \chi_{\max})^{a_2'} = a_1' \cdot {\chi_{\max}}^{a_2'} \cdot \alpha^{a_2'} = a_1 \cdot \alpha^{a_2}$$

χ and $\chi_{\max}$ are values of absolute crystallinity in %. It follows that $a_1 = a_1' \cdot {\chi_{\max}}^{a_2'}$ and that $a_2 = a_2'$. For $\chi_{\max}$, the initial crystallinities are used: $\chi_{\max,\mathrm{SG}} \approx 9.5\%$ and $\chi_{\max,\mathrm{cEVA}} \approx 17\%$ (see Table 4.5 and Table 4.6). The parameters of the LPT equation are shown in Table 5.6, and the course of the functions is shown in Figure 5.22.

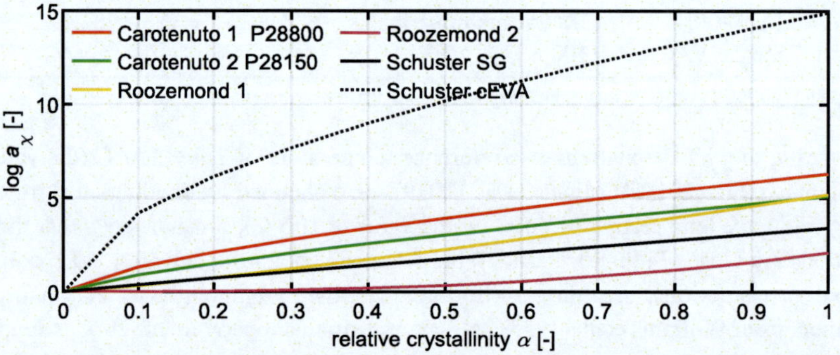

Figure 5.22 Comparison TCSP with literature, with α_{ref} = 0. Note that the here determined crystallinity shift factor functions (TCSP) were determined for absolute crystallinities of $\chi \leq 8\%$ for both materials, which corresponds to $\alpha_{\mathrm{SG}} \approx 0.84$ with $\chi_{\mathrm{initial,SG}}$ = 9.5% and $\alpha_{\mathrm{cEVA}} \approx 0.4$ with $\chi_{\mathrm{initial,cEVA}}$ = 17%.

Table 5.6 Comparison of TCSP data.

Author and material	**Test set-up and temp. prog.**	a_1 [-]	a_2 [-]	α_{ref} [-]
Carotenuto et al. (2019a), P28800, Repsol, 28 wt% VAC	DMTA plate-plate, [100:40]°C, ΔT = -10°C	6.3	0.68	0
Carotenuto et al. (2019a), P28150, Repsol, 28 wt% VAC	DMTA plate-plate, [100:40]°C, ΔT = -10°C	5.1	0.75	0
Roozemond et al. (2012), HDPE[1]	RheoDSC device, T = 124°C, different relative crystallinities	5.2	1.2	0
Roozemond et al. (2012), HDPE[2]	RheoDSC device, T = 124°C, different relative crystallinities	2.1	2.56	0
Schuster, SG6000, Kuraray	DMTA plate-plate, [100:-40]°C (TFS with ΔT = 5°C) and DSC C = +-3 K/min	3.4 [3]	0.96	0
Schuster, EV200, Glaast 28 wt% VAC	DMTA torsion strip, [100:+37]°C (FTS) and DSC C = +-2K/min,	14.9 [4]	0.55	0

[1] The constants are included in the comparison because they also consist of PE crystallites.
[2] In addition to the horizontal crystallinity shift factor, a vertical crystallinity dependent shift factor was applied: $\log b = b_1 \cdot \alpha^2 - b_2 \cdot \alpha$ with $b_1 = 0.22$, $b_2 = 1.01$ and α: relative crystallinity.
[3] The fit was initially performed in dependency on the absolute crystallinity χ. To get the parameters in dependency of the relative crystallinity α, a_1' has to be transformed to: $a_1 = 0.3932 \cdot 9.5^{0.9599} = 3.4$
[4] Modification from absolute crystallinity χ to relative crystallinity α: $a_1 = 3.1325 \cdot 17^{0.5508} = 14.9$

6 Determination of the linear viscoelastic behaviour and linearity limits for PVB

This chapter refers to standard and structural PVB. The Prony series and a time-temperature shift principle are determined based on chapter 2.3.1, chapter 2.4.1 and chapter 3. For PVB-based interlayers, this already corresponds to the state of the art, which is why the results are only shown briefly. However, this will be extended here to include the determination of the linearity limits according to chapter 2.3.2 and chapter 3. The linearity limits separate the linear from the nonlinear viscoelastic range.

All the experiments in this chapter were also carried out on the acoustic PVB. The results are shown exclusively in the appendix (chapter A.4.6). Detailed descriptions of parts of the investigations can be found in (Nitzling, 2018) and (Maddé, 2019). Furthermore, the main findings have already been published in (Schuster et al., 2021).

6.1 Linear viscoelastic material behaviour

6.1.1 Temperature-frequency sweeps

The experiments were performed on an Anton Paar MCR 302 rheometer using a plate-plate measuring system. The upper plate has a diameter of 8 mm and a textured surface. The convection oven was used in combination with liquid nitrogen for heating and cooling. After sample installation, the samples were briefly loaded at a temperature of 65°C with a normal force of 5 - 10 N, and unloaded (0 - 0.1 N). This should prevent slippage during the measurement. To relax the resulting stresses in the samples, they were then heated in a controlled manner to 100°C - 120°C and finally cooled down to the test temperature with a cooling rate of 5 K/min.

TFS was performed for both materials in a temperature range of [100:-40]°C (from warm to cold) with ΔT = -5°C in a frequency range of [0.1:10]Hz. The strain amplitude was in the linear viscoelastic range. The experimental parameters are shown in the appendix (see chapter A.24).

M. Schuster, *Characterization of Laminated Safety Glass Interlayers*,
Mechanik, Werkstoffe und Konstruktion im Bauwesen 66,
https://doi.org/10.1007/978-3-658-39821-7_6

6.1.2 Glass transition temperatures

The measurement data at a frequency of 1 Hz for all temperatures of the TFS were used to determine the glass transition temperature. The maximum of the loss factor led to $T_g \approx 30°C$ for standard PVB, see Figure 6.1 (a). For structural PVB, $T_g \approx 45°C$ results, see Figure 6.1 (b). However, it should be pointed out that the temperature increment was ΔT = -5°C.

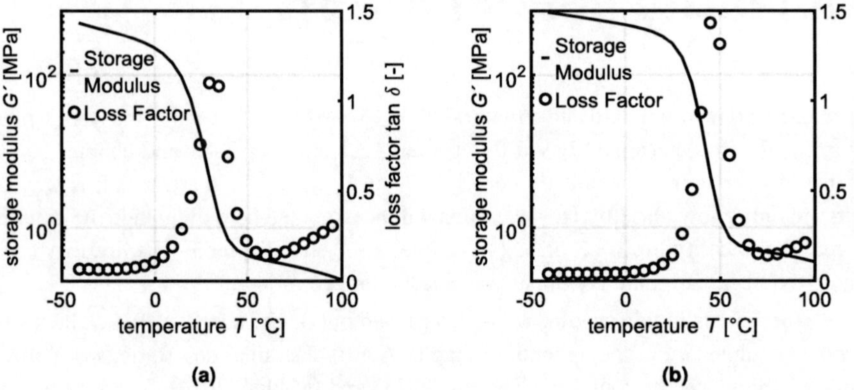

Figure 6.1 Temperature dependent material behaviour at f = 1 Hz, (a) standard PVB, (b) structural PVB.

6.1.3 Mastercurves and Prony parameters

Figure 6.2 and Figure 6.3 show the mastercurving process for the standard and structural PVB. The additional plots (mastercurves of storage modulus, loss modulus, loss factor, Cole-Cole and van-Gurp-Palmen plots) can be found in the appendix, see Figure A.13 and Figure A.14. The shift factors are listed in Table A.25 and A.26. A MATLAB® script, that incrementally shifts the individual complex modulus $|G^*|$ - frequency curves until a continuous curve is created is used for this purpose. Subsequently, the horizontal shift factors are mathematically approximated by means of time-temperature-superposition principle functions and the mastercurves of the storage G' and loss G'' modulus at its reference temperatures are mathematically approximated by a Prony series. The results are summarized in Table 6.1 and Table 6.2. Shorter Prony series with 10 Maxwell elements can be found in the appendix (Table A.27 and Table A.28). Comparisons of the the Prony series with the respective mastercurves are shown in Figure A.15 and A.16.
Looking at the shift factor - temperature curves, it can be seen that both WLF and Arrhenius do not optimally hit the incrementally determined shift factors in the entire temper-

Table 6.1 Prony series and TTSP parameters for standard PVB at $T_{ref} = 20\,°C$.

Time-Temperature-Superposition: WLF		
C_2	33.20	
C_2	212.42	°C

Time-Temperature-Superposition: Polynomial

$$\begin{aligned} \log a_T = & -2.24078 \cdot 10^{-9} \cdot (T - T_{ref})^5 \\ & -1.73583 \cdot 10^{-8} \cdot (T - T_{ref})^4 \\ & +2.52160 \cdot 10^{-5} \cdot (T - T_{ref})^3 \\ & +6.79443 \cdot 10^{-4} \cdot (T - T_{ref})^2 \\ & -0.21439 \cdot (T - T_{ref}) \end{aligned}$$

Prony series at $T_{ref} = 20°C$		
G_0	564.48	MPa
i	τ_i [s]	g_i
1	1.00E+09	2.2533E-05
2	1.00E+08	1.4395E-04
3	1.00E+07	2.0530E-04
4	1.00E+06	1.5643E-04
5	1.00E+05	5.6594E-05
6	1.00E+04	2.7530E-04
7	1.00E+03	5.2786E-04
8	1.00E+02	3.3805E-03
9	1.00E+01	1.5006E-02
10	1.00E+00	4.8825E-02
11	1.00E-01	5.7801E-02
12	1.00E-02	8.8516E-02
13	1.00E-03	7.3438E-02
14	1.00E-04	7.7868E-02
15	1.00E-05	6.5816E-02
16	1.00E-06	6.3514E-02
17	1.00E-07	7.2386E-02
18	1.00E-08	6.7712E-02
19	1.00E-09	3.7248E-02
20	1.00E-10	9.3140E-02
21	1.00E-11	5.1943E-02
22	1.00E-12	1.1400E-01
23	1.00E-13	6.7844E-02

Table 6.2 Prony series and WLF parameters for structural PVB at $T_{ref} = 20\,°C$.

Time-Temperature-Superposition: WLF		
C_1	196.91	
C_2	1139.40	°C

Time-Temperature-Superposition: Polynomial

$$\begin{aligned}\log a_T = &-4.91956 \cdot 10^{-9} \cdot (T - T_{ref})^5 \\ &+2.85095 \cdot 10^{-7} \cdot (T - T_{ref})^4 \\ &+3.77869e - 5 \cdot 10^{-5} \cdot (T - T_{ref})^3 \\ &-9.96994 \cdot 10^{-4} \cdot (T - T_{ref})^2 \\ &-0.24179 \cdot (T - T_{ref})\end{aligned}$$

Prony series at T_{ref} = 20°C		
G_0	714.95	MPa
i	τ_i [s]	g_i
1	1.00E+11	8.9574E-06
2	1.00E+10	3.0532E-04
3	1.00E+09	1.0889E-04
4	1.00E+08	1.0005E-05
5	1.00E+07	5.2886E-04
6	1.00E+06	4.4507E-03
7	1.00E+05	1.5092E-02
8	1.00E+04	6.1492E-02
9	1.00E+03	9.2061E-02
10	1.00E+02	1.0363E-01
11	1.00E+01	7.4996E-02
12	1.00E+00	5.3605E-02
13	1.00E-01	7.3131E-02
14	1.00E-02	3.8629E-02
15	1.00E-03	3.9315E-02
16	1.00E-04	3.9952E-02
17	1.00E-05	4.5781E-02
18	1.00E-06	4.1220E-02
19	1.00E-07	5.2191E-02
20	1.00E-08	4.9264E-02
21	1.00E-09	8.1482E-02
22	1.00E-10	1.7828E-04
23	1.00E-11	1.2922E-01
24	1.00E-12	2.9571E-03

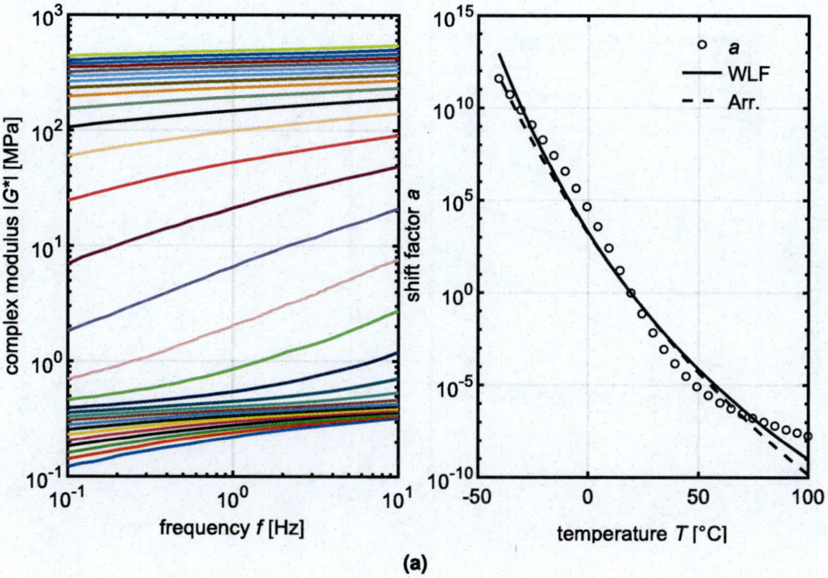

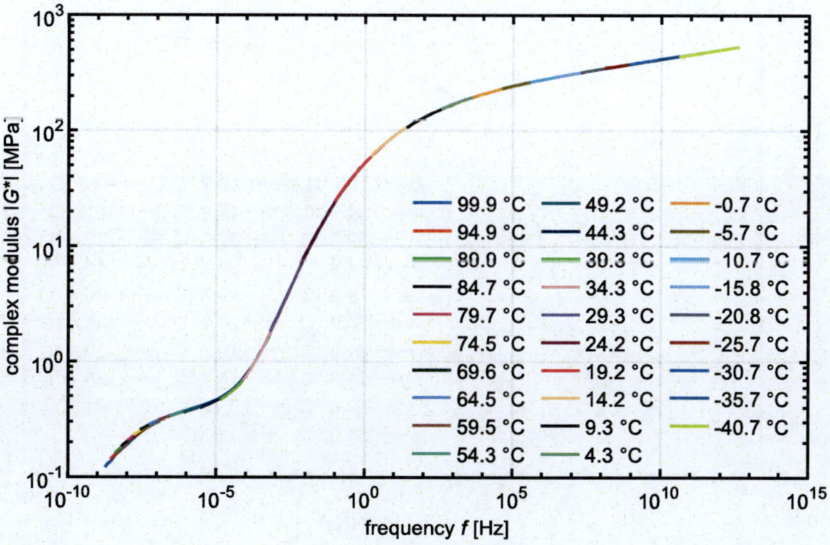

Figure 6.2 Standard PVB mastercurving process, (a) raw data and shift factors, (b) mastercurve at T_{ref} = 20℃.

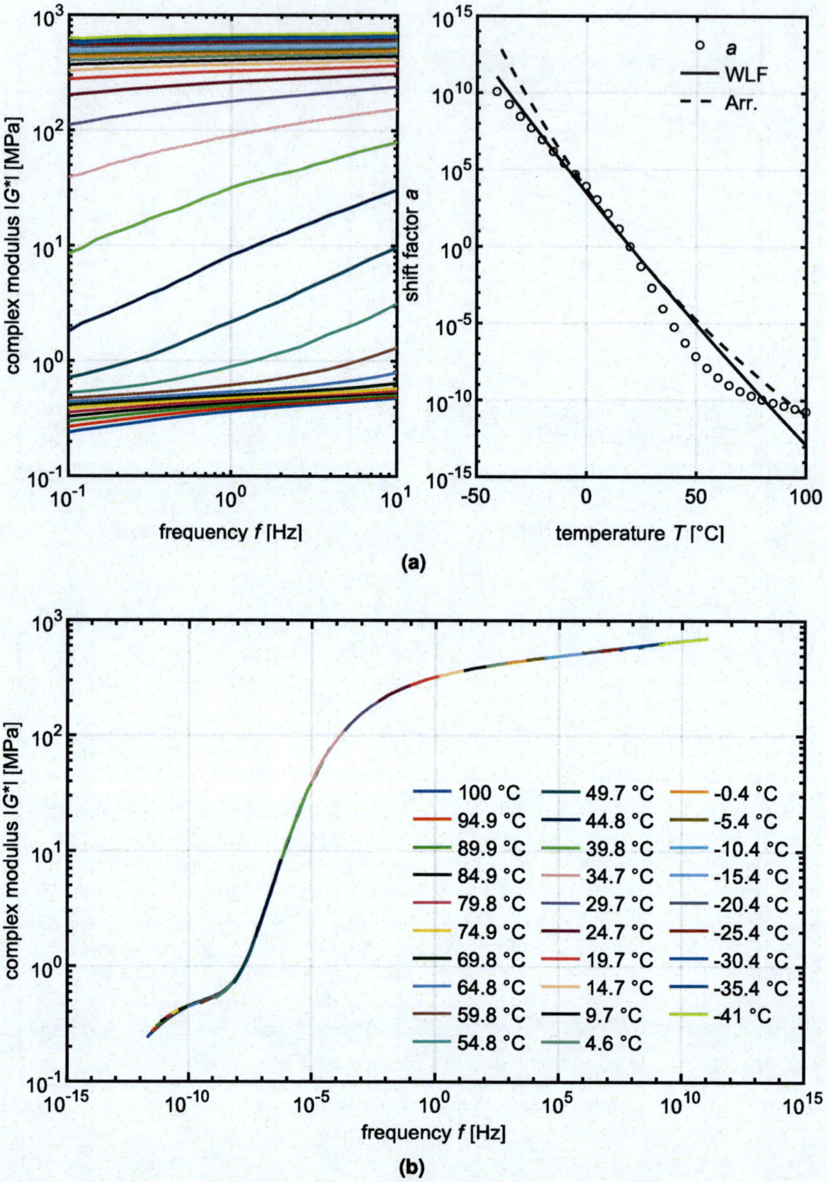

Figure 6.3 Structural PVB mastercurving process, (a) raw data and shift factors, (b) mastercurve at $T_{\mathrm{ref}} = 20\,°C$.

ature range. As an alternative TTSP, a polynomial of power 5 function is proposed for both materials, see Figure 6.4. The functions are only valid in the investigated temperature range [+100:-40]°C.

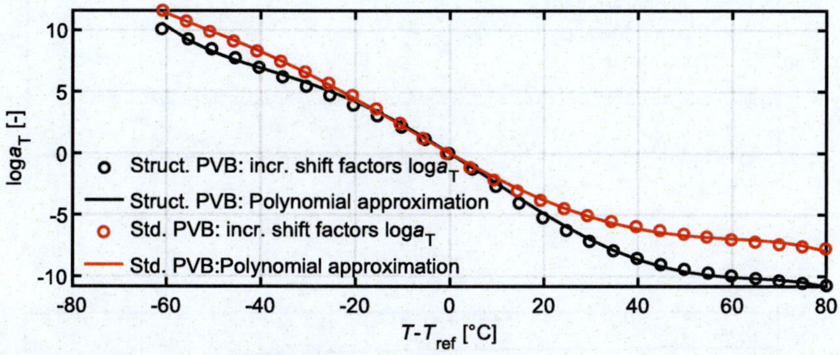

Figure 6.4 Polynomial of power 5 TTSP.

6.2 Linearity limits

The amplitude sweeps (AS) were performed with linearly increased strains for different temperatures T and different frequencies f. The most temperature levels were examined at a frequency of 1 Hz. For deviating frequencies (0.1 Hz or 10 Hz) only a few tests were carried out in different temperature ranges. The test parameters are summarized in Table A.29.

Figure 6.5 exemplary shows the determination of the linearity limit for the standard PVB interlayer at temperatures of 10°C and 40°C and a frequency of 1 Hz. Figure 6.6 shows the linearity limits of structural PVB at temperatures of 35°C and 60°C and a frequency of 1 Hz.

The linearity limit γ_{lin} was defined to the strain value at which the deviation of the storage modulus G' from the mean storage modulus in the linear viscoelastic range G'_{LVE} reaches 3%. To determine the mean storage modulus of the linear viscoelastic range G'_{LVE} under consideration of a certain measuring noise, the deviation of the actual measurement point from the mean value of all previous measured values is determined. G'_{LVE} is then calculated from the average of the values with a deviation less than 0.5%. The deviation for all measured values is then determined from this mean modulus. If the deviation reaches 3%, this point is defined as the linearity limit. To check that the stiffness decrease is not caused by slippage, the AS were driven both forward (from small to large strains) and backward (from large to small strains). If both curves lie on top of each other, it can be assumed that

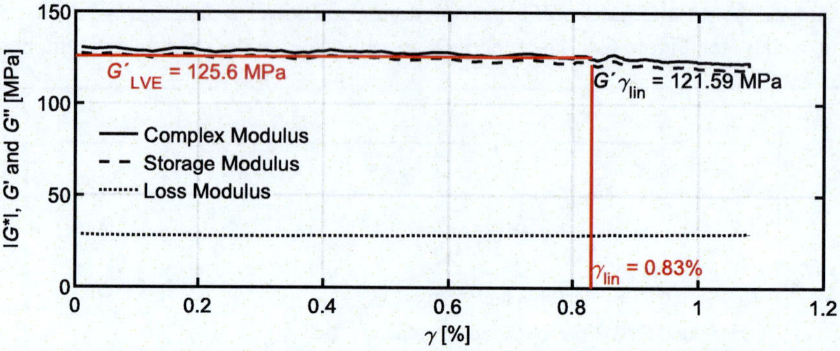

(a)

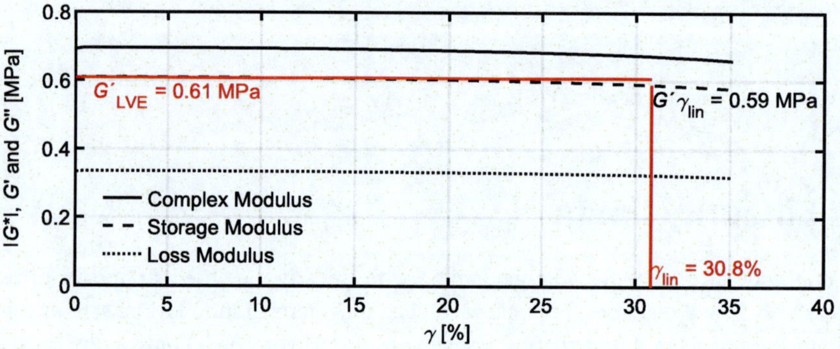

(b)

Figure 6.5 Standard PVB: exemplary determination of linearity limits at different temperatures and f = 1 Hz, (a) T = 10°C, (b) T = 40°C.

no slippage has occurred.

Figure 6.7 (a) shows the linearity limits of standard PVB determined as a function of temperature for two different frequencies (1 Hz and 0.1 Hz). At a frequency of 1 Hz, 14 different temperature levels were investigated. The selected temperature levels cover temperatures in the entropy-elastic as well as in the energy-elastic and the glass transition range. It becomes apparent that the linearity limit increases with increasing temperature. This corrodes to the literature references described before, see e.g. Figure 2.32. The linearity limit extends over a range of approx. 0.5% at low temperatures and 30-35% at high temperatures. At temperatures below 15°C, the linearity limit seems to increase again, but this could be due to the experimental setup. A new sample was used for each amplitude sweep, which means that there may be small differences in the sample preparation

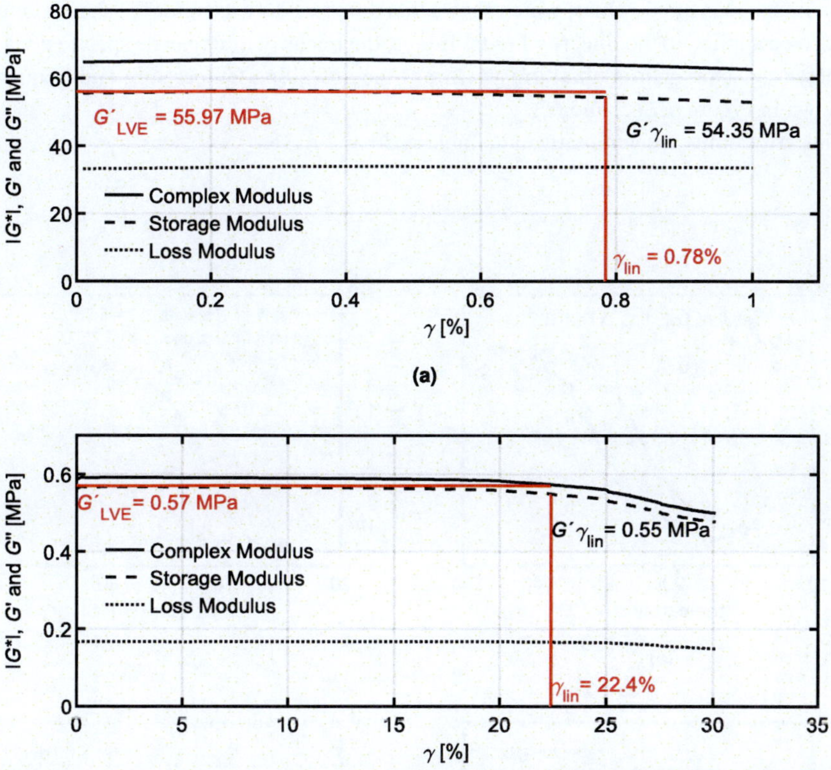

Figure 6.6 Structural PVB: exemplary determination of linearity limits at different temperatures and f = 1Hz, (a) T = 35 °C, (b) T = 60 °C.

and in the sample installation. At the frequency of 0.1 Hz, only two temperatures were investigated. The linearity limits are higher compared to the measurement data at 1 Hz. Remembering the thermorheology (time resp. frequency temperature analogy), a lower frequency results in a lower stiffness. Evaluating the TTSP with the parameters given in Table 6.1, the change in frequency from 1 Hz to 0.1 Hz leads to a shift factor of $\log a_{\mathrm{T}} = 1$, which in turn corresponds to a temperature decrease of ΔT = -6.21°C (WLF).
Figure 6.7 (b) shows analogously the linearity limits of structural PVB as a function of temperature for three different frequencies (0.1, 1 and 10 Hz). Using the TTSP parameters from Table 6.2, these frequency differences correspond to ΔT = +-5.75°C (WLF).

By shifting the black solid lines in Figure 6.7 (a) and in Figure 6.7 (b) with the calculated shift factors, the dashed and dotted lines are obtained. As can be seen this lines correlate

well with the measured values, which shows that the procedure used for the shear modulus can also be applied to the linearity limits. Hence, the linearity limit can be specified via the stiffness (e.g. storage modulus) irrespective of the prevailing temperature and frequency, as shown in Figure 6.7 (c) and (d).
The linearity limits are summarized in Table A.30 and A.31.

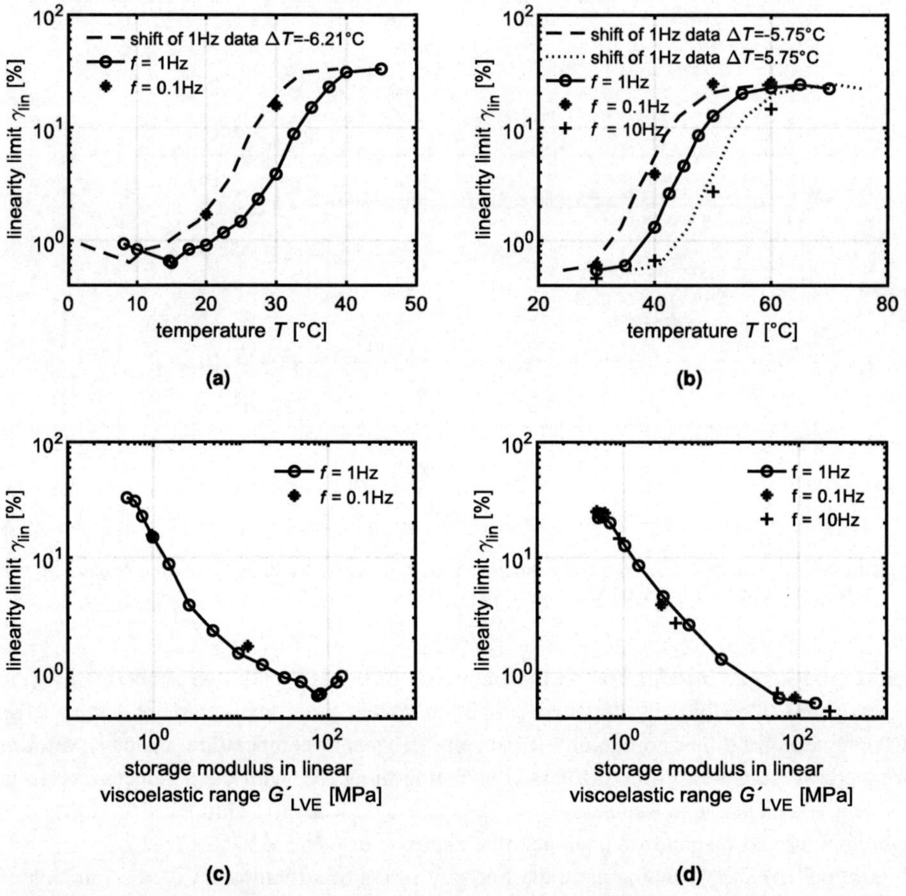

Figure 6.7 Linearity limits, (a) standard PVB in dependency of temperature and frequency, (b) structural PVB in dependency of temperature and frequency, (c) standard PVB in dependency of storage modulus, (d) structural PVB in dependency of storage modulus.

6.3 Comparison of two different PVBs

Comparing the two materials, many parallels can be seen. As mentioned earlier, the PVB types differ primarily in their plasticizer content. Figure 6.1 shows that both storage modulus G' curves have a similar shape. This becomes particularly clear by shifting the data of the structural PVB by the difference of the glass transition temperatures, Figure 6.8 (a). The loss factor and storage modulus of the structural PVB are slightly higher than those of standard PVB. Figure 6.8 (b) shows that all linearity limits, both for the standard and for structural PVB plotted as a function of the storage modulus, coincide and can be approximated with a single polynomial function (see Table 6.3). This can be explained by the fact that a variation of the plasticizer content affects the stiffness in the same way as a temperature change; the $G(t)$ curve shifts horizontally along the time axis.

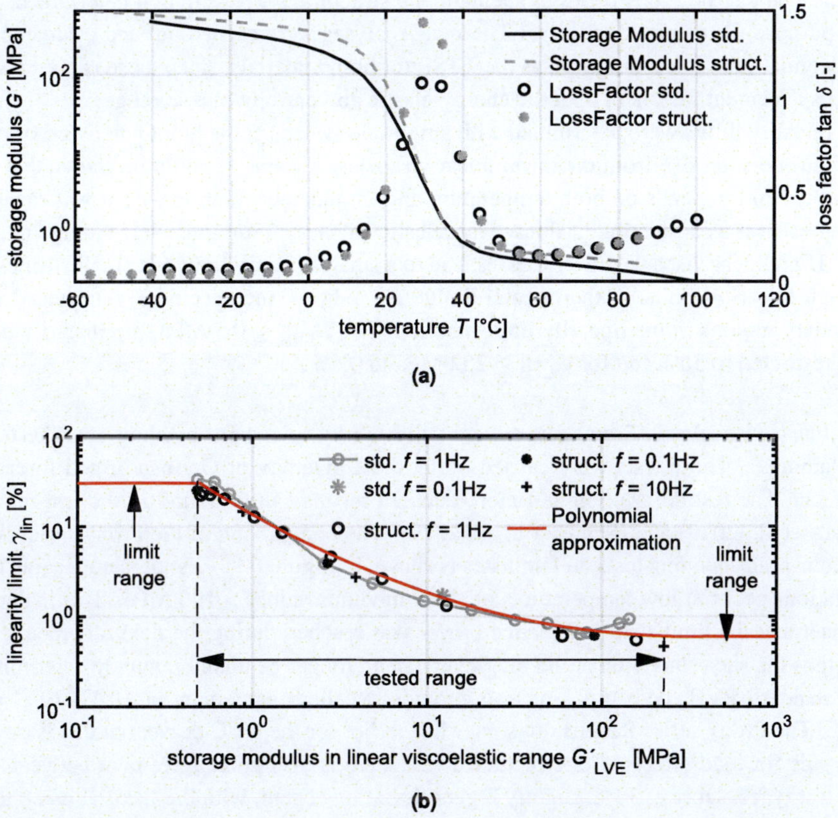

Figure 6.8 (a) temperature dependent material behaviour of standard and structural PVB at f = 1Hz. Structural PVB data shifted by ΔT = -15°C, (b) linearity limits.

Table 6.3 Approximation of linearity limits.

G'_{LVE} [MPa]	Approximation of γ_{lin} [%]
$G'_{LVE} \leq 0.5$	30
$G'_{LVE} \geq 252$	0.6
$0.5 < G'_{LVE} < 252$	$\log \gamma_{lin} = 2.1834 \cdot 10^{-1} \cdot x^2 - 1.0879 \cdot x + 1.1308$ with $x = \log G'_{LVE}$

6.4 Summary and comparison with literature

In this chapter, the Prony series of standard and structural PVB were first determined. For this purpose, Dynamic-Mechanical-Thermal-Analyses were carried out and evaluated. A polynomial function is proposed as the TTSP for both materials, since these can represent the displacement factors in both the energy-elastic and entropy-elastic ranges.
The linearity limits were determined with amplitude sweeps at the point where the storage modulus deviates 3% from the mean linear viscoelastic value. It could be shown that the linearity limit depends on both temperature and frequency, which is why it was decided to represent it as a function of linear viscoelastic stiffness (more precisely storage modulus). High linear viscoelastic stiffnesses lead to a small linearity limit and small stiffnesses to high linearity limits. Experimental limits (e.g. slip of the specimen) introduced two boundary regions in the linearity limit function. For $G'_{LVE} \leq 0.5$ MPa, the linearity limit was restricted to 30% and for $G'_{LVE} \geq 252$ MPa to 0.6%.

The linear viscoelastic Prony series were validated by four point bending creep tests on the laminate. These tests were carried out at the University of German armed forces in Munich. The test description and more detailed results can be found in the test reports (Kraus et al., 2019) and (Schuster et al., 2019). The comparison of the Prony series with the four-point-bending tests on laminates is shown in Figure 6.9. First, it is noticeable that deviations occur at low temperatures or shear modulus values > 100 MPa. This is due to the fact that the limit state *full shear transfer* was reached during the flexural creep tests, and thus the shear modulus of the interlayer can no longer be unambiguously determined. For standard PVB, both types of tests provide excellent agreement at 10°C, 20°C and 30°C. The Prony series flattens out somewhat earlier for the 40°C curve, but this is on the safe side for load effects. For structural PVB, there is also good agreement between the two test types. Only at 30°C slightly higher deviations occur, with the flexural creep tests providing lower shear modulus values.
For structural PVB, other Prony series and TTSP exist in literature (Z-70.3-230; Eastman Chemical Company, 2017) and are used in Figure 6.10 (a) for comparison. It can be ob-

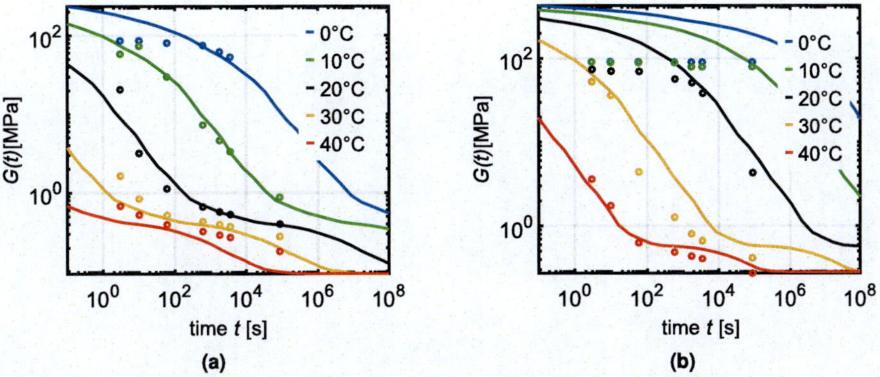

Figure 6.9 Comparison of Prony series with four-point bending creep tests, continuous lines correspond to the results obtained with the Prony series and TTSP (interpolation of incremental shift factors), the markers correspond to the mean results of four-point-bending creep tests (a) standard PVB, (b) structural PVB.

served that the Prony parameters in Z-70.3-230 are strongly on the safe side in comparison to the other results, which is noticeable in a shear modulus decreasing at earlier times. Figure 6.10 (b) compares the TTSP. At high temperatures, the WLF function specified by Eastman leads to a good agreement to the shift factors determined in this thesis. However, the deviations are enormous at temperatures below 15°C.

No literature was found on experimentally determined linearity limits of PVB. However, qualitative comparison with Figure 2.32 shows that the results are plausible.

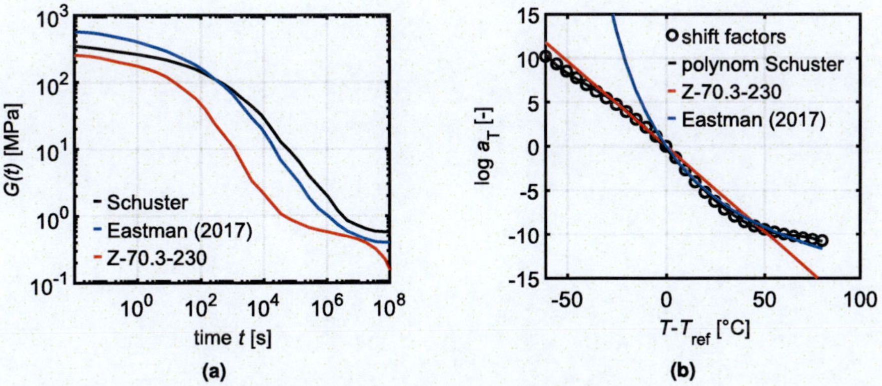

Figure 6.10 Structural PVB: Comparison with literature data, (a) Prony series at 20°C, (b) TTSP.

7 Characterization of the nonlinear viscoelastic behaviour of PVB

This chapter refers to standard and structural PVB. The relaxation behaviour at large deformations, well within the nonlinear viscoelastic range, is investigated. All the experiments in this chapter were also carried out with acoustic PVB. The results are shown exclusively in the appendix (chapter A.5.5). Detailed descriptions of parts of the investigations can be found in (Farr, 2018). Furthermore, some aspects have already been summarized in (Schuster et al., 2022) and submitted for publication.

7.1 Uniaxial tensile relaxation tests

The high strain relaxation behaviour of two different PVB interlayer types was investigated in uniaxial tensile mode. The tests were carried out with a universal testing machine in a climatic chamber and lasted between 1 and 2 hours (very few tests lasted 12 h). The Becker tension rod (Becker, 2009) was used as specimen geometry. Three different temperatures (T_1 in the energy-elastic area: 0°C for standard PVB and 20°C for structural PVB, T_2 in glass transition range: 20°C for standard PVB and 40°C for structural PVB and T_3 in entropy-elastic range: 40°C for standard PVB and 60°C for structural PVB) were investigated for each interlayer type at a constant relative humidity level of 50%rH. The influence of the strain level was investigated by carrying out tests with three different positions of the upper clamping device for each temperature and interlayer. For T_1, elongations of 2.5 cm, 5 cm and 7.5 cm were investigated. For T_2, the elongations corresponded to 5 cm, 10 cm and 12.5 cm and for T_3 to 5 cm, 10 cm and 15 cm. The displacement rate of the machine traverse to apply the deformation was identical for all tests (400 mm/min). The force was measured with a load cell and the applied deformation was controlled via the machine traverse. Since slippage may occur in the clamping devices, the strain in the tension rod was measured with DIC. An overview of the test settings and the performed tests are summarized in Table A.36 and Table A.38.
The time-dependent strain and stress measurements are represented in Figure A.24 and A.30 in the appendix. Figure 7.1 and 7.2 (a) show the relaxation functions at different strain levels (resp. crosshead displacements) for three different temperatures. They were calculated with the true stress and strain values at each time t, see eq. (3.26).

M. Schuster, *Characterization of Laminated Safety Glass Interlayers*,
Mechanik, Werkstoffe und Konstruktion im Bauwesen 66,
https://doi.org/10.1007/978-3-658-39821-7_7

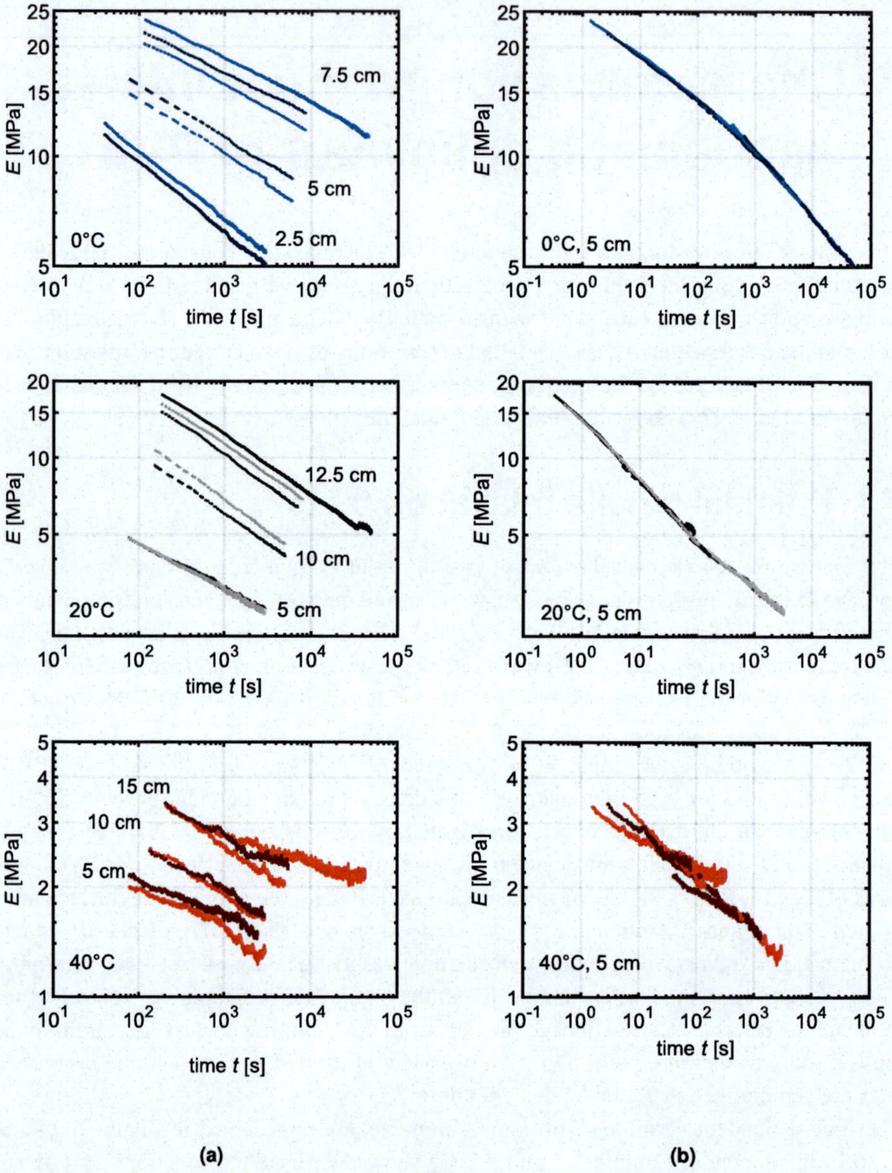

Figure 7.1 Standard PVB, (a) relaxation functions $E(t)$ for different temperatures and different strain levels. $E(t)$ was calculated from the true stress and strain values at each time t, (b) horizontally shifted relaxation functions. The reference displacement equals 5 cm in all cases.

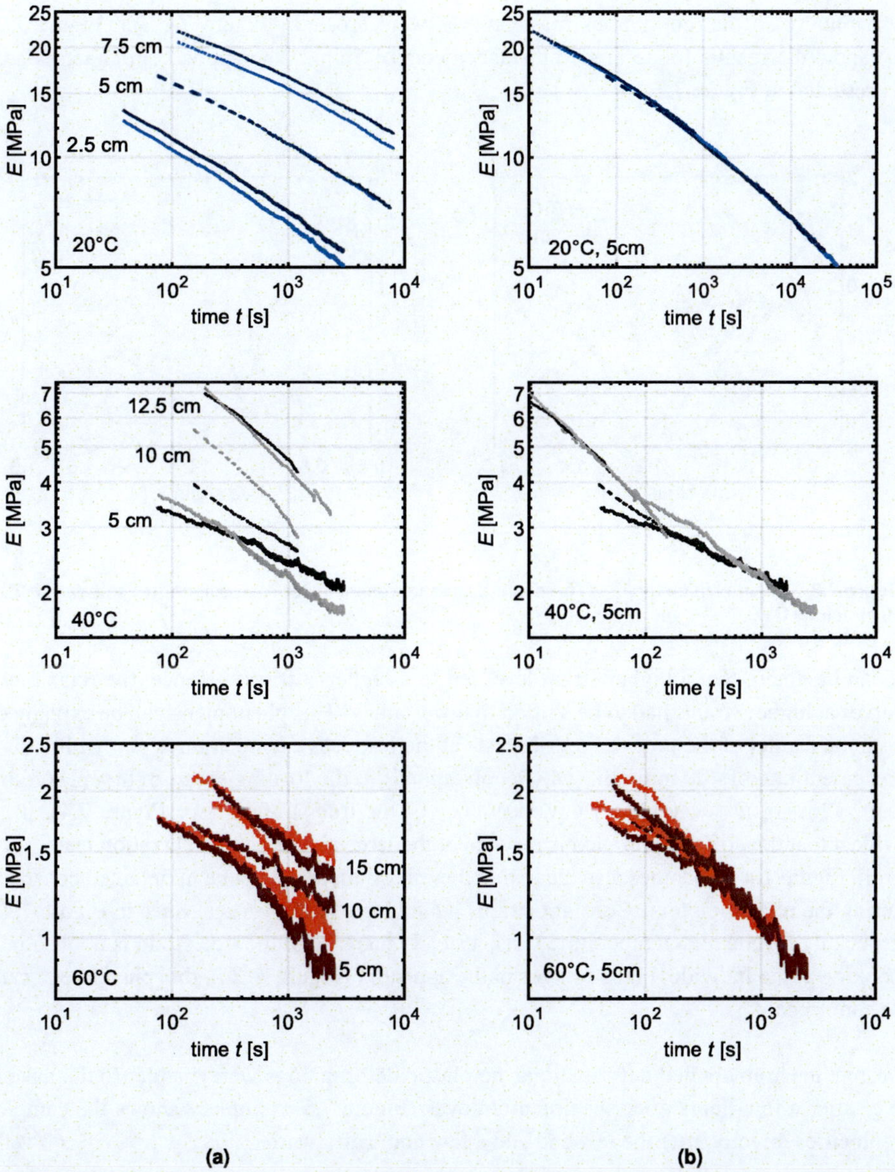

Figure 7.2 Structural PVB, (a) relaxation functions $E(t)$ for different temperatures and different strain levels. $E(t)$ was calculated from the true stress and strain values at each time t, (b) horizontally shifted relaxation functions. The reference displacement equals 5 cm in all cases.

Since the individual relaxation curves have overlapping areas on the y-axis, the time-strain-superposition according to chapter 2.4 was applied by shifting the individual curves purely horizontally so that continuous mastercurves were created. Figure 7.1 and Figure 7.2 (b) show the results for reference displacements of 5 cm. The applied shift factors are represented in Figure 7.3.

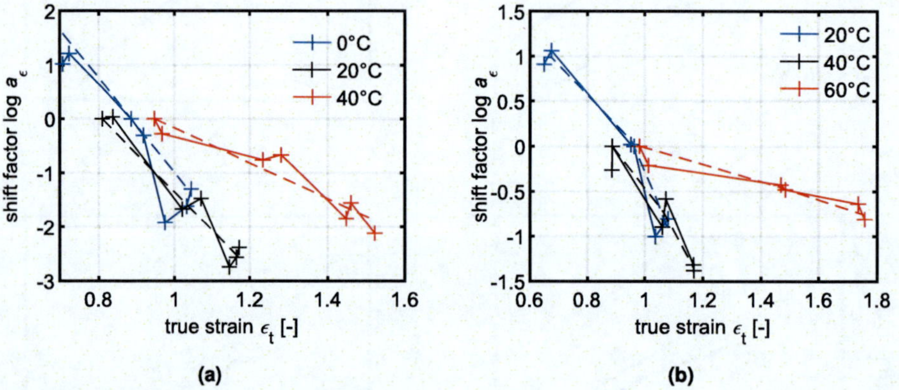

Figure 7.3 Strain shift factors a_ε. The dashed lines represent linear regressions. (a) standard PVB, (b) structural PVB.

It can be noted, that a higher strain level led to a higher stiffness. Hence, the relaxation curves at higher strains had to be shifted to lower times. This phenomena can be explained with the results of the uniaxial tensile tests at the respective temperatures and with a displacement rate of 400 mm/min (which corresponds to the loading speed in the relaxation test). Plotting $E = \sigma_t(t)/\varepsilon_t(t)$ in dependency of the true strain ε_t (see Figure 7.4), it is evident that the stiffening branch is reached at the large strains in the relaxation tests. From Figure 7.4 it becomes also clear, that the true strain could not be maintained constant during the relaxation text. Constant strains would have led to vertical lines in Figure 7.4. Especially for the high temperature tests, a steady increase of the true strain is noticeable. Looking at the individual DIC images in the appendix (Figure A.26), this phenomena can be confirmed.

Despite the high applied deformations, no plastic deformations were visible (to the naked eye) after a few hours after specimen removal. Figure 7.5 exemplary shows the sample geometries directly after the test end and a few hours afterwards.

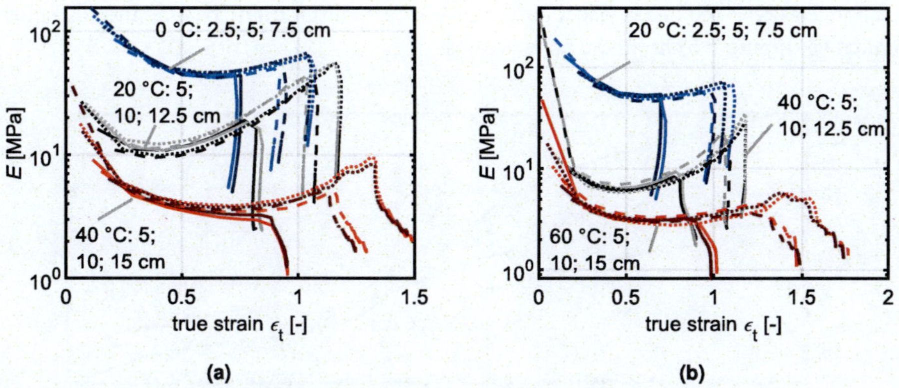

Figure 7.4 $E = \sigma_t(t)/\varepsilon_t(t)$ during the application of deformation with 400 mm/min and the relaxation tests (a) standard PVB, (b) structural PVB.

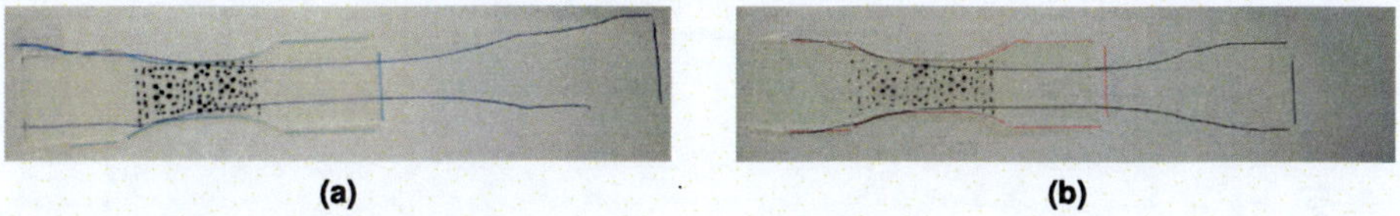

Figure 7.5 Sample geometries directly after the test and a few hours later (a) standard PVB, $T = 0$°C, 7.5 cm, (b) structural PVB, $T = 20$°C, 5 cm.

7.2 Comparison with Prony series

After mastercurves were created for all temperatures at a reference displacement of 5 cm, they were additionally shifted to a reference temperature of 20°C and compared with the linear viscoelastic Prony series. For this purpose, the TTSP determined by DMTA (in chapter 5, Figure 6.2 and 6.3) was applied.

It should be noted, however, that the reference displacement of 5 cm does not correspond to the exact same reference strain at all temperatures, see Figure 7.3, so that slight deviations are possible. In addition it has to be mentioned, that the Prony series was determined with DMTA in plate-plate mode and hence depict the shear modulus. In order to compare the results to the uniaxial tensile relaxation tests, the shear Prony series was transformed into a tensile Prony series by applying the approximate relation $E(t) \approx 3 \cdot G(t)$. This approximation is based on incompressible material behaviour with Poisson's ratio of $\mu = 0.5$. For PVB, however, the Poisson's ratio is time and temperature dependent and can have values between ≈ 0.3 - 0.5 (Kuntsche, 2015). Furthermore, the DMTA samples were stored in drying beads prior to testing while the Becker tensile rods were exposed to a relative humidity of 50%rH. Also for these reasons, the comparison of the large strain tensile relaxation tests with the linear viscoelastic Prony series must initially be classified as rather qualitative. Figure 7.6 shows the comparisons.

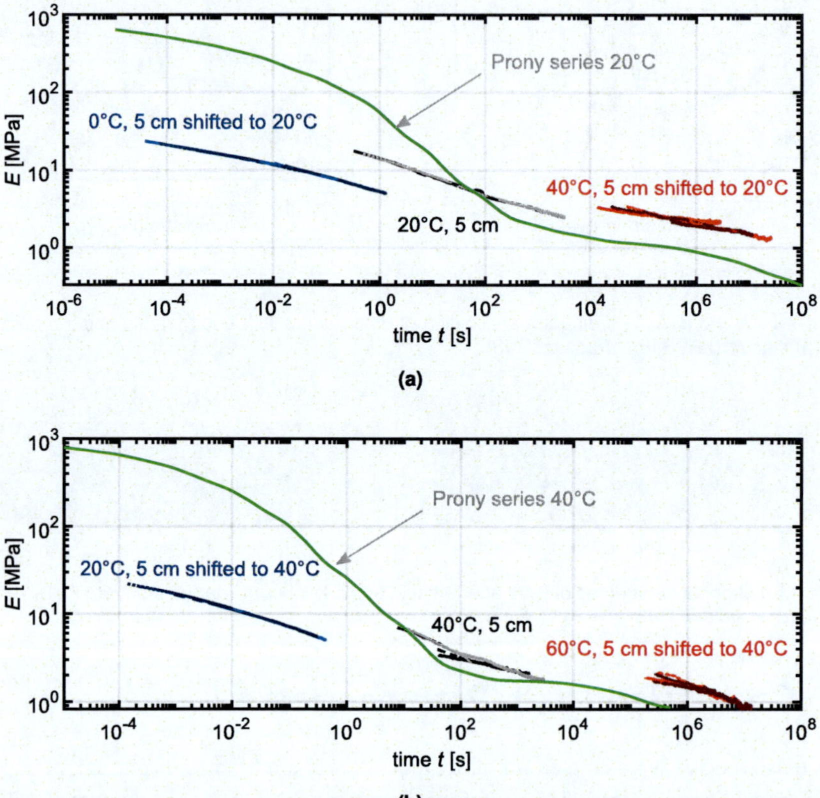

Figure 7.6 Comparison of linear viscoelastic Prony series with large strain uniaxial relaxation mastercurve. The horizontal temperature shift factors were determined with DMTA. The tensile Prony series was approximated with $E(t) \approx 3G(t)$. (a) standard PVB, T_{ref} = 20 °C and w_{ref} = 5 cm ($\varepsilon_{\mathrm{true}} \approx$ 80%), (b) structural PVB T_{ref} = 40 °C and w_{ref} = 5 cm ($\varepsilon_{\mathrm{true}} \approx$ 88%).

It can be observed, that no continuous mastercurves are obtained. It seems that in order to form continuous mastercurves, the individual temperature curves (after TSSP) have to be shifted vertically. Especially between T_1 (blue lines) and T_2 (black lines) a vertical offset is recognizable.

At low temperatures, the uniaxial tensile relaxation tests deliver much lower stiffness values than the linear viscoelastic Prony series. The shape of the relaxation curve, however, seems to be similar. In order to fit the Prony series, a vertical shift factor > 1 is necessary. At high temperatures, the uniaxial tensile tests deliver stiffness values that are in the range of those from the linear viscoelastic Prony series (note that the y-axis is in a logarithmic scale). At medium temperatures, the shape of the uniaxial tensile relaxation curve differs from that of the linear viscoelastic Prony series. With increasing time, the linear viscoelastic Prony series is approached. To explain this curve change, the degree of nonlinearity is introduced in the following section.

7.3 Introduction of the nonlinearity degree

As described in chapter 6, the linearity limit depends on the stiffness in the linear viscoelastic region (and hence on the prevailing temperature, loading duration, test frequency etc.). This has the following significance for relaxation experiments:

- Relaxation tests with a constant strain at a low test temperature are significantly further in the nonlinear viscoelastic range than the same tests at elevated temperature.
- A relaxation test that initially takes place far in the nonlinear viscoelastic range approaches the linear viscoelastic range over time.

To account for these effects, the degree of nonlinearity NL is introduced. It is calculated according to equation 7.1 from the quotient of the applied strain γ or ε to the linearity limit γ_{lin} or ε_{lin}. NL = 1 equals the linear viscoelastic range. Higher strain levels result in a vertical shift of the NL-curve. The linearity limit is calculated with the linearity limit function determined in chapter 6. The linear viscoelastic shear modulus values (Prony series) serve as input variables. The method described here is subjected to some uncertainties. On the one hand, the storage modulus is set equal to the shear modulus. Furthermore, it is assumed that the linearity limit for shear and tension are equal.

$$\text{NL} = \frac{\gamma}{\gamma_{\text{lin}}} = \frac{\varepsilon}{\varepsilon_{\text{lin}}} \tag{7.1}$$

Figure 7.7 shows the degree of nonlinearity for standard PVB as a function of time at an exemplary temperature of 20°C for different strain levels in the relaxation test. It can be seen that at an applied strain of 5%, the linear viscoelastic range is reached again after approx. 25 s, and therefore the Prony series is valid again from there on. With an elongation of 25%, on the other hand, the linear viscoelastic range is reached only after 1500 s (25 min). At strain levels higher than 30%, the Prony series is not reached at all. This is due to the definition of the nonlinearity limit, which has been assigned a maximum value of 30% (Figure 6.7).

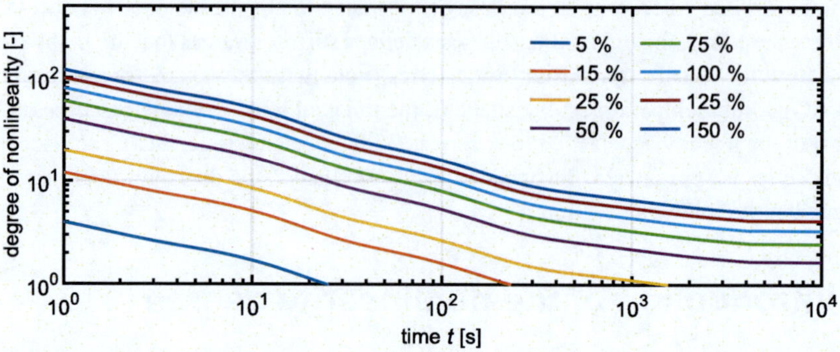

Figure 7.7 Degree of nonlinearity of standard PVB for different strain levels at a temperature of $T = 20\,°C$.

In order to determine the degree of nonlinearity in a tensile relaxation test, the following steps are performed:

- Shifting the Prony series (shear modulus) to the temperature at which the relaxation test takes place.
- Evaluating the Prony series at the measurement times of the relaxation test. As a result, the linear viscoelastic shear modulus values at each measurement time from the relaxation test are obtained. In order to determine the degree of nonlinearity of a strain- and temperature-based mastercurve, the Prony series has to be evaluated for the shifted times of the nonlinear viscoelastic mastercurve.
- Determination of the linearity limit with the linearity limit function in chapter 6 at each measurement time of the relaxation test resp. each shifted time of the nonlinear viscoelastic mastercurve. The linear viscoelastic shear modulus values serve as input variables.
- Determination of the degree of nonlinearity NL according to equation 7.1 for each measuring point of the relaxation test resp. for each shifted time of the nonlinear viscoelastic mastercurve. It is assumed that $\varepsilon_{\text{lin}} = \gamma_{\text{lin}}$.

Figure 7.8 shows the degrees of nonlinearity at T_{ref} and 5 cm along with the mastercurves at T_{ref} and 5 cm and the linear viscoelastic Prony series. As expected, the course of the degree of nonlinearity confirms that the change in the course of the relaxation function is related with a varying degree of nonlinearity.

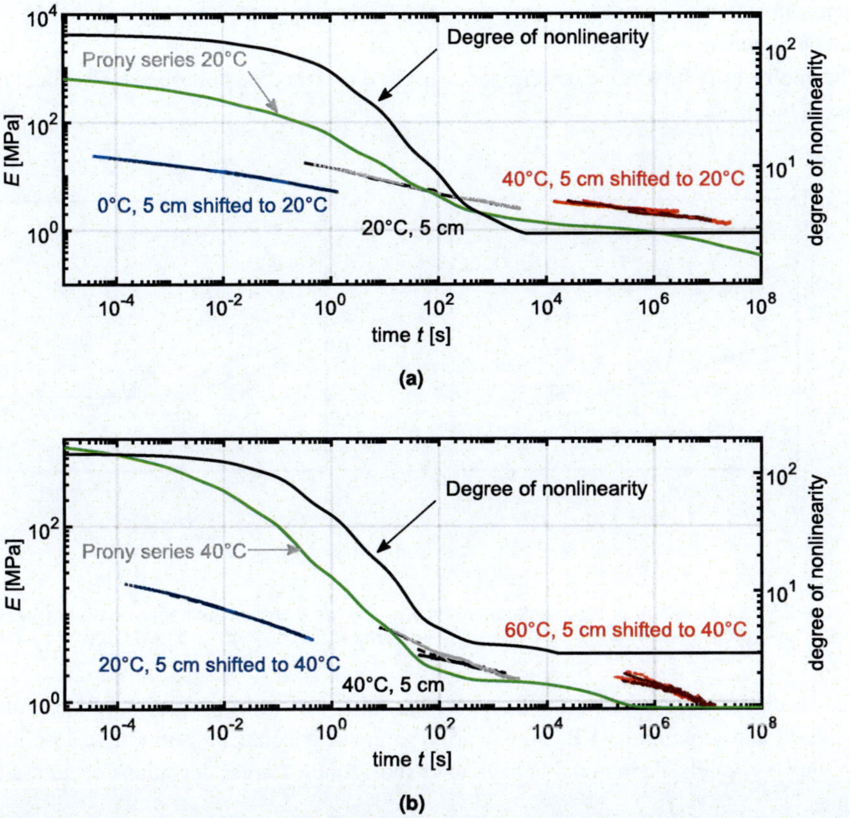

Figure 7.8 Comparison of linear viscoelastic Prony series with large strain uniaxial relaxation mastercurve for w_{ref} = 5 cm with indication of nonlinearity degree. (a) standard PVB, T_{ref} = 20 °C, (b) structural PVB, T_{ref} = 40 °C.

7.4 Determination of Schapery model parameters

In section 2.3.3, the Schapery model was introduced. Due to the various strain-dependent nonlinearity factors, this model is very adaptable. Figure 2.33 shows that by combining the nonlinearity factors with the Prony series, the required change in shape of the relaxation curve can be obtained. a_ε was used to create the mastercurve at 5 cm and at 20°C (standard PVB) resp. 40°C (structural PVB) and is therefore no longer considered for approximating the mastercurve with the Schapery model ($a_\varepsilon = 1$ is set). Note that this is analogous to the procedure of determining the Prony series in the linear viscoelastic range where first a_T is applied to construct the mastercurve and second, the Prony series is fitted to the mastercurve. Since the mastercurve is vertically compressed in the double logarithmic scale and

approximates the Prony series over time, $h_\infty = 1$ and $h_{1,2} < 1$, resp. $h_0 < 1$ and $h^*_{1,2} \neq h_0$ can be assumed.
The nonlinearity factors are determined with the direct method described by Brüller (1993), see also chapter 3.

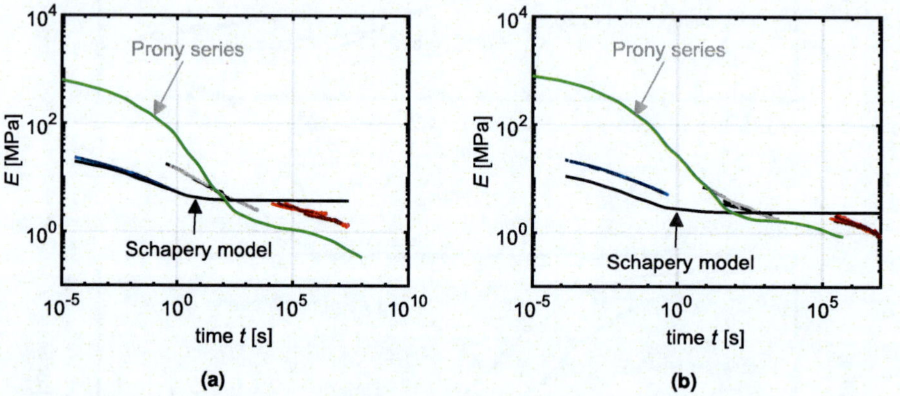

Figure 7.9 Comparison of the mastercurve for w_{ref} = 5 cm with the Schapery model, (a) standard PVB: T_{ref} = 20 °C, h_0 = 0.0169, $h^*_{1,2}$ = 0.0156, (b) structural PVB : T_{ref} = 40 °C, h_0 = 0.0358, $h^*_{1,2}$ = 0.0346.

The results of the fits using the Schapery model are shown in Figure 7.9. For both the standard and structural PVB, nonlinearity parameters could be found that describe the nonlinear viscoelastic material behaviour at short times. Larger deviations occur at higher times.

7.5 Summary and outlook

The relaxation behaviour of two PVB types was investigated at large strains and for three different temperatures in uniaxial tension. The applied strains were so large that a larger strain resulted in a stiffer material behaviour (stiffening branch). For all three temperatures, mastercurves for a reference displacement of 5 cm could be generated by purely horizontally shifting the individual relaxation curves. Subsequently, the 5 cm mastercurves were shifted to a reference temperature of 20°C (standard PVB) resp. 40°C (structural PVB) using the TTSP. Compared to the linear viscoelastic Prony series, it was found that the shape of the nonlinear viscoelastic relaxation curve differs significantly from the Prony series. At short times, a significantly smaller relaxation modulus is observed than in the linear case. However, it can also be seen that the nonlinear viscoelastic mastercurve approaches the linear viscoelastic Prony series with increasing time. This phenomenon could be explained by introducing a degree of nonlinearity. The vertical compression can theoretically

be represented by the Schapery model.
However, it is not yet clear why applying the TTSP to the individual 5 cm mastercurves does not produce a continuous curve. Especially between the T_1 (0°C resp. 20°C) and T_2 (20°C resp. 40°C) curves, there is a vertical offset which cannot be neglected. Here, further research is necessary in order to check the validity of the TTSP in the nonlinear viscoelastic range.
In addition, only very high strain levels have been considered so far. For a better understanding, preliminary tests at constant temperature (0°C) over a wide range of nonlinearity degrees have been conducted with standard PVB and will be presented here briefly. Those tests have been performed on rectangular samples (12 cm x 1.5 cm). The strain was measured with DIC. Similar to the previous tests, the deformations were applied with 400 mm/min. The following displacements have been investigated for $\approx$ 1h: 1 mm, 2 mm, 4 mm, 6 mm, 8 mm, 10 mm and 20 mm. Figure 7.10 (a) shows the results. The test settings and pictures from the test set up can be found in the appendix A.5.4. In Figure 7.10 (b), the previous results are represented for comparison (same y-scale than Figure 7.10 (a)). It can be seen that in the preliminary test results, higher elongation lead to lower stiffness. This is because small to medium strains were applied here. In addition, by comparison with the relaxation tests already presented, it can be seen that the relaxation curve at 20 mm fits in terms of magnitude with the high strain tests already presented.
Figure 7.10 (c) shows the result after applying a horizontal shift to the new, preliminary test data. The 1 mm curve was chosen as reference curve. In Figure 7.10 (d) the results from the large strain relaxation tests are also included. Due to the small reference elongation, it was expected that the Prony series is matched. However, if the mastercurve is compared with the Prony series, a certain similarity can be seen for short times only. For longer times, greater deviations occur. In addition, the shape of the large strain mastercurve is clearly different from the rest of the part. Figure 7.10 (e) and Table A.39 contain the horizontal shift factors in dependency of the mean true strain.
Since no continuous mastercurve arises through purely horizontal shifting of the individual relaxation curves, it is likely that in addition to the relaxation times, the spring stiffnesses of the generalized Maxwell model also behave variable as a function of the applied strain. This can be taken into account in the Schapery model with the nonlinearity factors h_0 or h_∞ and $h^*_{1,2}$ or $h_{1,2}$. In order to determine the nonlinearity factors, Schapery (1969a) generally suggests to split the relaxation behaviour into the time invariant part and the transient part and to consider both parts separately. By considering the time invariant part, h_0 or h_∞ can be determined and from the transient part $h^*_{1,2}$ or $h_{1,2}$ and a_ε.
The experiments are too short to adequately determine $E_\infty(\varepsilon)$ and hence h_∞, which is why it was chosen to do the further evaluation with $E_0(\varepsilon)$ and h_0. However, since the deformation cannot be applied suddenly and part of the relaxation already occurs during the application of the deformation, the determination of the true elastic fractions $E_0(\varepsilon)$ and hence h_0 is a major challenge.

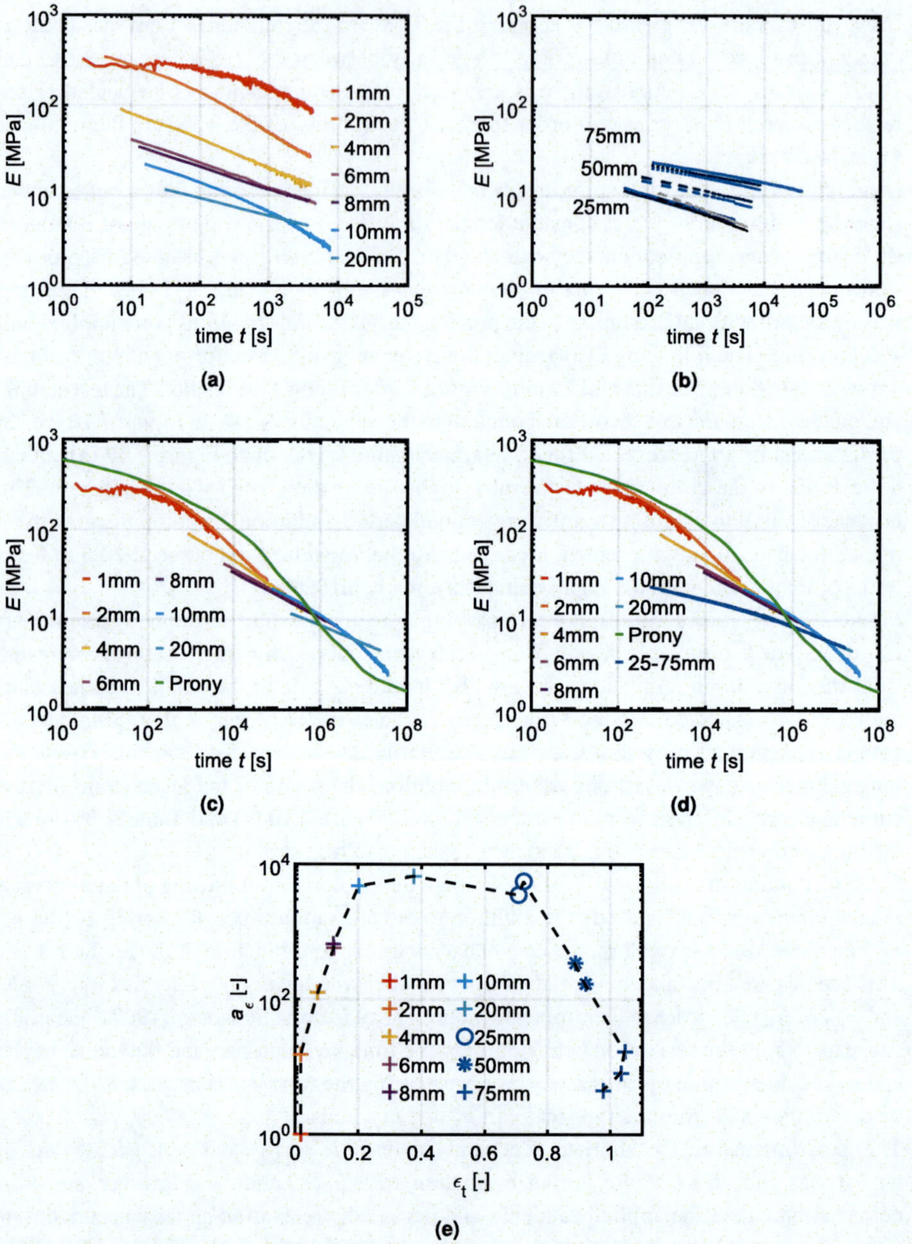

Figure 7.10 Standard PVB at 0℃: Uniaxial tensile relaxation tests (a) preliminary tests with small to medium strain deformations, (b) test results at high strain levels, (c) preliminary data horizontally shifted, 1 mm curve is the reference curve, (d) all data at 0℃, shifted horizontally, 1 mm curve is the reference curve, (e) shift factors.

Through a visual assessment of the individual curve slopes, it was found that a distinctly continuous mastercurve emerges as the relaxation curves (log-log curves) are shifted vertically upward and to the left as the strain increases. Since the course of the individual log-log curves remains the same, $h = h_0 = h^*_{1,2}$ is assumed. This also fits with the previous observation in Figure 7.8: At 0°C, the degrees of nonlinearity are nearly constant (see Figure A.27 in the appendix), so that the shape of the curve is preserved. Hence, after dividing the relaxation curves by h, a mastercurve is created only by horizontal shifting. Figure 7.11 (a) shows the manually determined vertical shift factors h, which were found through trial and error. In Figure 7.11 (b), E/h is represented. Figure 7.11 (c) contains the manually determined horizontal shift factors and in Figure 7.11 (d), the final mastercurve is presented. The shift factors are also summarized in Table A.40. Compared to Figure 7.10, a much more continuous mastercurve is obtained, however, large deviations from the linear viscoelastic Prony series are still present. The course of h shows similarities to $E_0(\varepsilon)$, which can be either defined as $E_0(\varepsilon) = E(\varepsilon, 11 \cdot t_0)$ or $E_0(\varepsilon) = E(\varepsilon, t_0)$, see Figure 7.11 (e), where t_0 is the time at which the respective relaxation experiment starts.

All in all, it can be stated that further research is needed to understand the mechanical behaviour of PVB under large deformations. The investigations carried out in this chapter are to be interpreted as preliminary test results and require validation. It is strongly believed that both the springs and dampers of the generalized Maxwell model are dependent on the applied strain. Initially, spring stiffnesses (defined here as secant moduli) decrease at small to medium strains, but increase at very large strains. This fits with the observations from uniaxial tensile tests. In the tests performed, horizontal displacement to shorter times with increasing strain gave the best result (most continuous mastercurve). This is in contradiction to the free volume theory according to Knauss and Emri (Kolařík and Pegoretti, 2006), which states that the free volume increases with increasing strain and the relaxation processes are faster with larger free volume. According to this, an increase in strain would cause the same effect as an increase in temperature and larger strain levels would have to be shifted to higher times.

It is recommended to further increase the experimental data base of the relaxation tests and to investigate a large number of strain levels at constant temperatures (analogous to the relaxation tests at 0°C shown here). Subsequently, individual tests can be repeated with other displacement rates. The following applies: the faster, the better, since E_0 can be better approximated in this way. In addition, it is suggested to investigate a number of temperatures for one strain level (e.g. in 2.5°C steps) to check the TTSP.
The influence of the degree of nonlinearity should also be taken into account.

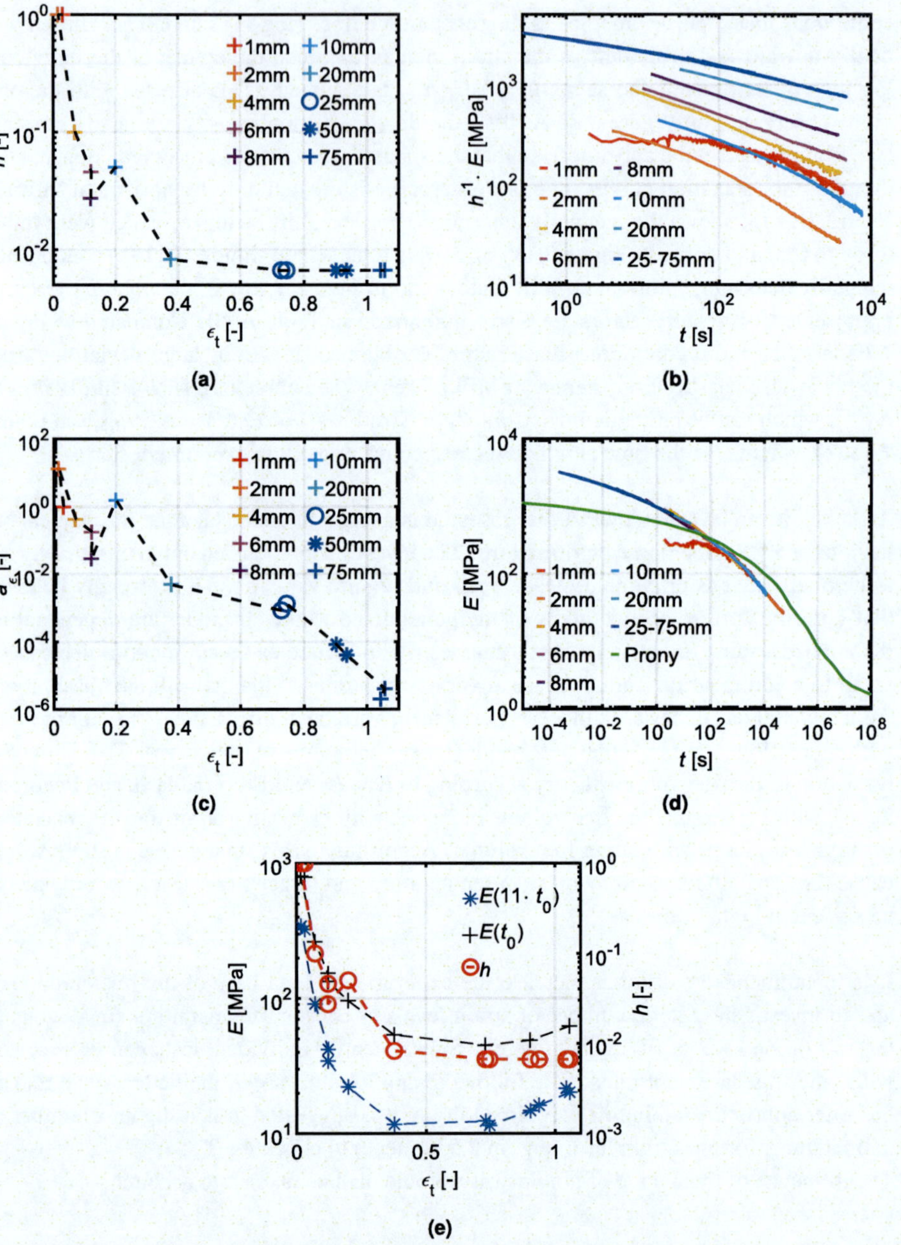

Figure 7.11 Standard PVB at 0 °C: Uniaxial tensile relaxation tests (a) vertical shift factor h, (b) measured E divided by h, (c) horizontal shift factor a_ε, (d) final mastercurve, (e) comparison of vertical shift factor h with $E(11 \cdot t_0)$ (first evaluation point) and $E(t_0)$ (begin of relaxation test).

Alternative tests to determine the nonlinear viscoelastic behaviour would be creep and creep recovery tests according to Schapery (Schapery, 1969a; Rand et al., 1996; Brinson et al., 2008). However, it must be ensured that a) sufficient space is available for the creep strain in the climatic chamber and b) the creep recovery can be measured.
In addition, uniaxial tensile tests with constant strain rate (and not constant displacement rate) can be helpful to determine nonlinearity factors, see e.g. (Woicke, 2006; Schmachtenberg, 1985).
Nonlinear viscoelastic material behaviour could also be studied using DMTA. For example, Golden et al. (1996, 1999); Strganac et al. (1995), describe a procedure in which a constant pre-strain or pre-stress is applied, which is then superimposed with a small amplitude harmonic vibration.
Another possibility to further investigate the nonlinear viscoelastic material behaviour would be to perform large amplitude oscillatory shear measurements where the amplitude of the sinusoidal oscillation exceeds the linear range. A response deviating from the sinusoidal curve is obtained and analyzed by means of Fourier transformation, (Hyun et al., 2002, 2011; Mezger, 2016; Farr, 2018). This measurement method is predominantly used for solutions and melts.

Finally, it should be noted that the studies to date have been limited to a one-dimensional consideration. Additionally, for a generalization, Mullins (Diani et al., 2009) or Payne effects (Lion et al., 2003; Yang et al., 2017), which frequently occur in polymers (especially in elastomers with fillers) at large strains, should be further investigated.

8 Application to engineering practice

Glass design in intact condition considering shear coupling through the interlayer requires shear modulus values. The following chapter will describe the experimental determination of the shear modulus values as a function of different interlayer types, with a special focus on semicrystalline materials. An approach to dealing with semicrystalline interlayers is also given with regard to the laminated glass design.

In the fractured state, the linearity limits of the materials are exceeded. Information on the experimental determination of the material behaviour when the linearity limits are exceeded is given.

Finally, this chapter contains specific design examples.

8.1 Linear viscoelastic material parameters and glass design

This section proposes a standard procedure for determining the linear viscoelastic material behaviour of different laminated glass interlayers. The consideration is limited to interlayer films. For this purpose, the information from EN 16613 and Kraus (2018) are supplemented by the experiences gathered by the author from the time at the institute, through discussions with interlayer manufacturers, existing standards, as well as new knowledge gained in this work are combined. The test guide contains information on samples, sample storage, testing, evaluation and validation. Additional information is provided for glass dimensioning.

General

In order to determine the temperature- and load duration-dependent material behaviour of polymeric interlayers undergoing small strains and stresses, Dynamic-Mechanical-Thermal-Analyses (DMTA) are recommended. In DMTA, the viscoelastic sample is excited at controlled temperature by means of harmonically oscillating strain or stress and the phase-

M. Schuster, *Characterization of Laminated Safety Glass Interlayers*,
Mechanik, Werkstoffe und Konstruktion im Bauwesen 66,
https://doi.org/10.1007/978-3-658-39821-7_8

shifted stress resp. strain response is measured. By varying the excitation frequency and temperature, the time and temperature-dependent stiffness characteristics can be derived. The test procedure to determine the linear viscoelastic material properties of polymeric interlayers may be three- or four-stage:

- Stage 1: Differential Scanning Calorimetry (DSC) measurements to determine thermo-mechanical phases (crystalline, amorphous) and characteristic temperatures (glass transition T_g, crystallization T_c and melting temperature T_m) is recommended generally. Alternatively, Dynamic-Mechanical-Thermal-Analysis temperature sweeps (DMTA-TS) can be used.
- Stage 2: Conduction of DMTA amplitude sweeps (DMTA-AS) to determine the linearity limits of the material under investigation and the choice of an excitation amplitude for Stage 4.
- Stage 3: Conduction of DMTA temperature-frequency sweeps (DMTA-TFS) with subsequent mastercurve generation to deduce Prony series and time-superposition principles.
- Stage 4: Conduction of large scale validation tests (bending creep or torsional relaxation) with glass laminates.

The testing procedure, evaluation methods and laminated glass design are dependent on the interlayer material type. The usual distinction for engineering purposes is taken up here as a function of the thermo-mechanical behaviour:

- Material type 1: Amorphous polymers
- Material type 2: Semi crystalline polymers
- Material type 3: Chemically crosslinked polymers
- Material type 4: Unknown material behaviour

The DSC test procedure and evaluation is described in EN ISO 11357-1 to EN ISO 11357-7 and will not be further discussed here.
The general test procedure for DMTA is described in EN 16613 (chapter 6), resp. EN ISO 6721-1. Different DMTA-test modes (e.g. tensile vibration ISO 6721-4, flexural vibration ISO 6721-5, shear vibration ISO 6721-6 resp. EN 16613, torsional vibration ISO 6721-7 or parallel plate oscillation ISO 6721-10) are available. Since due to glass bending shear is the main stress in the interlayer, it is recommended to perform the measurement in shear or oscillation mode and hence to avoid errors during the conversion of the tensile modulus $E(t,T)$ into a shear modulus $G(t,T)$. If this is not realizable, the shear modulus can be approximated with eq. (8.1). wherein E is the tensile modulus and μ is the Poisson's ratio.

This approximation is also included in EN 16613. Note: Non-isotropic interlayers, such as multilayers (which are frequently used for acoustic application) have to be tested in shear.

$$G = \frac{E}{2 \cdot (1+\mu)} \approx \frac{E}{3} \tag{8.1}$$

Requirements for the samples

The samples must be representative of normal manufacture. Since some interlayer materials change properties during lamination (e.g. shrinkage, crosslinking process, etc.), it is recommended that the samples have passed the lamination process before the test. Therefore, non-adhering materials (e.g. Teflon) can be placed between glass and interlayer during lamination. The geometry of the samples has to be adapted to the measuring device. Realistic thicknesses for construction applications should be used. Typical thicknesses for PVB interlayers are e.g. 0.76 mm and 1.52 mm.

Requirements for the samples storage

If the materials are hygroscopic, they should be handed over to the test center in vacuum-sealed or air-sealed bags. As soon as the sealed bag is opened, the manufacturer's instructions must be followed. Alternatively, in the context of this work, the storage of the samples in an airtight glass container filled with drying pearls has proven itself. Proper storage is important so that the moisture content within the film (and thus the linear viscoelastic material behaviour and the adhesion of the film) does not change.

Documentation

The determined Prony series and the time-superposition principles have to be documented in the form of equations. The application limits must be specified. This includes e.g. the temperature range in which the time-temperature-superposition principle is valid. If possible, the linearity limits or compliance limits for the linear viscoelastic material behaviour should be specified. Amplitude sweeps at different temperatures and frequencies can be performed for this purpose. The temperature and frequency dependence of the linearity limit can be summarized by specifying the linearity limit as a function of the linear viscoelastic stiffness (e.g. storage modulus G'_{LVE}).
In addition to the functions, the results have to be documented in form of a table containing shear modulus values for different load durations and different temperatures. The load durations to be considered are at least [3 s, 30 s, 5 min, 10 min, 30 min, 1 h, 6 h, 12 h, 24 h, 4 d, 5 d, 3 w, 30 d, 50 a]. The temperatures to be considered are at least [-20°C, 0°C,

20°C, 23°C, 30°C, 35°C, 40°C, 43°C, 60°C, 80°C]. Those values are based on EN 16613 and former national technical approvals e.g. Z-70.3-230.

Validation procedure

For validation of the generated time and temperature-dependent shear modulus values, creep or relaxation tests on the glass-interlayer-laminate are conducted. This can be done by bending creep or torsional relaxation tests, as described in Botz et al. (2018), whereby the torsional tests seem to be more complex in realization. The tests shall be performed according to the principles of EN 16613, appendix A1, resp. EN 1288-3 without breaking the test specimens. The measurement data of the creep or relaxation test are only to be evaluated for times $t > 10 \cdot t_0$, wherein t_0 is the time needed to apply the load. Shear modulus values can be back-calculated from the creep or relaxation tests.
Alternatively, the component test can be mapped numerically in FEM, taking into account the Prony series for the interlayer. The deformations or stresses between the model and the test can be compared. However, it should be noted that the measurable shear modulus resp. deformation and stress range is limited by the two limiting cases of full shear transfer and no shear transfer. Especially at high shear modulus values, deviations between DMTA and component tests may occur.
Acceptable deviations between the Prony series and the validation tests still have to be defined.

8.1.1 Material type 1: amorphous polymers

Testing Procedure

The proposed test procedure generally consists of different steps:

- Step 1: Amplitude sweeps in order to determine the strain or stress amplitude to be used for the temperature-frequency sweeps. The excitation amplitude of the TFS has to be in the linear viscoelastic range. To find an appropriate strain or stress amplitude, the sample is loaded with varying strain or stress amplitudes at constant temperature and excitation frequency. If the derived shear modulus is *constant*, it is ensured that the measuring range is within the linear viscoelastic range. A deviation of less than 3% is suggested as a constant value (this corresponds to the definition used here to determine the linearity limit). According to ISO 6721-11, the frequency is fixed to $f = 1$ Hz. However, it should be remembered that the linearity limit is very much dependent on the material stiffness (and therefore on temperature and frequency). If a constant excitation amplitude and a constant frequency range are to be investigated during the entire TFS, at least one amplitude sweep must be performed at T_{min} with f_{max} if the measurement is strain-controlled.

In this constellation the expected material stiffness is the highest, and therefore the linearity limit is the smallest. If the measurement is stress-controlled, the compliance limit is the lowest for T_{max} and f_{min}. If the parameters vary during the TFS, additional amplitude sweeps with the new boundary conditions are recommended. The amplitude sweep must be performed with the same device and in the same load mode as the temperature-frequency sweep (tension, shear, torsion,... stress- or strain-controlled).

- Step 2: Temperature-frequency sweeps in order to determine the time and temperature-dependent shear modulus. In contrast to EN 16613, the test frequencies must be adapted to the device and measuring system used. The measurements are repeated for each relevant temperature level: $T \in [T_g - 40 : T_g + 40] \subseteq [-40 : +80]$°C) wherein steps of 2-5°C are recommended. However, the temperature steps can be adjusted depending on the frequency range under investigation. The stiffness values measured between two adjacent temperature levels must have an overlapping range. It is recommended to perform the test from warm to cold to minimize effects from physical aging. As a result of these measurements, modulus-frequency curves (storage modulus $G'(f)$, loss modulus $G''(f)$, complex modulus $|G^*(f)|$) and loss factor-frequency curves $\tan\delta(f)$ are derived for different temperatures.

  Note: If the glass transition temperature T_g is not known, the temperature-frequency sweep can be performed in $T \in [-40 : +80]$°C. By evaluating the 1 Hz data, the glass transition temperature T_g is obtained. The accuracy depends on the selected temperature steps. Different evaluation methods are given in ISO 6721-11.

Evaluation

The evaluation consists of different steps:

- Step 1: Verification whether a purely horizontal displacement factor is sufficient to determine the mastercurves and whether thermorheologically simple material behaviour is present. For this purpose, three so called control plots *Wicket plot* (loss factor $\tan\delta$ vs. storage modulus G'), *Cole-Cole plot* (storage modulus G' vs. loss modulus G'') and *Van-Gurp Palmen plot* (loss angle δ or loss factor $\tan\delta$ vs. logarithmic amount of complex modulus $\log|G^*|$) can be created. If those so called control-plots result in smooth, temperature-independent curves, thermorheologically simple material behaviour can be supposed and solely horizontal shifts are sufficient to create mastercurves.

- Step 2: Creation of mastercurves at a reference temperature T_{ref} in the frequency domain. For the creation of mastercurves, the shift factors are determined incrementally. For this purpose, two neighboring isothermal curves are considered in each case.

There are different methods for this: Creating the mastercurves with storage modulus or complex modulus curves has the advantage that the curves are continuously increasing with increasing frequency, which simplifies the shifting process. Then, for example, two neighboring isothermal storage modulus frequency curves can be approximated by functions (e.g. splines), the overlapping area of both curves can be defined and finally the shift factor between both curves can be determined, which leads to the least square error. Another possibility is the "zigzag" connection of the measuring points of the two curves to an arc, whose length is then minimized (see further Bae et al. (2011)). The incremental displacement factor can be expressed either as a quotient of the frequencies of both curves $a_T(T_1, T_2) = f_1/f_2$ or in logarithmic notation as $\log a_T(T_1, T_2) = \log f_1 - \log f_2$.
Alternatively to the complex or storage modulus, the loss factor can also be used to create the mastercurves. The advantage of this quantity is that small errors, e.g. from the geometry and possibly vertical displacement factors from density differences are truncated. The disadvantage is that the loss factor does not have a monotonic course.

The incremental shift factors are multiplied or the logarithmic values are summed to determine the absolute displacement factors.

Regardless of which quantity was chosen for the mastercurving process, it is recommended to use the determined shift factors to examine the mastercurves of all remaining quantities (storage modulus, loss modulus, complex modulus and loss factor). In all cases, a continuous curve must be produced.

- Step 3: Approximation of shift factors with TTSP models. Contrary to the European standard EN 16613, the TTSP to generate the mastercurves should not be restricted to WLF. WLF works well especially at investigated temperatures in the glass transition region. In the energy-elastic state, however, the Arrhenius equation usually yields better results. Since most interlayer materials can be in the energy-elastic as well as in the glass transition-elastic or entropy-elastic range, for construction applications, it is recommended for practical engineering reasons to approximate the shift factors with other (physically insignificant) functions. The chosen function must approximate the experimentally determined displacement factors well, a permissible deviation must still be defined. Possible functions are e.g. polynomials of higher order, arctangent functions or others. The application limits (temperature range) must be specified.

- Step 4: Approximation of the mastercurves by a generalized Maxwell model and determination of a Prony series. Different procedures are accepted. Here, for example, the method of Kuntsche (2015) (two-step procedure with a genetic algorithm followed by a global search) and the method of Kraus (2018) (direct procedure, by setting up a matrix equation system) are referred to. Both procedures are described e.g. in (Kraus et al., 2017). In general, it is recommended to select one Maxwell element per time or frequency decade to be mapped. With the large temperature interval to be investigated, this often leads to very long Prony series with more than 20 Maxwell elements. Since this can be quite cumbersome in practice and, since for many common glass dimensions shear moduli of 10 MPa already lead to an almost full shear transfer, a shorter Prony series with 10 Maxwell elements can be determined for engineering purposes. In addition, for thermoplastics, the Prony series must be artificially set to zero for long times ($G_{\infty} = 0$).

8.1.2 Material type 2: semicrystalline polymers

Testing Procedure

If the melting and crystallization temperature as well as the degree of crystallization of the sample are unknown, DSC tests (EN ISO 11357-1) are recommended.
The characterization of the linear viscoelastic material behaviour can be determined in a simplified or detailed manner, whereby a distinction is to be made in the case of the simplified procedure depending on the application.

- Level 1: Simplified characterization in dependency on the application
  - Case 1: Applications where the melting range is always exceeded. The DMTA is only to be carried out for a temperature range above the melting point. Since the material is amorphous in the temperature range considered, the procedure for amorphous polymers applies (material type 1).
  - Case 2: Applications where the melting range is not reached by far. In this case, the crystallinity is assumed to be constant during the DMTA. The DMTA procedure is similar to that of the procedure for amorphous polymers.
  - Case 3: Applications where the melting range is not (or only rarely) reached. In this case, the DMTA can be performed from cold to warm. The degree of crystallization of the film in the installed state must be taken into account. If the melting range is exceeded during installation of the sample, the sample may be subjected to a physical aging process to restore crystallinity. In the context of this work, this was necessary for SG, for example. On the recommendation of the manufacturer, the sample was subjected to an aging process at 23°C for 5 days (whereby it was noticed that the equilibrium value had not yet been fully reached). If the melting

range is not far from the application temperature range, it is likely that temperatures exceeding the melting point will be measured during DMTA in order to determine the long-term behaviour. At these temperatures, the degree of crystallization changes (it decreases with higher temperatures) and the shift factor needed in mastercurve generation is a smeared value from temperature influence and degree of crystallization influence. It follows that the long-term behaviour is underestimated at temperatures not far from the melting range, which is on the safe side for design under load-induced loading. Hence, this can lead to deviations between the DMTA and the validation tests on the laminate, where the higher value from the component tests reflects the actual value. For such load cases, the DMTA can also be completely replaced by component tests on the laminate. The component test should be carried out at the temperature and for the load duration as defined in the load case. Care must be taken to cold crystallization and crystal perfection effects. For deformation-induced cases, full shear coupling is recommended during the glass dimensioning.

- Case 4: Applications where the temperature range is both below and above the melting point and the melting point is therefore regularly exceeded. It is recommended to perform the DMTA from warm to cold with a high cooling rate (in the case of DMTA-FTS, otherwise a high linearized cooling rate), The exact cooling rate should represent the maximum cooling rate that can occur in construction and still has to be defined. This provides a lower limit stiffness curve. The shear modulus values determined in this way are then on the safe side for the glass design with load-induced loads. For deformation-induced cases, it is recommended to take full shear coupling into account.
  Similarly to case 3, the horizontal shift factors are to be regarded as smeared factors, which are dependent on both the temperature and the degree of crystallization. The material behaviour for long times is therefore generally underestimated. During the component test, it is particularly important that the temperature history is known, realistic and comparable to the one in DMTA.

- Level 2: Detailed characterization by separating the effects of temperature and crystallinity. This allows the determination of the material relaxation behaviour for each temperature-crystallinity combination. Two DMTA experiments with different temperature histories must be carried out: e.g. one from warm to cold and one from cold to warm (after deleting the thermal prehistory), or heating DMTA on an unaged and one on an aged sample (secondary crystallization), or two cooling DMTA with different temperature rates.

The prerequisite is that the degree of crystallization must be known for both temperature programs. To determine this, DSC can be used, for example. In addition, large differences in crystallinity at the same temperature are desirable.
Mastercurves have to be generated for both temperature paths. The reference temperature should be the same in both cases and should be above the melting temperature. By comparing the shift factors of both temperature programs, the shift factor can be divided into a pure temperature-dependent shift factor and a pure crystallinity-dependent shift factor. Since the classical temperature-frequency sweep is a stair-stepped temperature program, the temperature rate can be simplified to be linearised by a straight line. However, this results in very small temperature rates, so that the determination of the degree of crystallization in DSC can become difficult. Frequency-temperature sweeps can solve this problem: In this case, many temperature sweeps are performed at different frequencies. The temperature rate can be set to match that in the DSC measurement.
For the design, a realistic consideration of the crystallization state must be made. However, since it is impossible to know the entire temperature history of a laminated glass, it is proposed to work with two limiting cases.

- First limiting case: few crystallites / low degree of crystallization. A relative degree of crystallization of 0 (or another small amount, that will always be achieved, still has to be defined) can be used.

- Second limiting case: many crystallites with a high degree of crystallization. A relative degree of crystallization of 1 can be used.

8.1.3 Material type 3: chemically crosslinked polymers

Chemically crosslinked interlayers with a constant degree of crosslinking (post crosslinking during testing and application is excluded) can be considered as material type 1 or 2, depending on whether they are otherwise amorphous or semicrystalline. Simplified, however, only the equilibrium modulus G_∞ can be used for the design (independent of the load duration and temperature of the load-induced cases). To determine this experimentally, a temperature sweep from warm to cold or cold to warm can be performed, for example. The basic elasticity is reached as soon as a plateau occurs in the entropy-elastic range or the stiffness increases again somewhat with increasing temperature. The basic elasticity can be validated, for example, by four-point bending creep tests on the laminate. To shorten the test time, a temperature from the entropy-elastic range is suggested. When a constant deformation level is reached, the basic elasticity can be back-calculated. Alternatively, torsional relaxation tests can be used.

8.1.4 Material type 4: unknown thermo-mechanical behaviour

If the material under investigation is new and unknown, DSC (EN ISO 11357-1) can be useful to detect thermal processes (transformations such as melting or crystallization, relaxations such as glass transitions or post crosslinking). Alternatively, DMTA-TS can already provide indications of the various processes. For this purpose, it is recommended to perform DMTA-TS from warm to cold and from cold to warm. Crystallizable materials tend to form hysteresis in the shear modulus temperature curves. In the case of post crosslinking, the first heating curve would differ from subsequent ones in that the stiffness in the first heating process would be lower. The subsequent heating and cooling curves would have to be congruent with each other due to irreversibility. In addition, it is helpful to run the DMTA-TS at different frequencies in order to distinguish frequency-dependent (e.g. glass transition) from frequency-independent processes (e.g. melting of crystallites). Depending on the DSC or DMTA-TS results, one of the already described procedures can be used for the different material types.

8.2 Nonlinear viscoelastic material behaviour

The nonlinear viscoelastic material behaviour of PVB could not be fully explored in this work. Nevertheless, first important findings were made, which may be helpful for further investigations. In the following, information is given on the nonlinear viscoelastic material behaviour, as well as recommendations for conducting and evaluating relaxation tests in uniaxial tensile mode.

General notes

Nonlinear viscoelastic material behaviour occurs, when the linearity limits are exceeded. In this work it was shown, that the linearity limit is a function of the linear viscoelastic stiffness. Low linear viscoelastic stiffnesses lead to high linearity limits, while high linear viscoelastic stiffnesses result in lower linearity limits. With respect to relaxation experiments, this means that:

- A relaxation test at a low temperature is much further in the nonlinear range than a relaxation test at the same strain level but higher temperature.
- A relaxation test that initially starts with a high degree of nonlinearity approaches the linear viscoelastic material behaviour with increasing time.

As a result of a variation of the degree of nonlinearity, the shape of the relaxation curve in the nonlinear region (at a constant strain) differs from that in the linear region. This change of shape can theoretically be depicted with the Schapery model. It is believed that

the shape of the relaxation curve is preserved when not the strain but the degree of nonlinearity is constant. To check this, it is recommended to create 3-dimensional plots from a large database of tests (x-axis: time, y-axis: degree of nonlinearity and z-axis: tensile relaxation modulus) and subsequently to consider sections at constant NL degrees.
The investigations at different temperatures at constant strain level suggest that the time-temperature-superposition TTSP is still applicable, at least in part. At low temperatures, however, a vertical offset occurs, the origin of which still has to be clarified. This could be further investigated with further relaxation experiments with smaller temperature steps.
The time-strain-superposition position TSSP leads to continuous mastercurves, but further investigation is needed for validation. For very large strains, a purely horizontal displacement was sufficient to obtain a continuous mastercurve for a constant temperature. For small / medium strain tests, additional vertical displacements were necessary. The course of the vertical displacement factors is similar to the course of the secant modulus in the uniaxial tensile test. It should be noted that this investigation has been carried out only at 0°C so far. Here, the NL degree during the relaxation tests is almost constant. If this is not given, it could be that the individual curves would have to be distorted (i.e. each measuring point would have to be shifted horizontally and vertically individually).

Recommendations for the relaxation tests

Before the relaxation tests, it is recommended to perform a uniaxial tensile test with a high deformation rate (the deformation rate should correspond to that of the relaxation test). The secant modulus strain diagram can be used to determine the strain levels to be investigated. In addition, it can be determined in advance at which point in the relaxation test it can be expected that the Prony series will be reached again. See for example Figure 7.7 for standard PVB at 20°C and Figure A.27 for standard PVB at 0°C.
At present, no standard is available for the performance and evaluation of relaxation tests on polymers. In principle, however, the procedure described in the withdrawn DIN 53441 can be used. There it is also recommended to use the data point at $t = 11 \cdot t_0$ as the first evaluation point, since all previous points may be influenced by the load application. In addition, true stress and strain data should be used for evaluation.

8.3 Laminated glass design examples

In this section, the influence of the findings on engineering practice will be presented. It is first shown how a semicrystalline interlayer can be used in the design. For this purpose, a more detailed method, which considers the influences of temperature and crystallization separately, and a simplified approach with a smeared time-superposition principle are considered.
Subsequently, the strains occurring in laminated glass are investigated and compared with

the linearity limits of PVB. Indications are given regarding which systems might exceed linearity limits even in intact laminated glass.

8.3.1 Linear viscoelastic behaviour of semicrystalline interlayers

In the following, the shear modulus values at temperatures of [-20°C, 0°C, 20°C, 23°C, 30°C, 35°C, 40°C, 43°C, 60°C, 80°C] at the time points [3 s, 30 s, 5 min, 10 min, 30 min, 1 h, 6 h, 12 h, 24 h, 4 d, 5 d, 3 w, 30 d, 50 a] are investigated exemplary on SG interlayer. Due to the lack of knowledge about the existing degree of crystallization of the interlayer in the laminated safety glass used in buildings, four different degrees of crystallization are examined as examples in the detailed procedure. The influence during glass design is discussed. The individual shift factors (temperature-dependent shift factor and crystallinity-dependent shift factor), as well as the combined shift factor, are given in each case.
In addition, the shear modulus values for the simplified procedure with smeared shift factors are given.

Detailed procedure: Influence of crystallinity on the relaxation function

First, the Prony series shown in Table 5.1 ($T_{\text{ref}} = 95°\text{C}$, $\chi_{\text{ref}} = 0$) is shifted to the temperatures [-20°C, 0°C, 20°C, 23°C, 30°C, 35°C, 40°C, 43°C, 60°C, 80°C]. The temperature-dependent shift factors can be evaluated with the TTSP in Table 5.1 or with Figure 5.16 (b) and are summarized in Table 8.1. To shift the Prony series to the new temperatures, the relaxation times τ_i are multiplied by a_T. The corresponding relaxation curves are shown in Figure 8.1. These relaxation curves can now be evaluated for different points in time. The results for the time points [3 s, 30 s, 5 min, 10 min, 30 min, 1 h, 6 h, 12 h, 24 h, 4 d, 5 d, 3 w, 30 d, 50 a] are shown in Table 8.4. It should be noted that the degree of crystallization so far still corresponds to the reference degree of crystallization of $\chi_{\text{ref}} = 0$.

Table 8.1 SG: $\log a_T$ for different temperatures, χ_{ref}=0.

T [°C]	-20	0	20	23	30	35	40	43	60	80
$\log a_T$	15.60	11.65	8.19	7.67	6.45	5.59	4.74	4.24	1.83	0.45

After the Prony series was shifted to the new temperatures, in a second step, the relaxation curves can be shifted to the prevailing degree of crystallization. To determine the degree of crystallization dependent shift factor, the TCSP from Table 5.1 or Figure 8.1 (a) can be used. In the following, a degree of crystallization of 0%, 5%, 10% and 15% are investigated as examples. The corresponding crystallinity shift factors are summarized

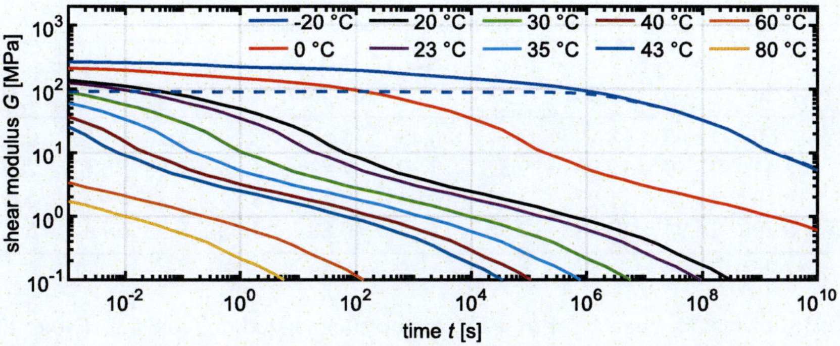

Figure 8.1 SG relaxation function at different temperature, $\chi_{\mathrm{ref}} = 0$. The dashed line corresponds to the evaluation of the short Prony series (Table A.23) at $T = 20$℃.

in Table 8.2. The influence of the different degrees of crystallization on the relaxation curve is shown in Figure 8.2 for a temperature of 20°C as an example. For this purpose, the relaxation times in Table 5.1 are multiplied by the temperature-dependent shift factor $a_{\mathrm{T}}(T = 20°\mathrm{C})$ and the crystallinity-dependent shift factors from Table 8.2.

Table 8.2 SG: $\log a_\chi$ for different crystallinities.

χ [%]	0	5	10	15
$\log a_\chi$	0	1.84	3.59	5.29

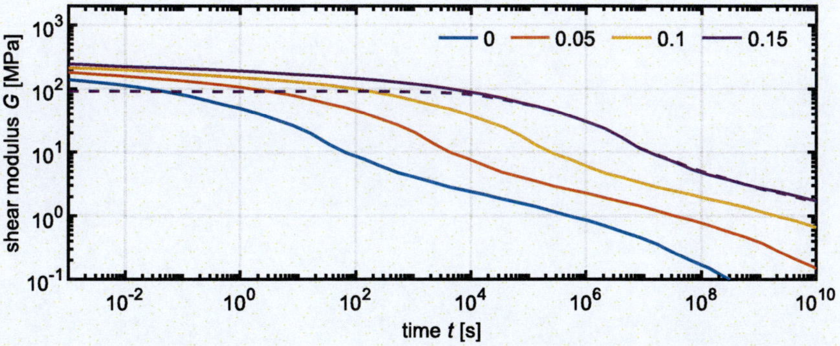

Figure 8.2 SG relaxation function at T = 20℃ and different degrees of crystallinity. The dashed line corresponds to the evaluation of the short Prony series (Table A.23) for a crystallinity of 15%.

Table 8.3 SG: $\log a_T + \log a_\chi$ for different temperatures and crystallinities.

χ [%]	T [°C] -20	0	20	23	30	35	40	43	60	80
0	15.60	11.65	8.19	7.67	6.45	5.59	4.74	4.24	1.83	0.45
5	17.44	13.49	10.03	9.51	8.30	7.43	6.58	6.08	3.67	2.30
10	19.19	15.24	11.77	11.25	10.04	9.17	8.32	7.83	5.41	4.04
15	0.89	16.94	13.48	12.96	11.75	10.88	10.03	9.53	7.12	5.75

The total shift factors ($\log a = \log a_T + \log a_\chi$) are summarized in Table 8.3. (Note: this corresponds to $a = a_T \cdot a_\chi$). The resulting shear modulus values at specific times are shown in Table 8.4. 1 month equals 30 days and 50 years equals 50 ·365 d.

Table 8.4 SG: $G(t,T,\chi)$...

T [°C]	χ [%]	3 s	30 s	5 min	10 min	30 min	1 h	6 h	12 h	1 d	4 d	3 w	1 m	50 a
-20	0	216.4	203.6	183.4	174.7	163.2	156.9	139.3	131.9	123.7	104.6	85.1	80.3	10.0
	5	249.9	231.7	214.4	210.4	204.5	199.7	183.0	174.3	166.5	151.2	137.5	133.6	50.7
	10	264.7	255.7	246.9	242.1	230.7	223.4	212.9	209.0	205.3	193.2	177.7	173.2	105.4
	15	282.7	270.9	262.2	259.6	255.1	252.9	244.7	238.9	231.4	217.9	210.0	208.1	148.5
0	0	137.2	111.0	79.9	69.3	54.5	46.8	26.9	20.5	14.1	7.5	4.6	4.2	1.0
	5	180.4	156.2	133.3	125.1	112.9	103.7	79.4	68.8	58.8	40.0	25.2	21.9	2.6
	10	211.7	197.3	172.9	165.5	155.4	148.2	130.2	122.0	114.5	92.7	73.1	67.5	7.6
	15	243.1	220.5	208.0	204.1	195.9	189.8	168.9	162.4	156.0	140.2	124.2	120.2	36.8
20	0	35.8	14.0	5.7	4.6	3.6	3.0	2.1	1.8	1.6	1.0	0.7	0.7	/
	5	91.3	58.7	31.9	25.6	15.3	10.7	5.7	4.6	3.9	2.7	2.0	1.8	0.3
	10	139.9	114.3	83.7	73.7	57.6	50.0	29.3	23.2	16.5	8.2	4.9	4.5	1.1
	15	179.9	155.9	132.9	124.7	112.4	103.2	78.9	68.2	58.3	39.5	24.9	21.6	2.5
23	0	24.3	8.6	4.1	3.5	2.7	2.4	1.6	1.4	1.2	0.8	0.5	0.5	/
	5	75.2	45.1	21.0	14.5	9.0	7.1	4.1	3.5	2.9	2.1	1.6	1.4	0.2
	10	126.8	98.4	65.9	56.5	44.1	35.5	18.4	12.5	9.4	5.3	3.7	3.4	0.8
	15	166.0	143.8	119.1	111.3	96.0	86.9	60.8	52.7	44.8	27.0	13.9	11.5	2.0
30	0	7.3	3.7	2.2	1.9	1.5	1.3	0.8	0.7	0.6	0.3	0.2	0.2	/
	5	38.9	16.2	6.3	4.9	3.8	3.2	2.2	1.9	1.6	1.1	0.8	0.7	/
	10	91.6	59.0	32.1	25.8	15.5	10.8	5.7	4.6	3.9	2.7	2.0	1.8	0.3
	15	139.3	113.7	82.9	72.8	56.9	49.3	28.8	22.6	16.0	8.1	4.8	4.4	1.0

continued . . .

...continued.

T [°C]	**χ** [%]	**t** 3s	30s	5min	10min	30min	1h	6h	12h	1d	4d	3w	1m	50a
35	0	4.0	2.4	1.5	1.2	0.9	0.8	0.5	0.4	0.3	0.1	/	/	/
	5	19.2	7.1	3.6	3.0	2.4	2.1	1.4	1.2	1.0	0.7	0.4	0.4	/
	10	63.2	35.5	13.8	10.0	6.9	5.3	3.4	2.8	2.5	1.8	1.3	1.2	0.1
	15	117.1	86.9	54.7	47.0	33.6	27.0	11.5	8.9	7.1	4.2	3.0	2.7	0.6
40	0	2.5	1.6	0.9	0.8	0.5	0.4	0.2	0.2	0.1	/	/	/	/
	5	8.0	3.9	2.4	2.0	1.6	1.4	0.9	0.7	0.6	0.4	0.2	0.2	/
	10	39.7	16.8	6.5	5.0	3.9	3.2	2.2	1.9	1.7	1.1	0.8	0.7	/
	15	91.3	58.7	31.9	25.6	15.3	10.7	5.7	4.6	3.9	2.7	2.0	1.8	0.3
43	0	2.0	1.2	0.7	0.5	0.4	0.3	0.1	/	/	/	/	/	/
	5	5.3	2.9	1.8	1.6	1.2	1.0	0.6	0.5	0.4	0.2	0.1	/	/
	10	27.5	9.6	4.5	3.8	2.9	2.5	1.7	1.5	1.3	0.9	0.6	0.5	/
	15	75.9	45.7	21.4	14.9	9.1	7.3	4.1	3.5	2.9	2.1	1.6	1.5	0.2
60	0	0.5	0.2	/	/	/	/	/	/	/	/	/	/	/
	5	1.5	0.9	0.4	0.3	0.2	0.2	/	/	/	/	/	/	/
	10	3.6	2.2	1.3	1.1	0.8	0.7	0.4	0.3	0.2	0.1	/	/	/
	15	12.9	5.4	3.0	2.6	2.1	1.8	1.2	1.0	0.9	0.5	0.3	0.3	/
80	0	0.1	/	/	/	/	/	/	/	/	/	/	/	/
	5	0.7	0.3	0.1	/	/	/	/	/	/	/	/	/	/
	10	1.8	1.0	0.6	0.5	0.3	0.2	0.1	/	/	/	/	/	/
	15	4.3	2.5	1.6	1.3	1.0	0.9	0.5	0.4	0.3	0.2	/	/	/

In order to demonstrate the influence of different degrees of crystallization on the structural design, the maximum glass stress σ_{glass} of a 1.0 m x 1.0 m laminated glass two sided linearly supported and loaded by two different load cases (load case 1: wind, load case 2: snow, dead load is neglected) are determined as an example (Figure 8.3). Two float glass panes of equal thickness $d_1 = d_2 = 3$ mm, 4 mm or 5 mm are investigated. The interlayer thickness is supposed to be 0.76 mm. The degrees of crystallization 0%, 5%, 10% and 15% are considered (it should be mentioned again that there have been no studies to date on the actual degrees of crystallization in the installed state).

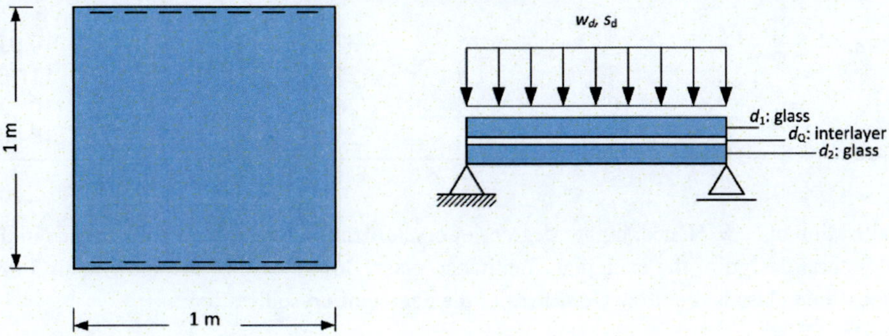

Figure 8.3 Structural system.

The first load case corresponds to the load case *wind storm load in mediterranean areas*. According to EN 16613, this is equivalent to a prevailing temperature of T = 35°C and a load duration of t = 3 s. The wind load is supposed to be w_d = 2.3 kN/m^2 (design value). As a second load case, *snow on a heated surface* is considered. In the general building approvals in Germany, a temperature of 23°C and a load duration of 30 days must be considered for this load case. A design snow load of s_d = 1.5 kN/m^2 is considered.
The maximum glass stresses are summarized in Table 8.5. The sandwich theory according to Wölfel (1987) serves as the basis for calculation (see also appendix A.1).
The glass design is done with the glass strength according to the German standard DIN 18008-1, see eq. (8.2), wherein $k_{mod,wind}$ = 0.7 resp. $k_{mod,snow}$ = 0.4 corresponds to the load duration factor, k_c = 1.8 to the construction factor, f_k = 45 MPa to the characteristic glass strength of float glass and γ_M = 1.8 to the partial safety factor. The use of laminated glass enables to increase the design value of the load-bearing resistance by 10% if the limit states *no shear coupling* or *full shear coupling* are considered. This is usually not allowed when the shear coupling effect is considered (see further existing general building authority approvals) and hence will not be respected here. Since the edge is under tensile stress in two-sided storage, the bending tensile strength must be reduced to 80%. Only the ultimate limit state (ULS) is considered.
For the wind load, the 4 mm glass thickness would fulfil the ULS from a degree of crys-

Table 8.5 SG: Maximum glass stress calculated with different interlayer crystallinities.

load case	χ [%]	G [MPa]	σ_{glass} [MPa]		
			$d_1 = 3$ mm	$d_1 = 4$ mm	$d_1 = 5$ mm
$w_{\text{d}} = 2.3\ \frac{\text{kN}}{\text{m}^2}$	0	1.2	43.64	26.87	18.38
	5	3.0	40.29	24.41	16.45
	10	10.0	38.57	23.09	15.39
	15	47.0	37.97	22.62	15.01
$s_{\text{d}} = 1.5\ \frac{\text{kN}}{\text{m}^2}$	0	0.5	32.66	20.41	14.05
	5	1.4	27.96	17.17	11.70
	10	3.4	26.09	15.78	10.62
	15	11.5	25.09	15.01	10.00

tallization of 5 %. Hence, higher degrees of crystallization would have no influence in the structural design of the exemplary laminated glass. In the case of the snow load, 5 mm thick float glass is required, regardless of the degree of crystallization.

$$\begin{aligned}\sigma_{\text{Rd}} &= 1.0 \cdot 0.8 \cdot \frac{k_{\text{mod}} \cdot k_{\text{c}} \cdot f_{\text{k}}}{\gamma_{\text{M}}} \\ \sigma_{\text{Rd,wind}} &= 25.2\text{MPa} \\ \sigma_{\text{Rd,snow}} &= 14.4\text{MPa}\end{aligned} \tag{8.2}$$

Simplified procedure: smeared temperature-crystallinity shift factors

Alternatively, a simplified shift factor smeared from temperature and degree of crystallization can be applied. However, it is important here that the characterization of the mechanical behaviour of the interlayer has been carried out with a suitable temperature history. See more on this in chapter 4 and the previous section 8.1.2.
At this point, an exemplary laminated safety glass application is considered, in which the melting point of the interlayer is rarely exceeded. Accordingly, it is believed that after the melting temperature is exceeded for a short time, there is sufficient time for crystallization, cold crystallization and crystal perfection to occur before high external stresses are applied at temperatures below the melting point. Under these assumptions, the results of the DMTA measurement from cold to warm, after 5d at room temperature can be used to derive shear modulus values.
In order to determine shear modulus values at the desired temperatures [-20°C, 0°C, 20°C, 23°C, 30°C, 35°C, 40°C, 43°C, 60°C, 80°C], the incrementally determined shift factors *a*

from Figure 5.9 can be used. They are summarized in Table 8.6. It should be noted again, that those shift factors are smeared values, which include effects of temperature change and crystallinity change. The crystallinity hence corresponds to the one during the measurement. The from Figure 5.17 estimated values are included in Table 8.6 for the sake of completeness.

Table 8.6 SG: smeared horizontal shift factors $\log a$ for different temperatures determined from DMTA TFS heating after 5 d at RT.

T **[°C]**	-20	0	20	23[1]	30	35	40	43[1]	60	80
log *a*	19.96	16.75	13.56	12.82	10.80	8.93	7.37	6.94	4.59	1.39
χ^2 **[%]**	11.00	14.00	14.60	13.80	11.80	8.90	6.70	7.00	7.30	2.20

[1] $\log a$ and χ determined through linear interpolation of the $\log a$ resp. χ values.
[2] determined from Figure 5.17.

The shear modulus values for specific temperatures and loading durations are summarized in Table 8.7. Comparing the values with those from Table 8.4, a high similarity to the values calculated for a crystallinity of χ = 10-15% can be seen, especially up to temperatures of $T \approx 35$°C. At higher temperatures, the values are more similar to those calculated with a crystallinity of χ = 5%. This result was expected from the Table 8.6.

8.3.1.1 Summary and conclusion

Two different methods for the design of laminated glass with semicrystalline interlayers were demonstrated using SG as an example. In the simplified method, a smeared shift factor is used, which contains effects from both temperature and the degree of crystallization. Therefore, it is very important in this method that the DMTA used to determine the Prony series and the smeared displacement factors is performed according to the laminated glass application (what temperature range and what degree of crystallization is expected in the application). In the example above, it was assumed that the melting point is only rarely exceeded.

The more detailed procedure separates the effects of temperature and degree of crystallization, so that the horizontal shift factor in theory is decomposed into two parts. This allows any temperature and degree of crystallization combination to be considered. It should be noted that the separation of the two effects is only theoretical and the two quantities temperature and degree of crystallization are physically related. This approach is particularly interesting for investigations still to be carried out on the degree of crystallization actually occurring in LSG: It must be further investigated which temperature histories represent the worst (but nevertheless realistic; lowest degree of crystallization to be expected at temperature T) and most favorable case (highest degree of crystallization at temperature T).

For this purpose, studies on physical aging, in particular cold crystallization, as well as crystallite perfection at isothermal and non-isothermal temperature loadings must also be taken into account. By annealing at high temperatures just before the melting temperature, small crystals melt and larger ones grow. The resulting shift of the melting point needs further research.

These outstanding research questions are limited to crystallinity studies. The resulting relaxation behaviour can then be predicted using the time-superposition principles already found. Of course, the results still need to be validated.

Table 8.7 SG: simplified $G(t,T)$. Assumption: Melting temperature is rarely exceeded.

T	***t***												
[°C]	**3 s**	**30 s**	**5 min**	**10 min**	**30 min**	**1 h**	**6 h**	**12 h**	**1 d**	**4 d**	**3 w**	**1 m**	**50 a**
-20	269.3	260.6	252.1	249.5	242.7	236.2	219.2	215.6	211.7	202.6	190.9	187.5	119.5
0	239.3	218.1	205.6	201.1	192.1	185.2	164.6	158.4	151.3	135.7	119.2	115.4	31.8
20	182.1	157.6	134.8	126.7	114.4	105.7	81.3	70.9	60.5	41.7	26.4	23.2	2.6
23	163.0	140.7	115.3	106.8	91.3	82.5	56.5	48.9	40.5	24.0	11.6	10.0	1.83
30	114.9	84.3	52.5	44.5	31.5	25.2	10.5	8.4	6.6	4.1	2.8	2.6	0.6
35	55.6	29.4	10.3	8.2	5.5	4.5	2.9	2.5	2.2	1.6	1.1	1.0	0.1
40	17.6	6.7	3.5	2.9	2.4	2.0	1.4	1.2	1.0	0.6	0.4	0.4	/
43	10.5	4.7	2.7	2.4	1.9	1.6	1.0	0.9	0.8	0.5	0.3	0.2	/
60	2.4	1.5	0.8	0.7	0.5	0.4	0.2	0.1	0.1	/	/	/	/
80	0.35	0.13	/	/	/	/	/	/	/	/	/	/	/

8.3.2 Linearity limit in intact laminated glass

In some literature, e.g. (Kuntsche, 2015), it is assumed that the interlayer in laminated glass is only subjected to small deformations and therefore linear viscoelastic material behaviour can be used for the calculation of maximum glass stresses. However, to the knowledge of the author of this thesis, there are only a few studies, on the actual strains occurring in the interlayers of intact laminated glass, (Schuler, 2003; Sackmann, 2008). Therefore, using a FEM model in *Ansys Workbench 2020R2* (ANSYS, Inc.) software, the surface loads that lead to shear deformations in the interlayer corresponding to the linearity limit are determined for a laminated glass supported linearly on two sides. As shown in chapter 6, the linearity limit depends on the linear viscoelastic stiffness. For this reason, different interlayer shear moduli between $0.1 \leq G \leq 30$ MPa are investigated. The linearity limits are based on the approximate equations in Table 6.3 and are summarized again in Table 8.8 for the investigated shear modulus values. Note that the linearity limits were determined with the storage modulus G' and that here $G' \approx G$ is assumed. The geometry of the glass under consideration was set to 1.35 m x 3 m. The following cross-sectional structures were investigated: 6 / 0.76 / 6 [mm], 6 / 1.52 / 6 [mm] and 8 / 1.52 / 8 [mm].

Figure 8.4 shows the three dimensional model build with *Solid 186* elements. Double symmetry was exploited. Linear elastic material behaviour with isotropic properties $E_{\text{glass}} = 70000$ MPa, $\mu_{\text{glass}} = 0.23$ was assigned to the glass. The PVB interlayer was also mapped linearly elastic, but with parameterized shear modulus values. The Poisson's ration was set to $\mu = 0.49$ (which corresponds to a simplification, as it also depends on temperature and loading duration, see e.g. (Kuntsche, 2015)). The three solids were defined as one component group with coincident nodes, which meant that no additional contact conditions between the film and the glass had to be defined. The laminated glass was linearly supported on the short sides, in that the lower edge of the bottom glass was non-displaceable in the z-direction. The surface loading was also parameterized.

The mesh size was set to 100 mm after a mesh study, see Table A.41 and the model was validated by a hand calculation according to Wölfel (1987) (Table A.42).

From the FEM model, the maximum elastic shear strain in the interlayer was then considered. Since increased values occur at the corner of the laminated glass, two mesh point samples were examined in more detail. The positions of the mesh point samples are shown in Figure 8.5. The analysis was performed in *large deformation mode*, which means that geometric nonlinearities were considered. The load was increased successively for certain shear modulus values until the linearity limit was reached in both mesh point samples. This means, that starting from the supported edge, 10 cm of the interlayer have exceeded the linearity limit in each case. The results are shown in Table 8.8.

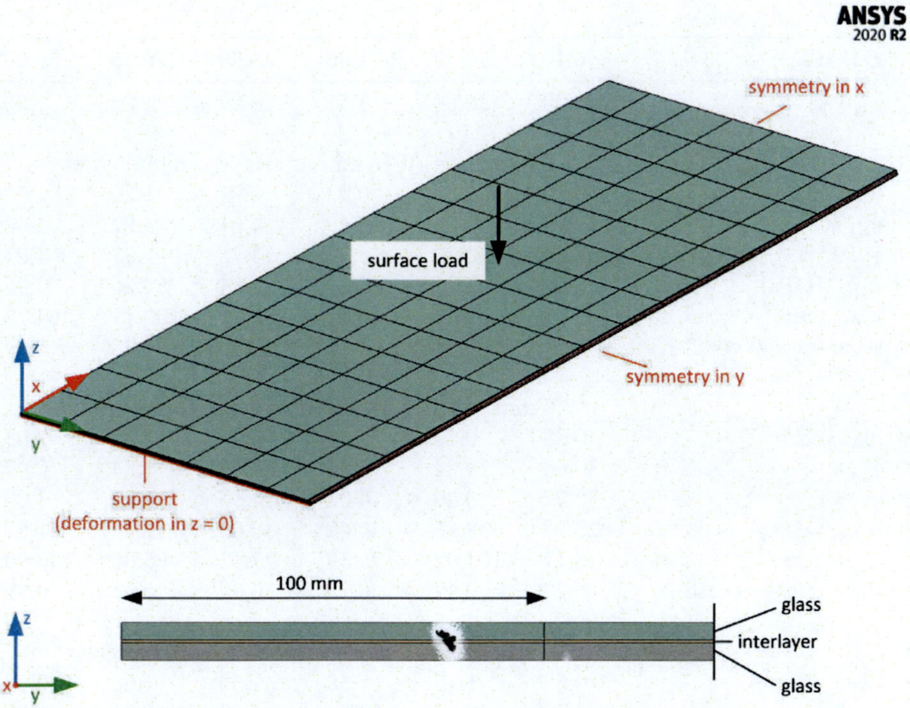

Figure 8.4 FEM model. Top: symmetries, support and load, (b) cross section and mesh size.

The characteristic tensile strength of heat strengthened glass is 70 MPa according to the German standard. Table 8.8 shows that the linearity limit of the interlayer can be reached for load heights realistic in civil engineering applications without the characteristic bending tensile stress of the glass being exceeded (hence, the glass is very likely to be in an intact state).

However, it should be noted that a very slim laminated glass with a large span width was deliberately selected here and that in all cases the occurring deflections $w_{\max}$ exceed the maximum permissible value according to serviceability ($w_{\mathrm{Rd}} = 3000/100 = 30$ mm).

Since the shear modulus of the interlayer initially decreases after the linearity limit is exceeded, the influence of a shear modulus that varies over the span was then investigated in the FEM model. For this purpose, the interlayer was divided into two areas: the 10 cm wide edge region that exceeded the linearity limit and the remaining, linearly viscoelastic region. A third mesh point sample was created ≈ 20 cm from the supported edge. In the linear viscoelastic region, an interlayer shear modulus of 3 MPa was specified.

Table 8.8 FEM analysis results.

G [MPa]	0.10	0.30	1.00	3.00	10.00	30.00
γ_{lin} [%]	30.00	30.00	13.51	4.59	1.82	1.00
	6 mm glass - 0.76 mm interlayer - 6 mm glass					
load [kN/m^2]	0.29	0.69	0.92	0.92	1.20	2.00
γ_2 [%]	29.98	30.06	13.49	4.64	1.82	1.00
γ_1 [%]	30.24	30.40	13.72	4.75	1.87	1.03
σ_{glass} [MPa]	16.67	33.76	42.24	41.84	54.26	87.96
w_{max} [mm]	49.53	82.64	88.12	81.21	102.11	164.18
shear transfer coeff. [-]	0.37	0.64	0.85	0.95	0.98	0.99
	6 mm glass - 1.52 mm interlayer - 6 mm glass					
load [kN/m^2]	0.38	0.80	1.05	0.99	1.30	2.05
γ_2 [%]	30.60	29.84	13.86	4.58	1.84	0.97
γ_1 [%]	30.77	30.09	14.02	4.66	1.88	1.00
σ_{glass} [MPa]	23.13	38.56	44.42	40.60	52.93	81.62
w_{max} [mm]	75.17	103.02	95.59	77.42	95.06	143.90
shear transfer coeff. [-]	0.23	0.47	0.75	0.90	0.97	0.99
	8 mm glass - 1.52 mm interlayer - 8 mm glass					
load [kN/m^2]	0.58	1.15	1.38	1.30	1.68	2.75
γ_2 [%]	30.92	30.67	13.58	4.60	1.81	1.00
γ_1 [%]	31.01	30.90	13.73	4.67	1.84	1.02
σ_{glass} [MPa]	22.43	34.98	35.60	32.02	40.84	66.72
w_{max} [mm]	57.28	75.25	61.79	49.07	58.17	91.48
shear transfer coeff. [-]	0.18	0.40	0.69	0.87	0.96	0.99

In the boundary region, a reduced shear modulus (2.5, 2, 1.5, 1 or 0.1 MPa) was investigated. Only the cross section 8 / 1.52 / 8 [mm] was considered. The surface load was 1.3 kN/m^2. Table 8.9 shows that the maximum glass stress and maximum deflection are only negligibly influenced by reduced interlayer shear modulus values in the edge. The occurring shear strains further increase. The higher the drop in the edge shear stiffness, the higher the shear strains. Hence, depending on the drop height in edge shear modulus, the excitation of the linearity limit may be restricted to the local edge area or may grow towards the center (see shear strains of the third mesh point sample in Table 8.9).
Other structures where at least local exceeding of the linearity limits is conceivable are, for example, the areas around the holes of a point-supported laminated glass or the areas around laminated supports.

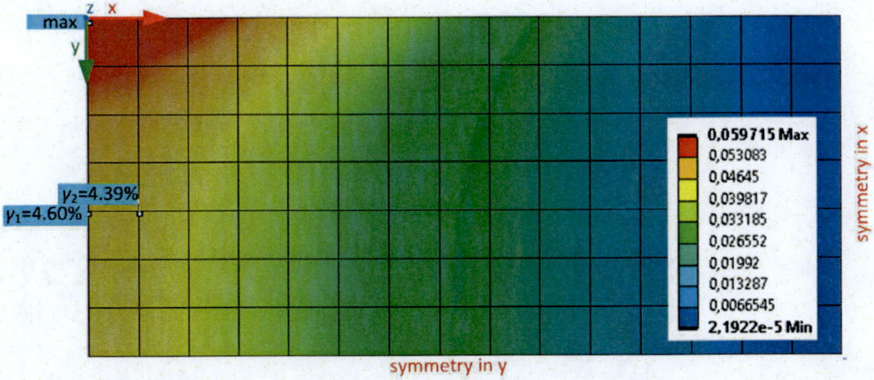

Figure 8.5 Maximum elastic shear of the interlayer. Cross sectional geometry: 8 mm glass /1.52 mm interlayer / 8 mm glass. The shear modulus value of the interlayer is G = 3 MPa. The first mesh point sample is located at $x \approx 0$ cm (supported edge), the second mesh point sample is located at $x \approx 10$ cm.

Conceivable applications where the linearity limit is exceeded in a larger area are, for example, thin glass laminates that are deliberately subjected to large deformations. Here it is advisable to take a closer look at the strains of the interlayer.
Irrespective of the thickness of the glass, also loads acting only on one glass pane, in the direction of the pane (load application via the glass edge), can lead to a global exceeding of the linearity limit. This would be thinkable e.g. by a slight edge offset of the glass panes of a glass sword due to the manufacturing process.

Table 8.9 Reduced shear modulus value at the edges. The edges are 10 cm width. γ_{lin} is the linearity limit, γ_i is the shear strain at the mesh point sample i, σ_{glass} corresponds to the maximum glass stress and w_{max} corresponds to the maximum deflection.

G [MPa]			3		
G_{edge} [MPa]	2.5	2	1.5	1	0.1
γ_{lin} [%]			4.59		
γ_3 [%]	4.55	4.74	4.96	5.22	5.84
γ_2 [%]	4.91	5.27	5.69	6.20	7.39
γ_1 [%]	5.06	5.52	6.06	6.70	8.05
σ_{glass} [MPa]	32.02	32.02	32.02	32.02	32.03
w_{max} [mm]	49.19	49.31	49.47	49.65	50.09

9 Summary and outlook

The present work deals with the mechanical characterization of different laminated glass interlayers. Three materials were considered: polyvinyl butyral (PVB), ethylene vinyl acetate (EVA resp. cEVA) and SentryGlas® (SG).

Since the state of the art for PVB based interlayers is much more advanced than for the other two materials, the molecular structure of uncrosslinked EVA, crosslinked cEVA and SG were first investigated in more depth. Using DSC, XRD and DMTA-TS, it was shown that the interlayers have crystalline regions. For SG an initial crystallinity (interlayer state as received from manufacturer, unknown thermal history) of about $\chi = 9.5$-10% and for EVA and cEVA of 28-29% could be determined. The melting and crystallization temperatures (T_m and T_c) were determined after elimination of the thermal history. For all materials, the melting temperature in the heating curve with a certain heating rate was higher than the crystallization temperature in the cooling curve with the same cooling rate. In addition, the crystallization temperatures were significantly more dependent on the cooling rate.

In contrast to EVA and cEVA, a flat exothermic peak was observed in the heating curve for SG, which was examined more closely with further DSC experiments and interpreted as cold crystallization. This means that during heating, the degree of crystallization initially continues to increase until finally the melting temperature is reached.

The glass transition temperature could be determined more easily with the DMTA-TS. Considering several DMTA-TS with different frequencies further facilitated the detection of the glass transition, since the maximum in T_g characteristic of $\tan\delta$ is frequency dependent.

For SG, the influence of physical aging through different storage times at room temperature was additionally investigated. It was found that a second endothermic peak forms in the DSC heating curves over time. The shape of the annealing peak depends on the storage time and annealing temperature, whereby the peak moves towards higher temperatures with increasing physical age (without exceeding the main melting point). At the same time, in DMTA-TS, a wider region with stiffer material behaviour is observed with increasing storage time.

An attempt was made to mathematically describe the crystallization kinetics of the cooling melt. For this purpose, both the peak areas between the DSC and baselines and the hystereses from the DMTA heating and cooling curves were considered. Due to a com-

M. Schuster, *Characterization of Laminated Safety Glass Interlayers*,
Mechanik, Werkstoffe und Konstruktion im Bauwesen 66,
https://doi.org/10.1007/978-3-658-39821-7_9

plex crystallization behaviour, only the Mo-Liu approach could approximately describe the crystallization process. To investigate the crystallization kinetics in more detail, it is recommended to divide the crystallization process into different phases.
Moreover, in the present work, only crystallization from the melt was considered in terms of crystallization kinetics. However, in the case of SG, cold crystallization and crystal perfection processes could also be mathematically described with respect to physical aging time and temperature.

After the influences from temperature and time history on the degree of crystallization and the stiffness (shear modulus) were investigated, DMTA-TFS have been performed. Three different temperature programs were investigated for SG: from warm to cold, from cold to warm after eliminating the thermal prehistory, and from cold to warm after 5 days of aging at RT. At the same time, DSC measurements were performed for these temperature histories, so that for each data point of the DMTA-TFS, the crystallinity was known. The creation of mastercurves at a temperature above the melting point of the crystallites by purely horizontally shifting the individual isothermal curves led to congruent mastercurves but different shift factors for equal temperatures. A comparison of the shift factors and the prevailing degrees of crystallinities, enabled the deduction of a time-crystallinity-superposition principle (TCSP), which has a similar effect to the well-known temperature-dependent shift factor. Hence, the total shift factors could be decomposed into two parts: the purely temperature-dependent shift factor and the purely crystallinity-dependent shift factor. It should be noted that the separation of the two effects is only theoretical and the two quantities temperature and degree of crystallization are physically related. The great advantage of the derived TCSP, however, is that by determining a Prony series with T_{ref} above the melting range and specifying the TTSP and TCSP, the material behaviour can be predicted for any combination of temperature and degree of crystallization. In addition, it was shown that the degree of crystallization of a material with known TCSP and TTSP can be back calculated.
Unfortunately, difficulties occurred obtaining reliable DMTA results with the cEVA interlayer. Various reasons for this are under debate.
The new knowledge gained in this work for determining the linear viscoelastic material properties of interlayers with semicrystalline structures was summarized in a guideline. For engineering application, further research is needed in order to determine realistic temperature histories and hence information on the degree of crystallinity in laminated glass applications.

Since the linear viscoelastic material behaviour of the most commonly used interlayer type PVB is already well studied, this work focused on the linearity limits and material behaviour at high strains (nonlinear viscoelasticity). In DMTA-AS, it was shown that the linearity limits are dependent on both temperature and frequency. Recalling the TTSP

provided the insight that the linearity limit can be expressed in terms of linear viscoelastic stiffness G'_{LVE}. In addition, it was shown, that the linearity limits of two different PVB types (standard and structural from the same manufacturer) can be described with the same function. This can be explained by the fact, that the two PVBs mainly differ in the amount of plasticizer, which shifts the glass transition temperature, without greatly influencing the shape of the relaxation curve.
At high linear viscoelastic stiffnesses, the linearity limit is small. With decreasing linear viscoelastic stiffness, the linearity limit increases. In the present study, the minimum value was determined to $\gamma_{\mathrm{lin}} = 0.6\%$, which occurred for linear viscoelastic storage modulus values of $G'_{\mathrm{LVE}} \geq 252$ MPa. The highest linearity limit was determined to $\gamma_{\mathrm{lin}} = 30\%$ for $G'_{\mathrm{LVE}} \leq 0.5$ MPa. The limit values occurred due to technical measurement limits (e.g. slippage of the sample). The linearity limits were determined in shear mode. Further research is needed in order to determine the linearity limits in tensile mode. Here, in a first approach, it was assumed that the linearity limits are independent on the loading mode. Supplementary, relaxation tests in shear and tensile mode should be conducted in order to validate the DMTA results.
In a small FEM example, it was shown that for a very slender laminated glass with a large span, the linearity limits in the edge region can be exceeded even in the intact state. However, since the deformations that occur in this case far exceed the permissible deformations according to SLS, it is assumed that linear viscoelastic material behaviour of the interlayer prevails for most intact standard laminated glass. Nevertheless, it is recommended to check the occurring strains in the interlayer for all glass designs, if an FEM model is available. The linearity limit can be exceeded, especially in the area of support (e.g. linear supports as in the FEM example shown here, around holes in point bearings or in laminated mounts). More global exceedances could occur e.g. with thin glass laminates or with laminated glass loaded in one glass plane. This can be further investigated in FEM studies.

At the latest in the post fracture state, the interlayer is subjected to large deformations, so that nonlinear viscoelastic material behaviour must be taken into account. Since the interlayer is subjected to tensile stress in the post fracture state, tensile relaxation tests were carried out on two PVB interlayer types as part of the present work. Three different temperatures were investigated ($T < T_g$, $T \approx T_g$ and $T > T_g$). The displacements, which were kept constant in the relaxation tests, were applied in a displacement-controlled manner (crosshead displacement speed 400 mm/min) and corresponded to true strains in a range of approx. 70% - 150%. The relaxation functions were evaluated by the quotient of true stress and true strain for each time t. For all three temperatures, as the applied displacement increased, stiffer material behaviour was observed. This can be explained by looking at the true stress-strain diagram during the application of the elongation: The applied strain is so high that it lies in the stiffening branch. By shifting purely horizontally, mastercurves could be generated for all three temperatures at a reference displacement of 5 cm. A higher

applied strain had to be shifted to shorter times. By applying the TTSP (from the master-curving process of linear viscoelastic DMTA), the relaxation curves for 5 cm displacement were shifted to 20°C. It could be shown, that the TTSP is at least in parts applicable in the nonlinear viscoelastic range. However, the origin of the vertical offset of the low temperature curves still has to be clarified, e.g. by performing more relaxation tests at the same strain level but with smaller temperature increments. The comparison with the linear viscoelastic Prony series shows a clear change in the shape of the curves: The initially very high deviation from the Prony series becomes smaller with increasing time. This is explained by the linearity limit depending on the linear viscoelastic stiffness. A relaxation test initially occurring far in the nonlinear viscoelastic range approaches the linear viscoelastic range over time. Hence the shape of the relaxation curve changes in comparison to the linear viscoelastic relaxation curve. In addition, a test at low temperature is much further in the nonlinear range than a test at higher temperature with the same amount of strain applied. To quantify this effect, the degree of nonlinearity NL was introduced, which describes the ratio of applied strain to linearity limit. By evaluating the degree of nonlinearity, the test duration can be determined at which the linear viscoelastic material behaviour should be regained. It is recommended to investigate whether the shape of the linear viscoelastic mastercurve is preserved at constant degree of nonlinearity (instead of constant strain). For this purpose, the NL degrees of all relaxation tests can be determined and then the data points with similar NL degrees can be further investigated.
The nonlinearity factors of the Schapery model have been determined for standard and structural PVB. Especially for short times a good agreement with the experimental data was achieved. The great advantage of this model is that the linear viscoelastic parameters (Prony series) remain unchanged and only the nonlinearity factors are added. Thus with one and the same model both linear and nonlinear viscoelastic material behaviour can be represented.
It is recommended to perform further tests at constant temperature and different strain levels from small to very large strains in order to check the time-strain-superposition principle TSSP. In this thesis, preliminary tests at 0°C have been performed at small to medium strains. A continuous mastercurve was created (with the smallest strain level as reference) by manually shifting the larger strain curves horizontally to shorter times and vertically to higher modulus values. Additionally it was shown, that the vertical shift factors is related to the initial true tensile modulus at the beginning of the relaxation test E_{t_0}.

All in all, new findings on the linear viscoelastic material behaviour of semicrystalline interlayers such as SG and cEVA or EVA, as well as important insights into the nonlinear viscoelastic behaviour of PVB, which becomes crucial especially in the post fracture behaviour, were obtained within the present work.

Bibliography

M. Abareshi, S. M. Zebarjad, and E. K. Goharshadi. Non-isothermal crystallization kinetics of polyethylene-clay nanocomposites prepared by high-energy ball milling. *Bulletin of Materials Science*, 37(5):1113–1121, 2014.

B. Adothu, S. Mallick, and P. Kartikay. Determination of crystallinity, composition, and thermal stability of ethylene vinyl acetate encapsulant used for pv module lamination. In *2019 IEEE 46th Photovoltaic Specialists Conference (PVSC)*, pages 0491–0494. IEEE, 2019.

K. Agroui, G. Collins, and J. Farenc. Measurement of glass transition temperature of crosslinked eva encapsulant by thermal analysis for photovoltaic application. *Renewable Energy*, 43:218–223, 2012.

C. Alter. *Nicht-lokale Versagensformulierung zur Simulation des spannungsratenabhängigen Bruchverhaltens von Verbundsicherheitsglas und ihre Anwendung beim Kopfaufprall auf Windschutzscheiben*. PhD thesis, Technische Universität Darmstadt, 2019.

L. Andreozzi, S. B. Bati, M. Fagone, G. Ranocchiai, and F. Zulli. Dynamic torsion tests to characterize the thermo-viscoelastic properties of polymeric interlayers for laminated glass. *Construction and Building Materials*, 65:1–13, 2014.

R. Androsch. Melting and crystallization of poly (ethylene-co-octene) measured by modulated dsc and temperature-resolved x-ray diffraction. *Polymer*, 40(10):2805–2812, 1999.

A. Arsac, C. Carrot, and J. Guillet. Rheological characterization of ethylene vinyl acetate copolymers. *Journal of applied polymer science*, 74(11):2625–2630, 1999.

AS 1288. Glass in buildings—selection and installation. Standard, 2006.

ASTM D2765. Standard test methods for determination of gel content and swell ratio of crosslinked ethylene plastics. Standard, ASTM International, 2016.

ASTM E1300. Standard practice for determining load resistance of glass in buildings. Standard, ASTM International, 2016.

M. Avrami. Kinetics of phase change. i general theory. *The Journal of chemical physics*, 7(12): 1103–1112, 1939.

M. Avrami. Kinetics of phase change. ii transformation-time relations for random distribution of nuclei. *The Journal of chemical physics*, 8(2):212–224, 1940.

M. Schuster, *Characterization of Laminated Safety Glass Interlayers*, Mechanik, Werkstoffe und Konstruktion im Bauwesen 66, https://doi.org/10.1007/978-3-658-39821-7

M. Avrami. Granulation, phase change, and microstructure kinetics of phase change. iii. *The Journal of chemical physics*, 9(2):177–184, 1941.

J.-E. Bae, K. S. Cho, K. H. Seo, and D.-G. Kang. Application of geometric algorithm of time-temperature superposition to linear viscoelasticity of rubber compounds. *Korea-Australia rheology journal*, 23(2):81–87, 2011.

J. Barredo, M. Soriano, M. Gómez, and L. Hermanns. Viscoelastic vibration damping identification methods. application to laminated glass. *Procedia Engineering*, 10:3208–3213, 2011.

D. Basu and A. Banerjee. Dynamic mechanical thermal analysis-crystallinity and tensile-strength of pbt/abs blends: Their interdependency and variations with abs content. 1994.

H. Batzer. *Polymere Werkstoffe: Chemie und Physik.* Polymere Werkstoffe I. Chemie und Physik. Thieme, 1985.

J. Bauch and R. Rosenkranz. Xrd-röntgenvielkristalldiffraktometrie. In *Physikalische Werkstoffdiagnostik*, pages 44–45. Springer, 2017.

F. Becker. *Entwicklung einer Beschreibungsmethodik für das mechanische Verhalten unverstärkter Thermoplaste bei hohen Deformationsgeschwindigkeiten.* PhD thesis, Martin-Luther-Universtiät Halle-Wittenberg, 2009.

J. S. Bergström and M. C. Boyce. Constitutive modeling of the large strain time-dependent behavior of elastomers. *Journal of the Mechanics and Physics of Solids*, 46(5):931–954, 1998.

H. Bertilsson and J.-F. Jansson. The limits of linear viscoelasticity in poly (methyl methacrylate) and poly (ethyl methacrylate). *Journal of Applied Polymer Science*, 19(7):1971–1978, 1975.

D. Beyer, W. Lautner, and H. Behrens. Zur feinstruktur von ethylen-vinylacetat-copolymeren. *Acta polymerica*, 35(6):466–469, 1984.

S. Bistac, P. Kunemann, and J. Schultz. Crystalline modifications of ethylene-vinyl acetate copolymers induced by a tensile drawing: effect of the molecular weight. *Polymer*, 39(20):4875–4881, 1998.

R. L. Blaine. Thermal applications note. *Polymer Heats of Fusion*, 2002.

M. Bonnet. *Kunststofftechnik - Grundlagen, Verarbeitung, Werkstoffauswahl und Fallbeispiele*, volume 3. Springer-Verlag, 2016.

M. Botz. *Beitrag zur versuchstechnischen und numerischen Beschreibung von Verbundglas mit PVB-Zwischenschicht im intakten und gebrochenen Zustand.* PhD thesis, München, Universität der Bundeswehr München, 2020.

M. Botz, G. Siebert, and M. A. Kraus. Experimental determination of the shear modulus of polymeric interlayers used in laminated glass. *Proceedings of GlassCon Global, Chicago*, pages 31–38, 2018.

K. Boutahar, C. Carrot, and J. Guillet. Crystallization of polyolefins from rheological measurements relation between the transformed fraction and the dynamic moduli. *Macromolecules*, 31(6):1921–1929, 1998.

H. F. Brinson, L. C. Brinson, et al. Polymer engineering science and viscoelasticity. *An introduction*, 2008.

O. Brüller. Zur charakterisierung des langzeitverhaltens von kunststoffen. *Archive of Applied Mechanics*, 63(6):363–376, 1993.

A. Burr and G. Harsch. Kunststoffe. In *Dubbel*, pages E76–E90. Springer, 2005.

C. Carotenuto, L. Grassia, L. P. Paduano, and M. Minale. Non-isothermal crystallization kinetics of an ethylene-vinyl-acetate. ii. time-temperature-crystallinity-superposition. *Polymer Engineering & Science*, 59(12):2550–2556, 2019a.

C. Carotenuto, L. P. Paduano, L. Grassia, and M. Minale. Viscoelasticity evolution of ethylene-vinyl-acetate copolymers during crystallization. *CHEMICAL ENGINEERING*, 74, 2019b.

CEN/TS 19100-2. Design of glass structures - part 2: Design of out-of-plane loaded glass components. Standard, 2021-11.

CNR-DT 210. Guide for the design, construction and control of buildings with structural glass elements. Standard, 2013.

A. Czanderna and F. Pern. Encapsulation of pv modules using ethylene vinyl acetate copolymer as a pottant: A critical review. *Solar energy materials and solar cells*, 43(2):101–181, 1996.

J. Dealy and D. Plazek. Time-temperature superposition—a users guide. *Rheol. Bull*, 78(2):16–31, 2009.

D. Delincé, D. Callewaert, J. Belis, and R. Van Impe. Post-breakage behaviour of laminated glass in structural applications. In *Conference on Architectural and Structural Applications of Glass*, pages 459–467. IOS Press, 2008.

P. D'Haene and W. Stevels. Modelling laminated glass based on multi-layer interlayers: Effective modulus approach. *Proceedings Glass Performance Days (Tampere Finland)*, pages 412–415, 2019.

A. Dhaliwal and J. Hay. The characterization of polyvinyl butyral by thermal analysis. *Thermochimica Acta*, 391(1-2):245–255, 2002.

M. Di Paola, L. Galuppi, and G. R. Carfagni. Fractional viscoelastic characterization of laminated glass beams under time-varying loading. *International Journal of Mechanical Sciences*, 196: 106274, 2021.

J. Diani, B. Fayolle, and P. Gilormini. A review on the mullins effect. *European Polymer Journal*, 45(3):601–612, 2009.

DIN 18008-1. Glass in building – design and construction rules – part 1: Terms and general bases. Standard, DIN Deutsches Institut für Normung e. V., 2020-05.

DIN 18008-2. Glass in building – design and construction rules – part 2: Part 2: Linearly supported glazings. Standard, DIN Deutsches Institut für Normung e. V., 2020-05.

DIN 18008-5. Glass in building – design and construction rules – part 5: Additional requirements for walk-on glazing. Standard, DIN Deutsches Institut für Normung e. V., 2013-07.

DIN 18008-6. Glass in building – design and construction rules – part 6: Additional requirements for walk-on glazing in case of maintenance procedures and for fall-through glazing. Standard, DIN Deutsches Institut für Normung e. V., 2018-02.

DIN 50035. Terms and definitions used on ageing of materials – polymeric materials. Standard, DIN Deutsches Institut für Normung e. V, 2012-09.

DIN 51810-2. Testing of lubricants – testing rheological properties of lubricating greases – part 2: Determination of flow point using an oscillatory rheometer with a parallel-plate measuring system, german version. Standard, DIN Deutsches Institut für Normung e. V, 2017-04.

DIN 53441. Testing of plastics; stress relaxation test. withdrawn standard, DIN Deutsches Institut für Normung e. V, 1984-01.

D.-I. H. Domininghaus. *Kunststoffe: Eigenschaften und Anwendungen*. Springer-Verlag, 2013.

A. K. Doolittle. Studies in newtonian flow. ii. the dependence of the viscosity of liquids on free-space. *Journal of Applied Physics*, 22(12):1471–1475, 1951.

A. V. Duser, A. Jagota, and S. J. Bennison. Analysis of glass/polyvinyl butyral laminates subjected to uniform pressure. *Journal of engineering mechanics*, 125(4):435–442, 1999.

Eastman Chemical Company. *Architectural Lamination Guide*, 2003.

Eastman Chemical Company. *Saflex® Q Series Advanced Acoustic Interlayer*, 2012.

Eastman Chemical Company. *Mechanical model for Saflex® DG structural interlayer*, 2017.

G. Ehrenstein. Polymer-werkstoffe: Struktur, eigenschaften. *Anwendung. Hanser*, 2011.

U. Eitner, S. Kajari-Schröder, M. Köntges, and R. Brendel. Non-linear mechanical properties of ethylene-vinyl acetate (eva) and its relevance to thermomechanics of photovoltaic modules. In *Proceedings of the 25th European Conference on Photovoltaic solar energy*, pages 4366–4368, 2010.

H.-G. Elias. *Makromoleküle: Physikalische Struktur und Eigenschaften*. John Wiley & Sons, 2009.

EN 1288-3. Glass in building - determination of the bending strength of glass - part 3: Test with specimen supported at two points (four point bending,; german version. Standard, DIN Deutsches Institut für Normung e. V, 2000-09.

EN 14770. Bitumen and bituminous binders – determination of complex shear modulus and phase angle using a dynamic shear rheometer (dsr), german version. Standard, DIN Deutsches Institut für Normung e. V, 2012-08.

EN 16612. Glass in building – determination of the lateral load resistance of glass panes by calculation, german version. Standard, DIN Deutsches Institut für Normung e. V., 2019-12.

EN 16613. Glass in building – laminated glass and laminated safety glass – determination of interlayer viscoelastic properties, german version. Standard, DIN Deutsches Institut für Normung e. V., 2020-01.

EN ISO 11357-1. Plastics – plastics - differential scanning calorimetry (dsc) - part 1: General principles, german version. Standard, DIN Deutsches Institut für Normung e. V, 2017-02.

EN ISO 11357-2. Plastics - differential scanning calorimetry (dsc) - part 2: Determination of glass transition temperature and step height, german version. Standard, DIN Deutsches Institut für Normung e. V, 2020-08.

EN ISO 11357-5. Plastics – differential scanning calorimetry (dsc) – part 5: Determination of characteristic reaction-curve temperatures and times, enthalpy of reaction and degree of conversion, german version. Standard, DIN Deutsches Institut für Normung e. V, 2014-07.

EN ISO 11357-7. Plastics – differential scanning calorimetry (dsc) – part 7: Determination of crystallization kinetics, german version. Standard, DIN Deutsches Institut für Normung e. V, 2015-12.

EN ISO 6427. Plastics - determination of matter extractable by organic solvents (conventional methods), german version. Standard, DIN Deutsches Institut für Normung e. V, 2014-08.

EN ISO 6721-1. Plastics – determination of dynamic mechanical properties – part 1: General principles, german version. Standard, DIN Deutsches Institut für Normung e. V, 2019-09.

F. Ensslen. *Zum Tragverhalten von Verbund-Sicherheitsglas unter Berücksichtigung der Alterung der Polyvinylbutyral-Folie*. PhD thesis, Ruhr-Universität Bochum, 2005.

U. Evans. The laws of expanding circles and spheres in relation to the lateral growth of surface films and the grain-size of metals. *Transactions of the Faraday Society*, 41:365–374, 1945.

Everlam. *EverlamQuiet Produktdatenblatt*.

evguard. *Handling guide*, 2014.

J. Farjas and P. Roura. Modification of the kolmogorov–johnson–mehl–avrami rate equation for non-isothermal experiments and its analytical solution. *Acta Materialia*, 54(20):5573–5579, 2006.

B. Farr. Investigations on the nonlinear viscoelastic material behaviour of polymeric laminated glass interlayers. Master's thesis, Technische Universität Darmstadt, 2018.

D. Fesko and N. Tschoegl. Time-temperature superposition in thermorheologically complex materials. In *Journal of Polymer Science Part C: Polymer Symposia*, volume 35, pages 51–69. Wiley Online Library, 1971.

Folienwerk Wolfen GmbH. *Technical data sheet - evguard® colorless transparent: EVA interlayer for laminated safety, security & architectural glass*, 2020.

J. Franz. *Untersuchungen zur Resttragfähigkeit von gebrochenen Verglasungen: Investigation of the residual load-bearing behaviour of fractured glazing*. PhD thesis, Technische Universität Darmstadt, 2015.

A. Frick and C. Stern. *DSC-Prüfung in der Anwendung*. Carl Hanser Verlag GmbH Co KG, 2013.

H. L. Friedman. Kinetics of thermal degradation of char-forming plastics from thermogravimetry. application to a phenolic plastic. In *Journal of polymer science part C: polymer symposia*, volume 6, pages 183–195. Wiley Online Library, 1964.

Y. Fung and R. Skalak. Biomechanics: mechanical properties of living tissues. 1981.

H. Gacem, Y. Chevalier, J.-L. Dion, M. Soula, and B. Rezgui. Long term prediction of nonlinear viscoelastic creep behaviour of elastomers: extended schapery model. *Mécanique & industries*, 9(5):407–416, 2008.

L. Galuppi and G. Royer-Carfagni. The effective thickness of laminated glass: Inconsistency of the formulation in a proposal of en-standards. *Composites Part B: Engineering*, 55:109–118, 2013.

L. Galuppi and G. Royer-Carfagni. The design of laminated glass under time-dependent bending and buckling. *Challenging Glass*, 4:431–438, 2014.

J. H. Gibbs and E. A. DiMarzio. Nature of the glass transition and the glassy state. *The Journal of Chemical Physics*, 28(3):373–383, 1958.

J. Göhler. *Das dreidimensionale Stoffverhalten im großen Temperatur-und Zeitbereich am Beispiel eines in der automobilen Klebtechnik verwendeten Epoxidharzklebstoffs*. PhD thesis, Technische Universität Dresden, 2010.

H. Golden, T. Strganac, and R. Schapery. A test and analysis protocol to rapidly characterize nonlinear viscoelastic properties of balloon materials. In *34th Aerospace Sciences Meeting and Exhibit*, page 572, 1996.

H. Golden, T. Strganac, and R. Schapery. An approach to characterize nonlinear viscoelastic material behavior using dynamic mechanical tests and analyses. 1999.

H. Gräf. *Untersuchungen zum Tragverhalten von Verbundsicherheitsglas bei linienförmiger und punktueller Lagerung*. PhD thesis, Technische Universität München, 2003.

W. Grellmann and S. Seidler. Mechanical and thermomechanical properties of polymers. *Landolt Börnstein*, 8, 2014.

R. M. Guedes. A viscoelastic model for a biomedical ultra-high molecular weight polyethylene using the time–temperature superposition principle. *Polymer testing*, 30(3):294–302, 2011.

G. H. Gómez, T. M. Gasparini, and S. V. Canevarolo. Solid-state morphology evolution of sodium neutralized poly(ethylene-ran-methacrylic acid) ionomer under dry and wet thermal annealing. *Materials Research*, 22(4), 2019.

A. Hammer. *UserCom31, Thermische Analyse von Polymeren; Teil 1: DSC an Thermoplasten.* Mettler-Toledo AG, Analytical, 2010.

T. Hána, M. Eliášová, M. Vokáč, and K. V. Machalická. Viscoelastic properties of eva interlayer used in laminated glass structures. In *IOP Conference Series: Materials Science and Engineering*, volume 800, page 012021. IOP Publishing, 2020.

M. Härth, S. J. Bennison, and S. Sauerbrunn. Determination of interlayer mechanical properties for use in laminated glass design. *Proceedings Glass Performance Days (Tampere Finland)*, pages 416–418, 2019.

P. He, W. Yu, and C. Zhou. Agglomeration of crystals during crystallization of semicrystalline polymers: A suspension-based rheological study. *Macromolecules*, 52(3):1042–1054, 2019.

A. M. Henderson. Ethylene-vinyl acetate (eva) copolymers: a general review. *IEEE Electrical Insulation Magazine*, 9(1):30–38, 1993.

E. Hirasawa, Y. Yamamoto, K. Tadano, and S. Yano. Formation of ionic crystallites and its effect on the modulus of ethylene ionomers. *Macromolecules*, 22(6):2776–2780, 1989.

C. Hirschl, M. Biebl-Rydlo, M. DeBiasio, W. Mühleisen, L. Neumaier, W. Scherf, G. Oreski, G. Eder, B. Chernev, W. Schwab, et al. Determining the degree of crosslinking of ethylene vinyl acetate photovoltaic module encapsulants—a comparative study. *Solar Energy Materials and Solar Cells*, 116:203–218, 2013.

C. Hirschl, L. Neumaier, W. Mühleisen, M. Zauner, G. Oreski, G. Eder, S. Seufzer, C. Berge, E. Rüland, and M. Kraft. In-line determination of the degree of crosslinking of ethylene vinyl acetate in pv modules by raman spectroscopy. *Solar Energy Materials and Solar Cells*, 152:10–20, 2016.

P. Hooper, B. Blackman, and J. Dear. The mechanical behaviour of poly (vinyl butyral) at different strain magnitudes and strain rates. *Journal of Materials Science*, 47(8):3564–3576, 2012a.

P. Hooper, R. Sukhram, B. Blackman, and J. Dear. On the blast resistance of laminated glass. *International Journal of Solids and Structures*, 49(6):899–918, 2012b.

K. Hyun, S. H. Kim, K. H. Ahn, and S. J. Lee. Large amplitude oscillatory shear as a way to classify the complex fluids. *Journal of Non-Newtonian Fluid Mechanics*, 107(1-3):51–65, 2002.

K. Hyun, M. Wilhelm, C. O. Klein, K. S. Cho, J. G. Nam, K. H. Ahn, S. J. Lee, R. H. Ewoldt, and G. H. McKinley. A review of nonlinear oscillatory shear tests: Analysis and application of large amplitude oscillatory shear (laos). *Progress in Polymer Science*, 36(12):1697–1753, 2011.

ISO 18437-6. Mechanical vibration and shock — characterization of the dynamic mechanical properties of visco-elastic materials — part 6: Time-temperature superposition. Standard, 2017-11.

ISO 6721-10. Plastics — determination of dynamic mechanical properties — part 10: Complex shear viscosity using a parallel-plate oscillatory rheometer. Standard, DIN Deutsches Institut für Normung e. V, 2015-09.

ISO 6721-11. Plastics - determination of dynamic mechanical properties - part 11: Glass transition temperature. Standard, DIN Deutsches Institut für Normung e. V, 2019-06.

ISO 6721-4. Plastics - determination of dynamic mechanical properties - part 4: Tensile vibration - non-resonance method, german version. Standard, DIN Deutsches Institut für Normung e. V, 2019-05.

ISO 6721-5. Plastics - determination of dynamic mechanical properties - part 5: Flexural vibration - non-resonance method. Standard, DIN Deutsches Institut für Normung e. V, 2019-04.

ISO 6721-6. Plastics - determination of dynamic mechanical properties - part 6: Shear vibration - non-resonance method. Standard, DIN Deutsches Institut für Normung e. V, 2019-04.

ISO 6721-7. Plastics - determination of dynamic mechanical properties - part 7: Torsional vibration - non-resonance method. Standard, DIN Deutsches Institut für Normung e. V, 2019-04.

J.-F. Jansson. The appearance of non-linear viscoelasticity in polyethyl methacrylate. *Die Angewandte Makromolekulare Chemie*, 37(1):27–34, 1974.

A. Jeziorny. Parameters characterizing the kinetics of the non-isothermal crystallization of poly (ethylene terephthalate) determined by dsc. *Polymer*, 19(10):1142–1144, 1978.

Z. Jing, J. Li, W. Xiao, H. Xu, P. Hong, and Y. Li. Crystallization, rheology and mechanical properties of the blends of poly (l-lactide) with supramolecular polymers based on poly (d-lactide)–poly (ε-caprolactone-co-δ-valerolactone)–poly (d-lactide) triblock copolymers. *RSC Advances*, 9(45): 26067–26079, 2019.

V. U. Johnsen and G. Nachtrab. Die kristallinität von äthylen-vinylacetat-copolymeren. *Die Angewandte Makromolekulare Chemie: Applied Macromolecular Chemistry and Physics*, 7(1):134–146, 1969.

Y. M. Joshi. Long time response of aging glassy polymers. *Rheologica Acta*, 53(5-6):477–488, 2014.

W. Kaiser. *Kunststoffchemie für Ingenieure - Von der Synthese bis zur Anwendung*. Carl Hanser GmbH&Co., 2011.

P. M. Kamath and R. W. Wakefield. Crystallinity of ethylene–vinyl acetate copolymers. *Journal of Applied Polymer Science*, 9(9):3153–3160, 1965.

I. Keridou, L. J. Del Valle, L. Funk, P. Turon, L. Franco, and J. Puiggalí. Non-isothermal crystallization kinetics of poly (4-hydroxybutyrate) biopolymer. *Molecules*, 24(15):2840, 2019.

Y. P. Khanna. Rheological mechanism and overview of nucleated crystallization kinetics. *macromolecules*, 26(14):3639–3643, 1993.

H. E. Kissinger et al. Variation of peak temperature with heating rate in differential thermal analysis. *Journal of research of the National Bureau of Standards*, 57(4):217–221, 1956.

E. Klompen and L. E. Govaert. Nonlinear viscoelastic behaviour of thermorheologically complex materials. *Mechanics of Time-Dependent Materials*, 3(1):49–69, 1999.

N. Koga. Ozawa's kinetic method for analyzing thermoanalytical curves. *Journal of thermal analysis and calorimetry*, 113(3):1527–1541, 2013.

M. Kohzaki, Y. Tsujita, A. Takizawa, and T. Kinoshita. The crystallization and formation of cluster of ethylene ionomer during physical aging. *Journal of applied polymer science*, 33(7):2393–2402, 1987.

J. Kolařík and A. Pegoretti. Non-linear tensile creep of polypropylene: Time-strain superposition and creep prediction. *Polymer*, 47(1):346–356, 2006.

M. Kothe. *Alterungsverhalten von polymeren Zwischenschichtmaterialien im Bauwesen*. PhD thesis, Saechsische Landesbibliothek-Staats-und Universitaetsbibliothek Dresden, 2013.

M. A. Kraus. *Machine Learning Techniques for the Material Parameter Identification of Laminated Glass in the Intact and Post-Fracture State*. PhD thesis, München, Universität der Bundeswehr München, 2018.

M. A. Kraus, M. Schuster, J. K. Kuntsche, G. Siebert, and J. Schneider. Parameter identification methods for visco-and hyperelastic material models. *Glass Structures & Engineering*, 2(2):147–167, 2017.

M. A. Kraus, M. Schuster, J. Schneider, and G. Siebert. *Report – Work Package A: Linear Viscoelasticity, A.3.: Four-Point Bending Tests*, 2019.

J. K. Kuntsche. *Mechanisches Verhalten von Verbundglas unter zeitabhängiger Belastung und Explosionsbeanspruchung: Mechanical behaviour of laminated glass under time-dependent and explosion loading*. PhD thesis, Technische Universität Darmstadt, 2015.

J. K. Kuntsche, M. Schuster, J. Schneider, and S. Langer. Viscoelastic properties of laminated glass interlayers–theory and experiments. *Glass Performance Days*, pages 143–147, 2015.

J. K. Kuntsche, M. Schuster, and J. Schneider. Engineering design of laminated safety glass considering the shear coupling: a review. *Glass Structures & Engineering*, 4(2):209–228, 2019.

Kuraray. *Kuraray Glass Laminating Solutions SENTRYGLAS® IONOPLAST INTERLAYER TECHNICAL MANUAL*, 2014.

Kuraray Trosifol World of Interlayers. *Verarbeitung von TROSIFOL PVB-Folie*, 2012.

Kuraray Trosifol World of Interlayers. *TROSIFOL SentryGlas*, 2019.

Kuraray Trosifol World of Interlayers. *Product portfolio Architectural Glazing*, 2021.

M. Kutterer. Verbundglasplatten–schubverbund und membrantragwirkung–teil i. *Stahlbau*, 74(1): 39–46, 2005.

G. Lamberti, G. Peters, and G. Titomanlio. Crystallinity and linear rheological properties of polymers. *International Polymer Processing*, 22(3):303–310, 2007.

M. D. Lechner, K. Gehrke, and E. H. Nordmeier. *Makromolekulare Chemie*. Springer, 2014.

P. Ledergerber, T. Balmer, C. Bernet, C. Brändli, and S. Schaible. Basic study on the evaluation of thermoplastic polymers as hot-melt adhesives for mixed-substrate joining. *Open Journal of Applied Sciences*, 6:579–592, 2016.

T. Lee, F. Boey, and K. Khor. X-ray diffraction analysis technique for determining the polymer crystallinity in a polyphenylene sulfide composite. *Polymer composites*, 16(6):481–488, 1995.

A. Lion, C. Kardelky, and P. Haupt. On the frequency and amplitude dependence of the payne effect: Theory and experiments. *Rubber chemistry and technology*, 76(2):533–547, 2003.

T. Liu, Z. Mo, S. Wang, and H. Zhang. Nonisothermal melt and cold crystallization kinetics of poly (aryl ether ether ketone ketone). *Polymer Engineering & Science*, 37(3):568–575, 1997.

Y. Long, R. A. Shanks, and Z. H. Stachurski. Kinetics of polymer crystallisation. *Progress in Polymer Science*, 20(4):651–701, 1995.

R. Longworth and D. J. Vaughan. Physical structure of ionomers. *Nature*, 218(5136):85–87, 1968.

Y.-L. Loo, K. Wakabayashi, Y. E. Huang, R. A. Register, and B. S. Hsiao. Thin crystal melting produces the low-temperature endotherm in ethylene/methacrylic acid ionomers. *Polymer*, 46 (14):5118–5124, 2005.

W. MacKnight and T. R. Earnest Jr. The structure and properties of ionomers. *Journal of Polymer Science: Macromolecular Reviews*, 16(1):41–122, 1981.

L. Maddé. Investigations on the linearity limits of viscoelastic laminated glass interlayers using amplitude-sweeps and relaxation tests. Master's thesis, Technische Universität Darmstadt, 2019.

J. Málek. The kinetic analysis of non-isothermal data. *Thermochimica acta*, 200:257–269, 1992.

J. Málek. Kinetic analysis of crystallization processes in amorphous materials. *Thermochimica Acta*, 355(1-2):239–253, 2000.

C. L. Marx and S. L. Cooper. The crystallinity of ionomers. *Journal of Macromolecular Science, Part B: Physics*, 9(1):19–33, 1974.

C. L. Marx, D. F. Caulfield, and S. L. Cooper. Morphology of ionomers. *Macromolecules*, 6(3): 344–353, 1973.

V. Mathot and M. Pijpers. Heat capacity, enthalpy and crystallinity for a linear polyethylene obtained by dsc. *Journal of Thermal Analysis and Calorimetry*, 28(2):349–358, 1983.

J. Matthews, H. Peiser, and R. Richards. The x-ray measurement of the amorphous content of polythene samples. *Acta Crystallographica*, 2(2):85–90, 1949.

B. R. C. d. Menezes, T. M. B. Campos, T. L. d. A. Montanheiro, R. G. Ribas, L. d. S. Cividanes, and G. P. Thim. Non-isothermal crystallization kinetic of polyethylene/carbon nanotubes nanocomposites using an isoconversional method. *Journal of Composites Science*, 3(1):21, 2019.

T. G. Mezger. *Das Rheologie Handbuch: 5.* FARBE UND LACK, 2016.

M. Minale, C. Carotenuto, L. P. Paduano, and L. Grassia. Nonisothermal crystallization kinetics of an ethylene-vinyl-acetate: I calorimetry versus rheology. *Polymer Engineering & Science*, 59 (12):2557–2563, 2019.

Z. Mo and H. Zhang. The degree of crystallinity in polymers by wide-angle x-ray diffraction (waxd). *Journal of Macromolecular Science, Part C: Polymer Reviews*, 35(4):555–580, 1995.

M. Moneke. *Die Kristallisation von verstärkten Thermoplasten während der schnellen Abkühlung und unter Druck.* PhD thesis, Technische Universität Darmstadt, 2001.

H. Münstedt. *Rheological and morphological properties of dispersed polymeric materials: filled polymers and polymer blends.* Carl Hanser Verlag GmbH Co KG, 2016.

N. Murthy and H. Minor. General procedure for evaluating amorphous scattering and crystallinity from x-ray diffraction scans of semicrystalline polymers. *Polymer*, 31(6):996–1002, 1990.

MVV TB. Muster verwaltungsvorschrift technische baubestimmungen. Report, Deutsches Institut für Bautechnik DIBt, 2020-01.

K. Nagamatsu, T. Takemura, T. Yoshitomi, and T. Takemoto. Effect of crystallinity on the viscoelastic properties of polyethylene. *Journal of Polymer Science*, 33(126):515–518, 1958.

O. Narayanaswamy. A model of structural relaxation in glass. *Journal of the American Ceramic Society*, 54(10):491–498, 1971.

T. Naumann. *Beitrag zur Beschreibung des mechanischen Langzeitdeformationsverhaltens von thermoplastischen Kunststoffen.* PhD thesis, Universität des Saarlandes, 2012.

NBN S 23-002-2. Glaswerk - deel 2 : Berekening van de glasdikte. Standard, 2020.

NEN 2608. Vlakglas voor gebouwen- eisen en bepalingsmethode (glass in building - requirements and determination method). Standard, 2014.

NETZSCH-Gerätebau GmbH. *Thermal Characterization of Polymers Thermoplastics, Thermoplastic Elastomers, Elastomers and Thermosets. Accessed March 11 2021.*

L. Nitzling. Investigations on the linearity limits of viscoelastic laminated glass interlayers. Master's thesis, Technische Universität Darmstadt, 2018.

NS 3510. Sikkerhetsruter i byggverk - krav til prosjektering og-klasser for ulike bruksområder (safety glass in constructionworks requirements for design and classes in various application areas). Standard, 2015.

P. Ohmen. Experimental investigations on the mechanical behaviour of polymeric interlayers of laminated glass under uniaxial tensile stress. Bachelor's thesis, Technische Universität Darmstadt, 2018.

ÖNORM B 3716-1. Glas im bauwesen—konstruktiver glasbau teil 1:grundlagen (glass in building—structural glass construction—part1: Basic principles). Standard, 2015.

T. Ozawa. Kinetics of non-isothermal crystallization. *Polymer*, 12(3):150–158, 1971.

T. Ozawa. Estimation of activation energy by isoconversion methods. *Thermochimica acta*, 203: 159–165, 1992.

T. Ozawa and T. Koto. A simple method for estimating activation energy from derivative thermoanalytical curves and its application to thermal shrinkage of polycarbonate. *Journal of thermal analysis*, 37(6):1299–1307, 1991.

M. Pander, S. Dietrich, S.-H. Schulze, U. Eitner, and M. Ebert. Thermo-mechanical assessment of solar cell displacement with respect to the viscoelastic behaviour of the encapsulant. In *2011 12th Intl. Conf. on Thermal, Mechanical & Multi-Physics Simulation and Experiments in Microelectronics and Microsystems*, pages 1–6. IEEE, 2011.

R. Pantani, V. Speranza, and G. Titomanlio. Simultaneous morphological and rheological measurements on polypropylene: Effect of crystallinity on viscoelastic parameters. *Journal of Rheology*, 59(2):377–390, 2015.

G. Papageorgiou, D. N. Bikiaris, and K. Chrissafis. A different approach for the study of the crystallization kinetics in polymers. key study: poly (ethylene terephthalate)/sio2 nanocomposites. *Polymer international*, 59(12):1630–1638, 2010.

G. Papanicolaou and S. Zaoutsos. Viscoelastic constitutive modeling of creep and stress relaxation in polymers and polymer matrix composites. In *Creep and fatigue in polymer matrix composites*, pages 3–59. Elsevier, 2019.

J. Pelfrene. *Numerical analysis of the post-fracture response of laminated glass under impact and blast loading*. PhD thesis, Ghent University, 2016.

E. Piorkowska and A. Galeski. Overall crystallization kinetics. *Handbook of polymer crystallization*, pages 215–236, 2013.

A. Pipkin and T. Rogers. A non-linear integral representation for viscoelastic behaviour. *Journal of the Mechanics and Physics of Solids*, 16(1):59–72, 1968.

N. V. Pogodina, H. H. Winter, and S. Srinivas. Strain effects on physical gelation of crystallizing isotactic polypropylene. *Journal of Polymer Science Part B: Polymer Physics*, 37(24):3512–3519, 1999.

N. Pour-Moghaddam. *On the fracture behaviour and the fracture pattern morphology of tempered soda-lime glass*. PhD thesis, Technische Universität Darmstadt, 2019.

K. Puller. *Untersuchung des Tragverhaltens von in die Zwischenschicht von Verbundglas integrierten Lasteinleitungselementen*. PhD thesis, Universität Stuttgart, 2013.

J. L. Rand, J. K. Henderson, and D. A. Grant. Nonlinear behavior of linear low-density polyethylene. *Polymer Engineering & Science*, 36(8):1058–1064, 1996.

M. Reading, D. M. Price, and H. Orliac. Measurement of crystallinity in polymers using modulated temperature differential scanning calorimetry. In *Materials Characterization by Dynamic and Modulated Thermal Analytical Techniques*. ASTM International, 2001.

W. Retting. Viskoelastisches verhalten bei zügig wachsender spannung und verformung. In G. Schreyer, editor, *Konstruieren mit Kunststoffen: Grundlagen und Eigenschaften, Konstruktionsprinzipien und Anwendungsbeispiele:[in zwei Teilen]*. C. Hanser, 1972.

P. C. Roozemond, V. Janssens, P. Van Puyvelde, and G. W. Peters. Suspension-like hardening behavior of hdpe and time-hardening superposition. *Rheologica acta*, 51(2):97–109, 2012.

J. Rösler, H. Harders, and M. Bäker. *Mechanisches Verhalten der Werkstoffe*. Springer-Verlag, 2012.

P. E. Rouse Jr. A theory of the linear viscoelastic properties of dilute solutions of coiling polymers. *The Journal of Chemical Physics*, 21(7):1272–1280, 1953.

A. Rühl. *On the time and temperature dependent behaviour of laminated amorphous polymers subjected to low-velocity impact*. PhD thesis, Technische Universität Darmstadt, 2017.

V. Sackmann. *Untersuchungen zur Dauerhaftigkeit des Schubverbunds in Verbundsicherheitsglas mit unterschiedlichen Folien aus Polyvinylbutyral*. PhD thesis, Technische Universität München, 2008.

S. Schaible, C. Bernet, P. Ledergeber, T. Balmer, and C. Brändli. Basic study on the evaluation of thermoplastic polymers as hot-melt adhesives for mixed-substrate joining. *Open Journal of Applied Sciences*, 06(08):579–592, 2016.

R. A. Schapery. *A theory of non-linear thermoviscoelasticity based on irreversible thermodynamics*. American Society of Mechanical Engineers New York, NY, 1966.

R. A. Schapery. On the characterization of nonlinear viscoelastic materials. *Polymer Engineering & Science*, 9(4):295–310, 1969a.

R. A. Schapery. *Further development of a thermodynamic constitutive theory: stress formulation*. Purdue University School of Aeronautics, Astronautics and Engineering Sciences, 1969b.

C. Schick. Differential scanning calorimetry (dsc) of semicrystalline polymers. *Analytical and bioanalytical chemistry*, 395(6):1589–1611, 2009.

E. Schmachtenberg. *Die mechanischen Eigenschaften nichtlinear viskoelastischer Werkstoffe*. PhD thesis, Rheinisch-Westfälische Technische Hochschule, 1985.

J. Schneider. Next generation ionoplast for improved lamination. In *Glass performance days 2019*, pages 61–63. Glass performance days, 2019.

J. Schneider, J. K. Kuntsche, S. Schula, F. Schneider, and J.-D. Wörner. *Glasbau: Grundlagen, Berechnung, Konstruktion*. Springer-Verlag, 2016.

M. Schubnell. Bestimmung der kristallinität bei polymeren aus dsc-messungen. *Metter Toledo UserCom*, 1, 2001.

C. Schuler. *Einfluß des Materialverhaltens von PVB auf das Tragverhalten von Verbundsicherheitsglas in Abhängigkeit von Temperatur und Belastung*. PhD thesis, Technische Universität München, 2003.

M. Schuster, M. A. Kraus, J. Schneider, and G. Siebert. Investigations on the thermorheologically complex material behaviour of the laminated safety glass interlayer ethylene-vinyl-acetate. *Glass Structures & Engineering*, 3(2):373–388, 2018.

M. Schuster, M. A. Kraus, J. Schneider, and G. Siebert. *Report – Work Package A: Linear Viscoelasticity A.5.: Correlation of DMTA and Large Scale Testing*, 2019.

M. Schuster, K. Thiele, and J. Schneider. Investigations on the viscoelastic material behaviour and linearity limits of pvb. In *Engineered Transparency Conference*. Engineered Transparency Conference, 2021.

M. Schuster, K. Thiele, and J. Schneider. Consideration of the nonlinear viscoelasticity of pvb through a time-strain superposition. submitted, not yet published. In *Glass performance days 2021*, page to appear. Glass performance days, 2022.

F. R. Schwarzl. *Polymermechanik: Struktur und mechanisches Verhalten von Polymeren*. Springer-Verlag, 2013.

V. Sessini, J. M. Raquez, J. M. Kenny, P. Dubois, and L. Peponi. Melt-processing of bionanocomposites based on ethylene-co-vinyl acetate and starch nanocrystals. *Carbohydr Polym*, 208:382–390, 2019.

J. Šesták and G. Berggren. Study of the kinetics of the mechanism of solid-state reactions at increasing temperatures. *Thermochimica Acta*, 3(1):1–12, 1971.

B. K. Sharma, U. Desai, A. Singh, and A. Singh. Effect of vinyl acetate content on the photovoltaic-encapsulation performance of ethylene vinyl acetate under accelerated ultra-violet aging. *Journal of Applied Polymer Science*, 137(2):48268, 2020.

F. Shehzad, S. P. Thomas, and M. A. Al-Harthi. Non-isothermal crystallization kinetics of high density polyethylene/graphene nanocomposites prepared by in-situ polymerization. *Thermochimica Acta*, 589:226–234, 2014.

L. Spieß, R. Schwarzer, H. Behnken, and G. Teichert. Methoden der röntgenbeugung. In *Moderne Röntgenbeugung*, pages 147–210. Springer, 2005.

F. J. Stadler, T. Takahashi, and K. Yonetake. Lattice sizes, crystallinities, and spacing between amorphous chains-characterization of ethene-/α-olefin copolymers with various comonomers and comonomer contents measured by wide angle x-ray scattering. *e-Polymers*, 9(1):040, 2009.

K. Stamm and H. Witte. *Sandwichkonstruktionen: Berechnung, Fertigung, Ausführung*, volume 3. Springer-Verlag, 2013.

A. Staverman and F. Schwarzl. Linear deformation behaviour of high polymers. In *Theorie und molekulare Deutung technologischer Eigenschaften von hochpolymeren Werkstoffen*, pages 1–125. Springer, 1956.

S. A. Steineck. Comparison of different approaches for laminate- and sandwich-elements in glass and steel constructions. Master's thesis, Technische Universität Darmstadt, 2018.

P. Stephan, K. Schaber, K. Stephan, and F. Mayinger. Thermodynamisches gleichgewicht und stabilität. *Thermodynamik: Grundlagen und technische Anwendungen–Band 2: Mehrstoffsysteme und chemische Reaktionen*, pages 153–177, 2010.

M. Stommel, M. Stojek, and W. Korte. *FEM zur Berechnung von Kunststoff-und Elastomerbauteilen*. Carl Hanser Verlag GmbH Co KG, 2018.

T. W. Strganac, A. Letton, D. F. Payne, and B. A. Biskup. Characterization of nonlinear viscoelastic behavior using a dynamic mechanical approach. *AIAA journal*, 33(5):904–910, 1995.

L. Struik. The mechanical and physical ageing of semicrystalline polymers: 1. *Polymer*, 28(9): 1521–1533, 1987.

Y. T. Sung, C. K. Kum, H. S. Lee, J. S. Kim, H. G. Yoon, and W. N. Kim. Effects of crystallinity and crosslinking on the thermal and rheological properties of ethylene vinyl acetate copolymer. *Polymer*, 46(25):11844–11848, 2005.

C. Suryanarayana and M. G. Norton. *X-ray diffraction: a practical approach*. Springer Science & Business Media, 2013.

K. Thiele, M. Schuster, M. A. Kraus, and J. Schneider. Fractional viscoelastic constitutive modelling of multi-layered acoustic pvb. *unpublished*, 2022.

G. Titomanlio, G. Spadaro, and F. La Mantia. Stress relaxation of a polyisobutylene under large strains. *Rheologica Acta*, 19(4):477–481, 1980.

A. Tobolsky and J. McLoughlin. Viscoelastic properties of crystalline polymers: Polytrifluorochloroethylene. *The Journal of Physical Chemistry*, 59(9):989–990, 1955.

E. Trösch. *Tragverhalten von überlappend laminierten Verbundglasträgern für grosse Spannweiten*. PhD thesis, ETH Zurich, 2015.

Y. S. Urzhumtsev. Time—temperature superposition for thermorheologically complex materials. *Polymer Mechanics*, 10(2):180–185, 1974.

M. van Drongelen, P. Roozemond, E. Troisi, A. Doufas, and G. Peters. Characterization of the primary and secondary crystallization kinetics of a linear low-density polyethylene in quiescent- and flow-conditions. *Polymer*, 76:254–270, 2015.

M. Van Gurp and J. Palmen. Time-temperature superposition for polymeric blends. *Rheol. Bull*, 67 (1):5–8, 1998.

S. Vyazovkin and N. Sbirrazzuoli. Estimating the activation energy for non-isothermal crystallization of polymer melts. *Journal of thermal analysis and calorimetry*, 72(2):681–686, 2003.

K. Wakabayashi and R. A. Register. Morphological origin of the multistep relaxation behavior in semicrystalline ethylene/methacrylic acid ionomers. *Macromolecules*, 39(3):1079–1086, 2006.

K. Wang and Q. Deng. The thermal and mechanical properties of poly(ethylene-co-vinyl acetate) random copolymers (peva) and its covalently crosslinked analogues (cpeva). *Polymers (Basel)*, 11(6), 2019.

J.-L. Wietor, D. J. M. van Beek, G. W. Peters, E. Mendes, and R. P. Sijbesma. Effects of branching and crystallization on rheology of polycaprolactone supramolecular polymers with ureidopyrimidinone end groups. *Macromolecules*, 44(5):1211–1219, 2011.

M. L. Williams, R. F. Landel, and J. D. Ferry. The temperature dependence of relaxation mechanisms in amorphous polymers and other glass-forming liquids. *Journal of the American Chemical society*, 77(14):3701–3707, 1955.

N. Woicke. *Viskoelastizität von Polypropylen im Glasübergang*. PhD thesis, Universtiät Stuttgart, 2006.

N. Woicke, M. Keuerleber, B. Hegemann, and P. Eyerer. Three-dimensional thermorheological behavior of isotactic polypropylene across glass transition temperature. *Journal of applied polymer science*, 94(3):877–880, 2004.

E. Wölfel. Nachgiebiger verbund. eine näherungslösung und deren anwendungsmöglichkeiten. *STAHLBAU, DER*, 56(6), 1987.

B. Wunderlich. *Thermal analysis of polymeric materials*. Springer Science & Business Media, 2005.

B. Wunderlich. The application of mtdsc to polymer melting. In *Modulated Temperature Differential Scanning Calorimetry*, pages 217–319. Springer, 2006.

R. Yang, Y. Song, and Q. Zheng. Payne effect of silica-filled styrene-butadiene rubber. *Polymer*, 116:304–313, 2017.

I. Yannas. Transition from linear to nonlinear viscoelastic behavior. iii. linearity below and above tg. *Rubber Chemistry and Technology*, 45(6):1623–1629, 1972.

Z-70.3-170. Verbund-sicherheitsglas aus sentryglas® sg5000 mit schubverbund. Allgemeine bauaufsichtliche zulassung, Deutsches Institut für Bautechnik DIBt, 2016-08.

Z-70.3-197. Verbund-sicherheitsglas mit einer verbundfolie der produktfamilie evasafe mit ansatz eines schubverbundes. Allgemeine bauaufsichtliche zulassung, Deutsches Institut für Bautechnik DIBt, 2016-10.

Z-70.3-230. Verbund-sicherheitsglas aus der produktfamilie saflex dg mit schubverbund. Allgemeine bauaufsichtliche zulassung, Deutsches Institut für Bautechnik DIBt, 2016-08.

Z-70.3-236. Verbund-sicherheitsglas mit der pvb-folie trosifol es mit schubverbund. Allgemeine bauaufsichtliche zulassung, Deutsches Institut für Bautechnik DIBt, 2016-10.

Z-70.3-256. Verglasungen aus verbund-sicherheitsglas mit der pvb-folie trosifol extra stiff b130. Allgemeine bauartengenehmigung, Deutsches Institut für Bautechnik DIBt, 2020-05.

P. Zadrapa. *The Morphology and Properties of Selected Filler/Poly (Ethylene-Co-Methacrylic Acid) Copolymer Systems*. PhD thesis, Tomas Bata University, Czech Republic, 2011.

G. Zografi and A. Newman. Interrelationships between structure and the properties of amorphous solids of pharmaceutical interest. *Journal of pharmaceutical sciences*, 106(1):5–27, 2017.

Appendix

M. Schuster, *Characterization of Laminated Safety Glass Interlayers*,
Mechanik, Werkstoffe und Konstruktion im Bauwesen 66,
https://doi.org/10.1007/978-3-658-39821-7

Appendix A

Additional data

A.1 Laminated safety glass design with Wölfel's sandwich theory

A vertical 1.0 m x 1.0 m large, two-side linearly supported laminated glass loaded by wind (design value w_d = 2.3 kN/m^2) is considered. The interlayer thickness is d_Q = 0.76 mm and the glass thickness is $d_1 = d_2$ = 6 mm, see Figure A.1.

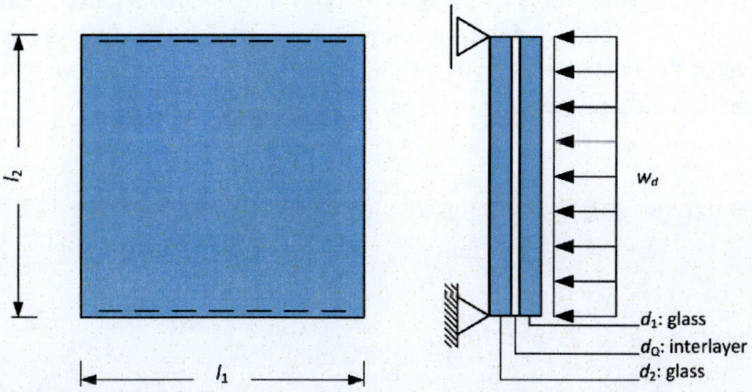

Figure A.1 Structural system.

The shear modulus value of the interlayer is assumed to be $G = 2$ MPa. The limiting case of *full shear transfer* is considered first. A maximum glass stress of 11.98 MPa is obtained for this example, see eq. (A.1). If the limit case *no shear transfer* is considered, the maximum glass stress is 23.96 MPa, see eq.(A.2), which is significantly greater than in the limit state of *full shear transfer*.

$$\sigma_{\text{glass}} = \frac{3 \cdot w_\mathrm{d} \cdot l_2 \cdot l_1^2}{4 l_2 \cdot (d_1 + d_2)^2} = \frac{3 \cdot 2.3 \frac{\mathrm{kN}}{\mathrm{m}^2} \cdot (1\mathrm{m})^2}{4 \cdot (12\mathrm{mm})^2} = 11.98 \text{ MPa.} \tag{A.1}$$

M. Schuster, *Characterization of Laminated Safety Glass Interlayers*, Mechanik, Werkstoffe und Konstruktion im Bauwesen 66,
https://doi.org/10.1007/978-3-658-39821-7

$$\sigma_{\text{glass}} = \frac{3 \cdot 0.5 \cdot w_{\text{d}} \cdot l_2 \cdot l_1^2}{4 l_2 \cdot d_1^2} = \frac{3 \cdot 0.5 \cdot 2.3 \frac{\text{kN}}{\text{m}^2} \cdot (1\text{m})^2}{4 \cdot (6\text{mm})^2} = 23.96 \text{ MPa}. \tag{A.2}$$

Coupling effects that are between the two limit cases of *full shear* and *no shear* can be respected during the glass design by the commonly known sandwich theories, e.g. (Kutterer, 2005; Stamm and Witte, 2013; Wölfel, 1987). A good summary of different theories can be found, for example, in (Steineck, 2018). Widely used in glass construction, e.g. in (Gräf, 2003; Schuler, 2003; Ensslen, 2005) is the approximation solution for a flexible composite according to Wölfel (1987), which was originally developed for steel sandwich panels and is only valid for statically determinate systems and linear elastic materials. In 2009, it was also implemented into the ASTM E1300 , where the original equations were reformulated into equivalent thicknesses for calculating the maximum deformation (see eq. A.6) and for calculating the maximum glass stress (see eq. A.7). Wölfel's method can be summarized as splitting the acting bending moment into two parts: On the one hand, there is the contribution of the glass panes, which carry a part through their bending stiffness. On the other hand, there is the composite part, in which the bending moment is transformed into a pair of forces with an inner lever arm, so that a tensile normal force is generated in the lower glass pane and a compressive normal force in the upper glass pane. The distance of the centers of gravity of the glass panes is called d_s (eq. (A.3)) and the area moment of inertia of the composite part is called I_s (eq. (A.4))

$$d_s = 0.5 \cdot (d_1 + d_2) + d_Q = 0.5 \cdot (6 + 6) + 0.76 = 6.76 \text{ mm} \tag{A.3}$$

$$I_s = \frac{d_1 \cdot d_2 \cdot d_s^2}{d_1 + d_2} = \frac{6 \cdot 6 \cdot 6.76^2}{6 + 6} = 137.09 \text{ mm}^3 \tag{A.4}$$

The composite part varies depending on the shear transfer (expressed by the shear transfer coefficient Γ, see eq. (A.5) or by $k = \Gamma^{-1} - 1$ as written in Wölfel (1987)) between the limit state *no shear transfer* and the limit state *full shear transfer*. The latter is achieved by high Γ values approaching 1 (resp. low k values approaching 0). Besides the dependency of Γ (resp. k) on the geometry (glass thickness, interlayer thickness, spans,..) and on the material properties, e.g. Young's modulus of the glass E_{glass} and shear modulus of the interlayer G, there is also a dependency on the loading type and structural system (e.g. uniformly distributed loads on a single span beam or single load in midspan,..) which is taken into account with the value α in eq. (A.5). For uniformly distributed loads on a single span beam $\alpha = 9.6$ end e.g. for a single load in midspan $\alpha = 12$. In the example presented here, an interlayer shear modulus of 2 MPa results in a shear transfer coefficient $\Gamma = 0.566$

(see eq. (A.5)), an equivalent thickness for deformations calculations of $d_w = 11.09$ mm (see eq.(A.6)) and an equivalent thickness for stress calculations of $d_\sigma = 11.78$ mm (eq. A.7)). This results in a glass stress of 12.43 MPa (see eq. (A.8)), which is much closer to the value of the limit state *full shear transfer* than to the limit state *no shear transfer*.

$$\Gamma = \frac{1}{1+\alpha\frac{E_{\text{glass}}\cdot I_s \cdot d_Q}{G\cdot(d_s\cdot l_1)^2}} = \frac{1}{1+9.6\cdot\frac{70000\cdot 137.09\cdot 0.76}{2\cdot(6.76\cdot 1000)^2}} = 0.566 \tag{A.5}$$

$$d_w = \sqrt[3]{d_1^3 + d_2^3 + 12\Gamma I_s} = 11.09 \text{ mm} \tag{A.6}$$

$$d_{\sigma,i} = d_{\sigma,1} = d_{\sigma,2} = \sqrt{\frac{d_w^3}{d_i + 2\Gamma d_{s,i}}} = 11.78 \text{ mm} \tag{A.7}$$

$$\sigma_{\text{glass}} = \frac{3w_d \cdot l_2 \cdot l_1^2}{4l_2 \cdot d_{\sigma,i}^2} = \frac{3\cdot 2.3\frac{\text{kN}}{\text{m}^2}\cdot(1\text{m})^2}{4\cdot(11.78 \text{ mm})^2} = 12.43 \text{ MPa}. \tag{A.8}$$

A.2 Semicrystalline structure

A.2.1 DSC and XRD test settings

Table A.1 Test matrix: DSC tests.

	SG	cEVA	EVA	standard PVB
test equipment	DWDK Netzsch Type 214 Polyma	DWDK Netzsch Type 214 Polyma	Mettler Toledo DSC [1]	Mettler Toledo DSC [1]
temperature range [°C]	[-60:140]	[-60:100]	[-60:120]	[-60:120]
number of cycles [-] x temperature rates [K/min]	1st heating: 1x20 other cycles: 2x20, 2x10, 2x3[2]	1st heating: 1x20 other cycles: 2x20, 2x10, 2x2	1st heating: 1x10 other cycles: 1x20, 1x10, 1x2	1st heating: 1x10 other cycles: 1x10, 1x20, 1x2
tested samples	1	1	1	1
sample mass [mg]	4.1	7.0	9.6	9.9

[1] measurement conducted by the department of macromolecular and paper chemistry TU Darmstadt.
[2] after the last cycle, the sample was left for 24 h at 23°C in the DSC and then finally reheated from [23:50]°C with 3 K/min.

Table A.2 Test matrix: XRD tests.

	SG	cEVA	EVA	standard PVB
test equipment	Bruker D2 Phaser[1]			StadiP, Fa. Stoe and Cie. GmbH[2]
temperature [°C]	room temperature			room temperature
range of 2Θ [°]	[5:70]			[3:65]
tested samples	1	1	1	1
sample geometry [cm]	⌀ = 3			no information

[1] measurement conducted by the department of construction and building materials TU Darmstadt.
[2] measurement conducted by the department of Eduard-Zintl-Institute for inorganic and physical chemistry TU Darmstadt.

Table A.3 Test matrix: DSC test to check cold crystallization.

	SG
test equipment	DWDK Netzsch Type 214 Polyma
temperature range [°C]	1st and final cycle: [-60:140] 2nd and 3rd cycle: [-60:60]
temperature rate [K/min]	20
tested samples	1
sample mass [mg]	5.7

Table A.4 Test matrix: DSC test to check influence of heating rates.

	SG
test equipment	DWDK Netzsch Type 214 Polyma
temperature range [°C]	[-60:140]
temperature rates [K/min]	cooling: -10 heating: 0.3, 3, 10, 20
tested samples	1
sample mass [mg]	6.0

Table A.5 Test matrix: DMTA-TS.

	SG	cEVA	EVA
test equipment	Anton Paar MCR 302, PP configuration	Anton Paar MCR 302, SRF configuration	Anton Paar MCR 302, PP configuration
sample geometry [mm]	⊘ = 8	50 x 10	⊘ = 8
	0.89 thick	0.8 thick	0.8 thick
temperature range [°C]	[-40:120]	[-60:100]	[-60:100]
temperature rate [K/min]	3	2	2
tested samples	1	1	1
frequency [Hz]	1	1	1
excitation amplitude [%]	0.025	0.1	0.025

Table A.6 Test matrix: DSC test to validate kinetic model.

	SG
test equipment	DWDK Netzsch Type 214 Polyma
temperature range [°C]	[-60:120]
temperature rates [K/min]	20, 5, 0.3
tested samples	1
sample mass [mg]	5.4

Table A.7 Test matrix: DSC to investigate physical aging.

	SG
test equipment	DWDK Netzsch Type 214 Polyma
temperature range [°C]	[-60:100]
temperature rates [K/min]	3[1], 5, 10, 20
tested samples	1
sample mass [mg]	6.1

[1] only the 3 K/min data is used for further evaluation. The remaining data is used for comparison with the other DSC measurements.

Table A.8 Test matrix: DMTA-TS to investigate physical aging of SG.

	SG	
physical aging procedure	elimination of thermal history (age 0)	as received[1]
	24 h at RT	
	4 d at RT	
test equipment	Anton Paar MCR 302, PP configuration	Anton Paar MCR 302, SRF configuration
sample geometry [mm]	∅ = 8	50 x 10
	0.89 thick	0.89 thick
temperature range [°C]	age 0: [-40:100]	[-40:80][2]
	other: [23:100]	
temperature rate [K/min]	3	3
tested samples	1	1
frequency [Hz]	1	1
excitation amplitude [%]	0.025	0.1

[1] as received indicates that the thermal prehistory has not been removed.
[2] the sample was installed at -40°C.

A.2.2 Crystallization kinetics with DMTA

Table A.9 Determination of the Arrhenius equations for the storage modulus in the totally molten and totally crystallized state: Parameters of G'_m and G'_s.

	SG	cEVA	EVA
G'_s (heating curve)			
k_s [MPa]	2.4067e-05	0.2004	1.6936e-10
$E_{a,s}$ [J/mol]	2.9378e+04	3.8001e+03	5.7441e+04
Fitting range [°C]	[60:100]	[50:100]	[60:100]
Heating rate [K/min]	3	2	2
G'_m (cooling curve)			
k_m [MPa]	1.2795e-08	0.0011	1.0579e-04
$E_{a,m}$ [J/mol]	5.5664e+04	2.0520e+04	2.5502e+04
fitting range [°C]	[20:30]	[0:20]	[70:80]
cooling rate [K/min]	-3	-2	-2

A.2.3 Relative crystallinities

Table A.10 SG: relative crystallinities at different temperature rates. Values above 100% in the heating curves represent cold crystallization.

	α at					
T [°C]	-20 K/min	-10 K/min	-2 K/min	2 K/min	10 K/min	20 K/min
≤ -21	100.00%	100.00%	100.00%	100.00%	100.00%	100.00%
20	99.87%	100.00%	100.00%	100.00%	100.00%	100.00%
24	94.48%	98.87%	99.33%	100.00%	100.00%	100.00%
28	79.77%	91.66%	95.33%	100.08%	100.00%	100.00%
32	56.39%	75.39%	87.28%	101.65%	100.55%	100.02%
36	30.32%	48.98%	72.09%	105.26%	102.93%	100.98%
40	10.85%	19.75%	47.06%	110.56%	107.81%	104.11%
44	1.79%	3.27%	16.78%	117.80%	114.60%	109.57%
48	0.00%	0.05%	2.38%	124.82%	122.30%	116.46%
52	0.00%	0.00%	0.30%	130.71%	129.15%	123.51%
56	0.00%	0.00%	0.00%	133.52%	133.45%	129.07%
60	0.00%	0.00%	0.00%	131.11%	133.03%	131.00%
64	0.00%	0.00%	0.00%	123.09%	126.47%	127.00%
68	0.00%	0.00%	0.00%	108.30%	112.52%	115.40%
72	0.00%	0.00%	0.00%	88.06%	91.19%	95.97%
76	0.00%	0.00%	0.00%	63.62%	64.91%	70.70%
80	0.00%	0.00%	0.00%	38.27%	37.67%	43.84%
84	0.00%	0.00%	0.00%	16.39%	15.16%	21.03%
88	0.00%	0.00%	0.00%	3.87%	2.82%	6.71%
92	0.00%	0.00%	0.00%	0.16%	0.04%	1.04%
≥ 96	0.00%	0.00%	0.00%	0.00%	0.00%	0.00%

Table A.11 cEVA: relative crystallinities at different temperature rates.

	α at					
T [°C]	-20 K/min	-10 K/min	-2 K/min	2 K/min	10 K/min	20 K/min
≤ -21	100.00%	100.00%	100.00%	100.00%	100.00%	100.00%
-22	100.00%	100.00%	100.00%	100.00%	100.00%	100.00%
-18	99.70%	99.95%	99.98%	99.93%	99.90%	100.00%
-14	98.82%	99.44%	99.64%	99.43%	99.29%	100.00%
-10	97.40%	98.40%	98.85%	98.53%	98.24%	100.00%
-6	95.47%	96.88%	97.63%	97.24%	96.76%	99.80%
-2	93.01%	94.89%	95.96%	95.50%	94.90%	99.07%
2	89.94%	92.41%	93.91%	93.30%	92.66%	97.84%
6	86.22%	89.43%	91.35%	90.75%	90.06%	96.16%
10	81.88%	85.87%	88.22%	87.84%	87.17%	94.03%
14	76.83%	81.69%	84.61%	84.62%	83.91%	91.44%
18	70.77%	76.79%	80.48%	81.03%	80.32%	88.42%
22	63.48%	71.11%	75.71%	77.03%	76.40%	84.97%
26	54.46%	64.34%	70.08%	72.80%	72.19%	81.11%
30	43.22%	56.08%	63.40%	68.23%	67.70%	76.84%
34	29.36%	45.72%	55.52%	63.47%	62.97%	72.20%
38	11.02%	32.01%	45.89%	58.57%	57.98%	67.18%
42	0.12%	7.88%	35.46%	53.45%	52.82%	61.85%
46	0.00%	0.02%	9.23%	48.17%	47.50%	56.21%
50	0.00%	0.00%	0.01%	42.65%	41.99%	50.27%
54	0.00%	0.00%	0.00%	36.26%	36.09%	43.94%
58	0.00%	0.00%	0.00%	28.36%	29.18%	36.95%
62	0.00%	0.00%	0.00%	18.93%	20.91%	28.83%
66	0.00%	0.00%	0.00%	9.35%	11.77%	19.37%
70	0.00%	0.00%	0.00%	2.89%	4.32%	9.92%
74	0.00%	0.00%	0.00%	0.34%	0.87%	3.42%
78	0.00%	0.00%	0.00%	0.00%	0.16%	0.79%
82	0.00%	0.00%	0.00%	0.00%	0.05%	0.14%
86	0.00%	0.00%	0.00%	0.00%	0.00%	0.02%
≥ 91	0.00%	0.00%	0.00%	0.00%	0.00%	0.00%

Table A.12 EVA: relative crystallinities at different temperature rates.

	α at					
T [°C]	-20 K/min	-10 K/min	-2 K/min	2 K/min	10 K/min	20 K/min
≤ -23	100.00%	100.00%	100.00%	100.00%	100.00%	100.00%
-19	99.92%	99.92%	100.00%	100.00%	100.00%	100.00%
-15	99.38%	99.44%	100.00%	100.00%	100.00%	100.00%
-11	98.39%	98.55%	99.74%	100.00%	100.00%	99.92%
-7	96.95%	97.23%	98.94%	100.00%	99.95%	99.50%
-3	95.08%	95.53%	97.62%	100.00%	99.56%	98.71%
1	92.79%	93.41%	95.80%	100.00%	98.79%	97.56%
5	90.04%	90.88%	93.47%	100.00%	97.63%	96.05%
9	86.83%	87.92%	90.61%	99.90%	96.08%	94.16%
13	83.10%	84.49%	87.21%	99.42%	94.14%	91.87%
17	78.76%	80.53%	83.25%	98.51%	91.79%	89.19%
21	73.76%	75.95%	78.68%	97.18%	89.04%	86.09%
25	68.12%	70.70%	73.45%	95.40%	85.85%	82.56%
29	62.09%	64.81%	67.49%	93.14%	82.20%	78.60%
33	56.16%	58.66%	60.86%	90.34%	78.06%	74.25%
37	50.25%	52.75%	54.07%	86.89%	73.38%	69.56%
41	43.12%	46.94%	47.58%	82.51%	68.20%	64.56%
45	28.74%	40.67%	41.35%	76.86%	62.57%	59.25%
49	5.60%	27.19%	35.35%	69.79%	56.53%	53.64%
53	0.28%	4.28%	27.19%	61.84%	50.12%	47.75%
57	0.00%	0.20%	7.12%	53.52%	43.57%	41.76%
61	0.00%	0.00%	0.65%	45.08%	36.74%	35.51%
65	0.00%	0.00%	0.01%	35.64%	28.97%	28.45%
69	0.00%	0.00%	0.00%	25.08%	20.45%	20.66%
73	0.00%	0.00%	0.00%	14.40%	11.89%	12.68%
77	0.00%	0.00%	0.00%	5.65%	4.72%	5.66%
81	0.00%	0.00%	0.00%	1.27%	1.07%	1.56%
85	0.00%	0.00%	0.00%	0.04%	0.11%	0.22%
89	0.00%	0.00%	0.00%	0.00%	0.02%	0.01%
≥ 93	0.00%	0.00%	0.00%	0.00%	0.00%	0.00%

A.2.4 DSC and XRD results for standard PVB

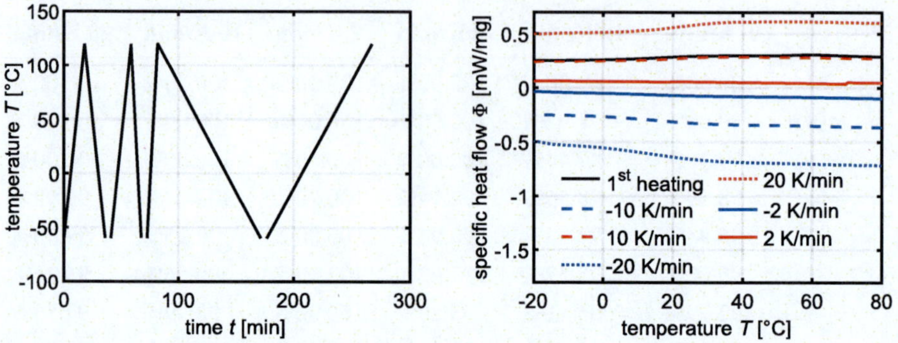

(a) (b)

Figure A.2 DSC test with standard PVB (a) temperature program, (b) results. The glass transition temperature at a cooling rate of -20 K/min is determined to $T_{g,cool}$ = 19.49 °C and at a heating rate of 20 K/min to $T_{g,heat}$ = 17.87 °C by the inflection point method. No crystallization or melting peaks can be observed.

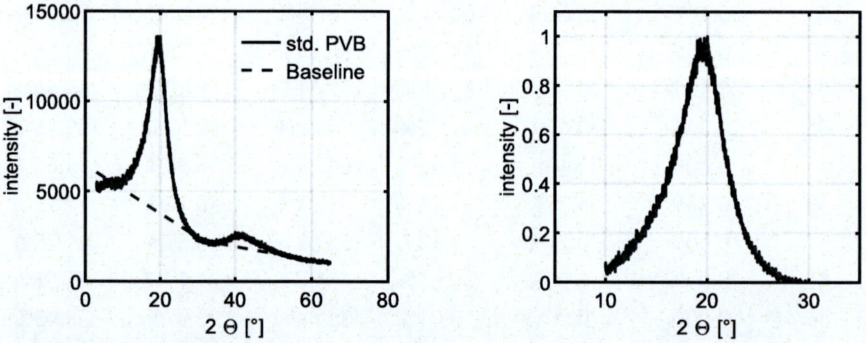

(a) (b)

Figure A.3 (a) XRD results for standard PVB, (b) normalized results. The tests were performed in transmission geometry. Machine: StadiP, (Fa. Stoe & Cie. GmbH, TU Darmstadt, institute of chemistry, inorganic chemistry) Ge[111]-Monochromator, CuKα1-radiation, λ = 1.54060 Å, detector: MYTHEN1K (Fa. Dectris, Baden, Switzerland). Only one amorphous halo can be seen in the region of 2Θ = 10-30°. No Bragg reflections can be detected.

A.3 Linear viscoelastic behaviour of SG and cEVA

A.3.1 Test settings

Table A.13 Test matrix: DMTA-TFS SG.

	SG	
test equipment	Anton Paar rheometer MCR 302, PP configuration	
sample preparation	punched with hole punch iron	
sample geometry [mm]	⌀ = 8	⌀ = 8
	0.89 thick	0.89 thick
contact normal force [N]	0.1 pressure	0.1 pressure
tested samples	1	2
	Time sweep	
temperature [°C]	-	23
frequency [Hz]	-	1
excitation amplitude [%]	-	0.025
duration [d]	-	5[1]
	Amplitude sweeps	
temperatures [°C]	-40, 100	-40, 100
frequencies [Hz]	0.1, 1, 10	0.1, 1, 10
excitation amplitude range [%]	[0.01:0.05]	[0.01:0.05]
	Temperature-frequency sweeps	
temperature range [°C]	heating: [-40:105], ΔT = 5	[-40:105], ΔT = 5
	cooling: [100:20], ΔT = -5	[100:20], ΔT = -5
	convection oven + nitrogen	convection oven + nitrogen
frequency range [Hz]	[0.1:10]	[0.1:10]
excitation amplitude [%]	0.025 in T = [-40:65]	0.025 in T = [-40:65]
	0.04 in T = [70:105]	0.04 in T = [70:105]

[1] 1 data point/hour.

Table A.14 Test matrix: DMTA-TFS cEVA.

	cEVA	cEVA thick
test equipment	Anton Paar rheometer MCR 302, SRF configuration	
sample preparation	cut with cutter knife	
approx. sample geometry [mm]	50 x 10	50 x 10
	0.8 thick	1.6 thick
contact normal force [N]	0.25 traction	0.25 traction
tested samples	1	1
	Amplitude sweeps	
temperatures [°C]	100, -50	100
frequencies [Hz]	0.1, 1, 10	0.1, 1, 10
excitation amplitude range [%]	[0.01:0.15]	-
	Temperature-frequency sweeps	
temperature range [°C]	[100:-50] with ΔT = -2.5	[100:22.5] with ΔT = -2.5
	convection oven and nitrogen	convection oven and nitrogen
frequency range [Hz]	[0.1:10]	[0.1:10]
excitation amplitude [%]	0.1	0.1

Table A.15 Test matrix cEVA: DMTA-FTS.

	cEVA
test equipment	Anton Paar MCR 302, SRF configuration
sample preparation	cut with cutter knife
approx. sample geometry [mm]	50 x 10 1.6 thick
contact normal force [N]	0.25 traction
tested samples	1
temperature range [°C]	[37.5:100] with 2[2] K/min convection oven
frequencies [Hz]	0.1, 0.121, 0.147, 0.261, 0.383, 0.562, 0.681, 0.825, 1, 1.21, 1.47, 1.78, 2.15, 2.61, 3.16, 3.83, 4.64, 5.62, 6.81, 8.25, 10
incitation amplitude [%]	0.1

[1] 1 data point/hour.

[2] the measured heating rate corresponds to 2.05 K/min, the measured cooling rate corresponds to -1.57 K/min.

A.3.2 Mastercurves, control plots and shift factors: SG

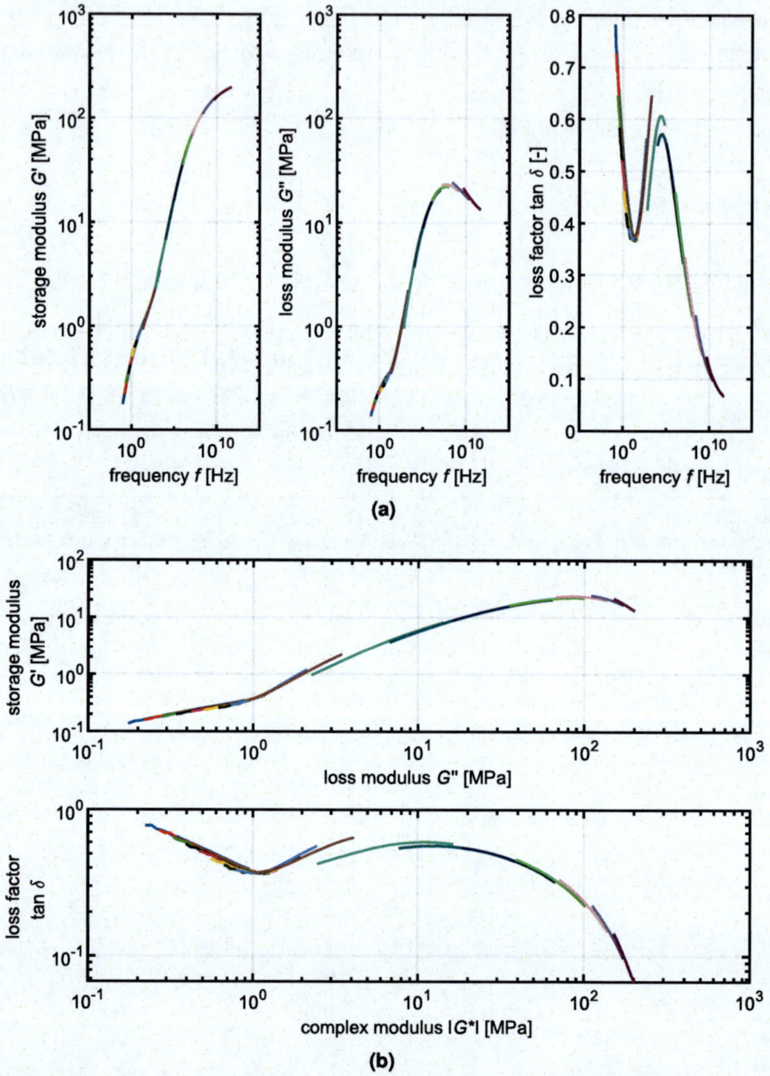

Figure A.4 SG TFS cooling data (a) mastercurves at T_{ref} = -95℃, (b) control plots: top: Cole-Cole bottom: Van-Gurp Palmen.

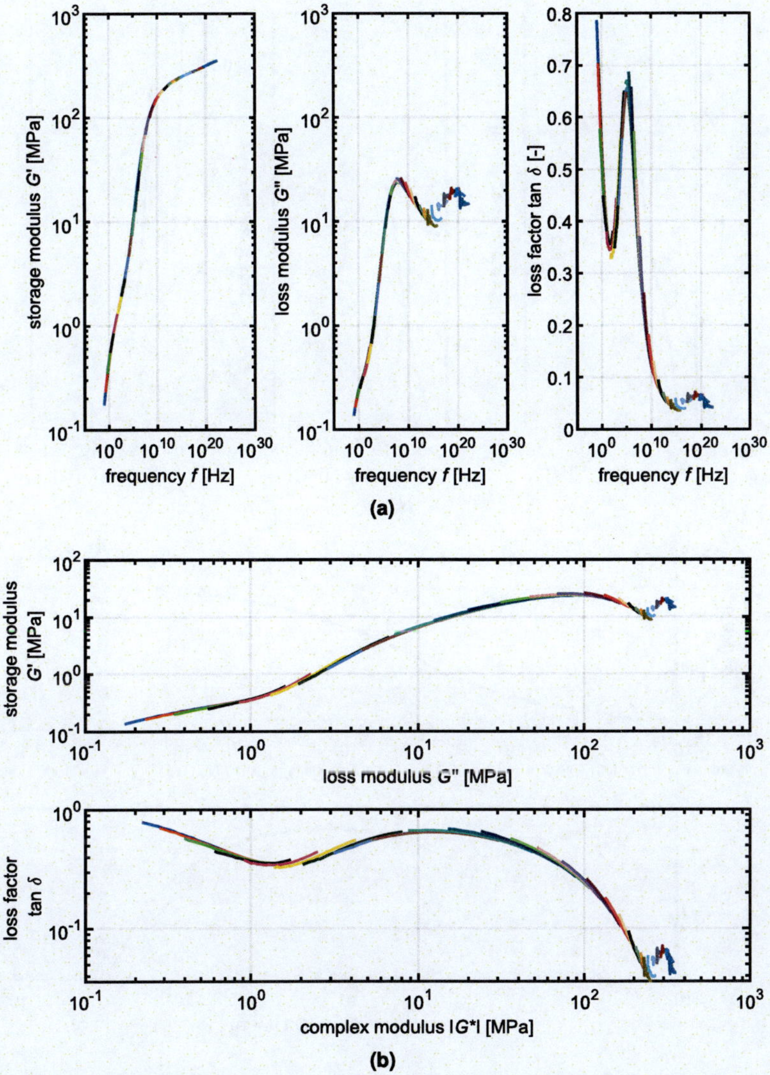

Figure A.5 SG TFS heating data (a) mastercurves at T_{ref} = -95°C, (b) control plots: top: Cole-Cole bottom: Van-Gurp Palmen.

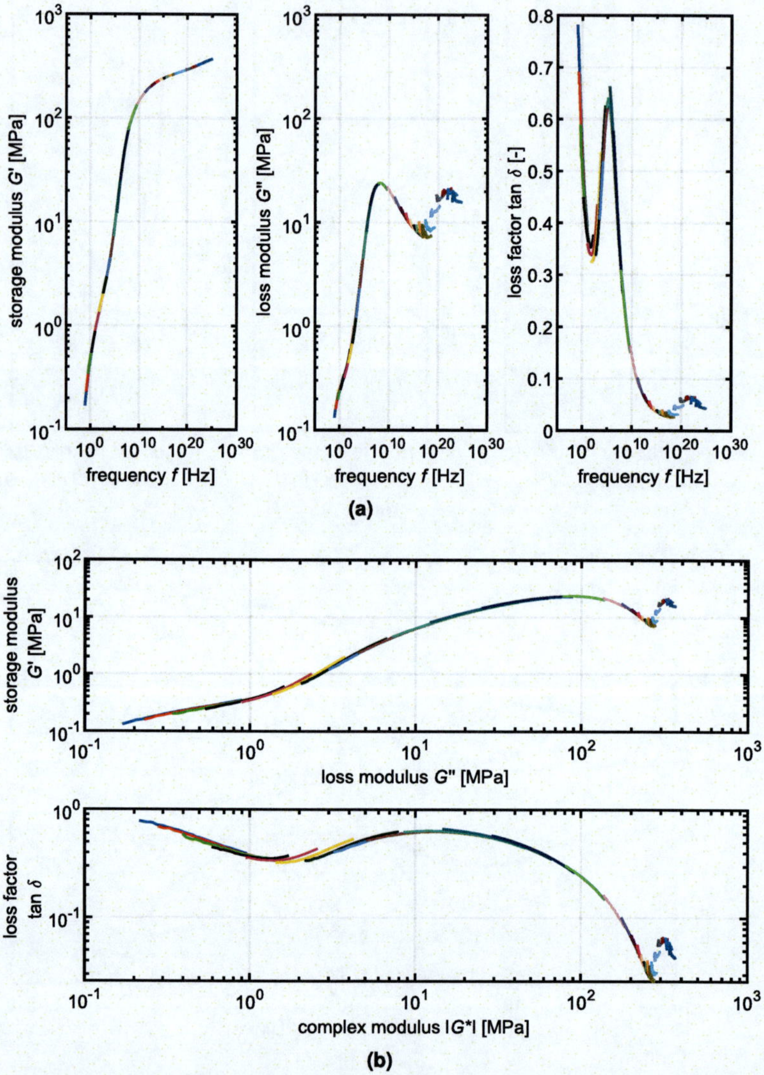

Figure A.6 SG TFS heating after 5d at T = 23°C (a) mastercurves at T_{ref} = -95°C, (b) control plots: top: Cole-Cole bottom: Van-Gurp Palmen.

Table A.16 Incrementally determined shift factors for SG: TFS cooling after 5d at 23 °C...

T [°C]	95.0	90.0	85.0	80.0	75.0	70.0	65.0	60.0	55.0
log a_T	0.00	0.19	0.38	0.60	0.82	1.07	1.35	1.68	2.30

... continuation.

50.0	44.9	39.9	34.9	29.9	24.9	19.9
3.81	4.97	5.99	7.04	8.18	9.40	10.68

Table A.17 Incrementally determined shift factors for SG: TFS heating ...

T [°C]	95.2	90.2	85.3	80.1	75.0	69.9	64.9	60.0	55.0
log a_T	0.00	0.26	0.65	1.23	1.96	2.78	3.59	4.37	5.15

... continuation...

50.0	45.0	40.0	35.0	30.0	25.0	20.0	15.1	10.0	5.0
5.83	6.46	6.99	7.51	8.06	8.68	9.48	10.40	11.38	12.36

... continuation.

0.0	-5.0	-10.0	-15.0	-20.0	-25.0	-30.0	-35.0	-40.0
13.28	14.10	14.87	15.80	16.96	18.18	19.39	20.55	21.41

Table A.18 Incrementally determined shift factors for SG: TFS heating after 5d at 23 °C...

T [°C]	95.0	90.0	85.0	80.0	74.9	70.0	65.0	60.0	55.0
log a_T	0.00	0.24	0.63	1.22	1.99	2.82	3.64	4.42	5.15

... continuation...

50.0	45.0	40.0	34.9	30.0	25.0	20.0	15.0	10.0	5.0
5.83	6.47	7.19	8.76	10.62	12.14	13.38	14.38	15.23	15.94

... continuation.

0.0	-5.0	-10.0	-15.0	-20.0	-25.0	-30.0	-35.0	-40.0
16.58	17.17	17.75	18.44	19.52	20.69	21.91	22.96	23.98

A.3.3 Mastercurves, control plots and shift factors: cEVA

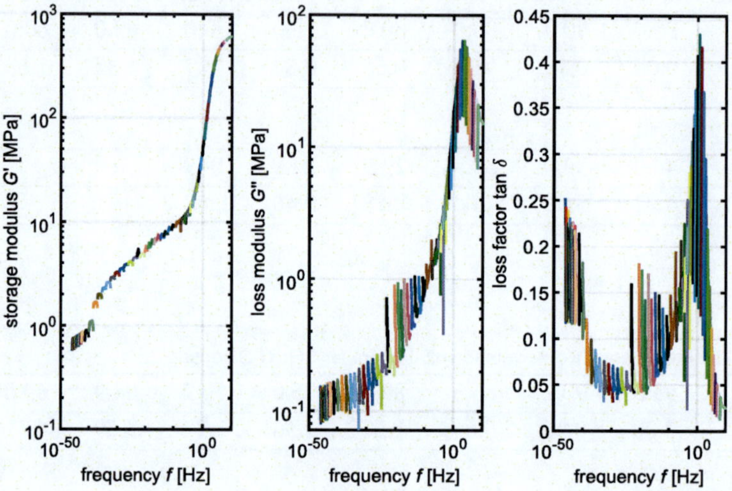

(a)

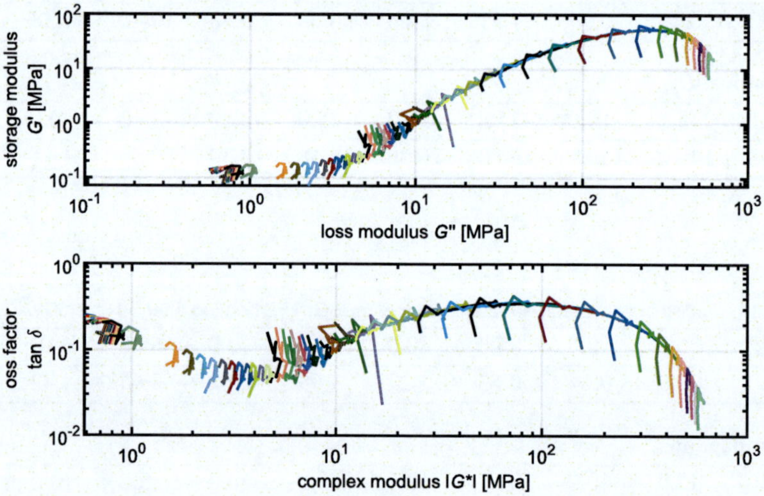

(b)

Figure A.7 cEVA TFS cooling (a) mastercurves at T_{ref} = -20℃, (b) control plots: top: Cole-Cole bottom: Van-Gurp Palmen.

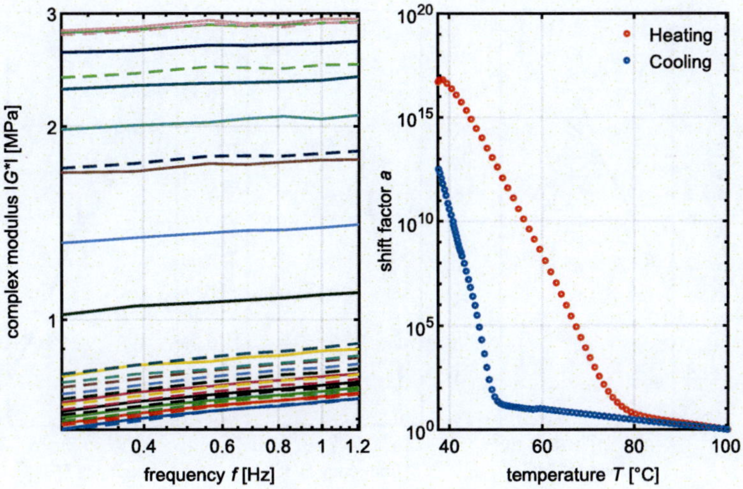

(a)

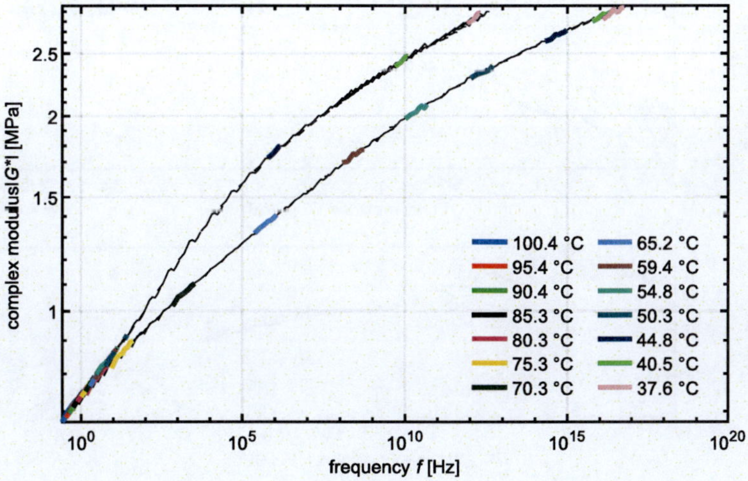

(b)

Figure A.8 cEVA: comparison heating and cooling FTS, (a) raw data and shift factors, (b) master-curves.

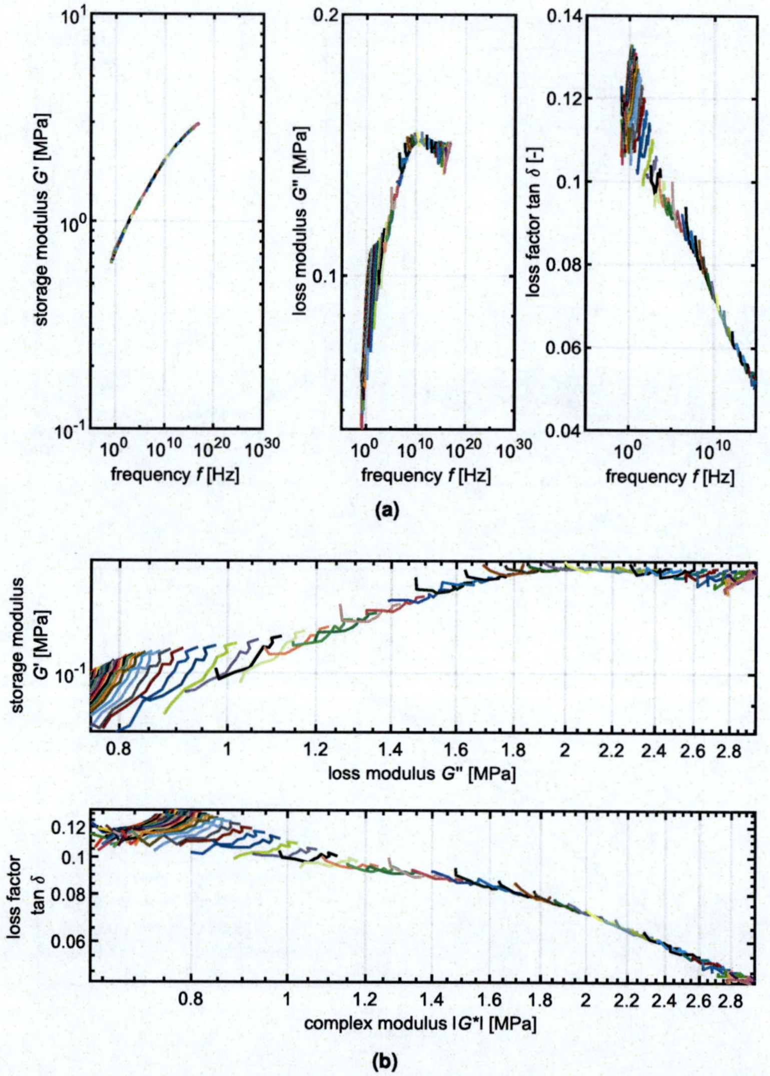

Figure A.9 cEVA FTS heating (a) mastercurves, (b) control plots: top: Cole-Cole bottom: Van-Gurp Palmen.

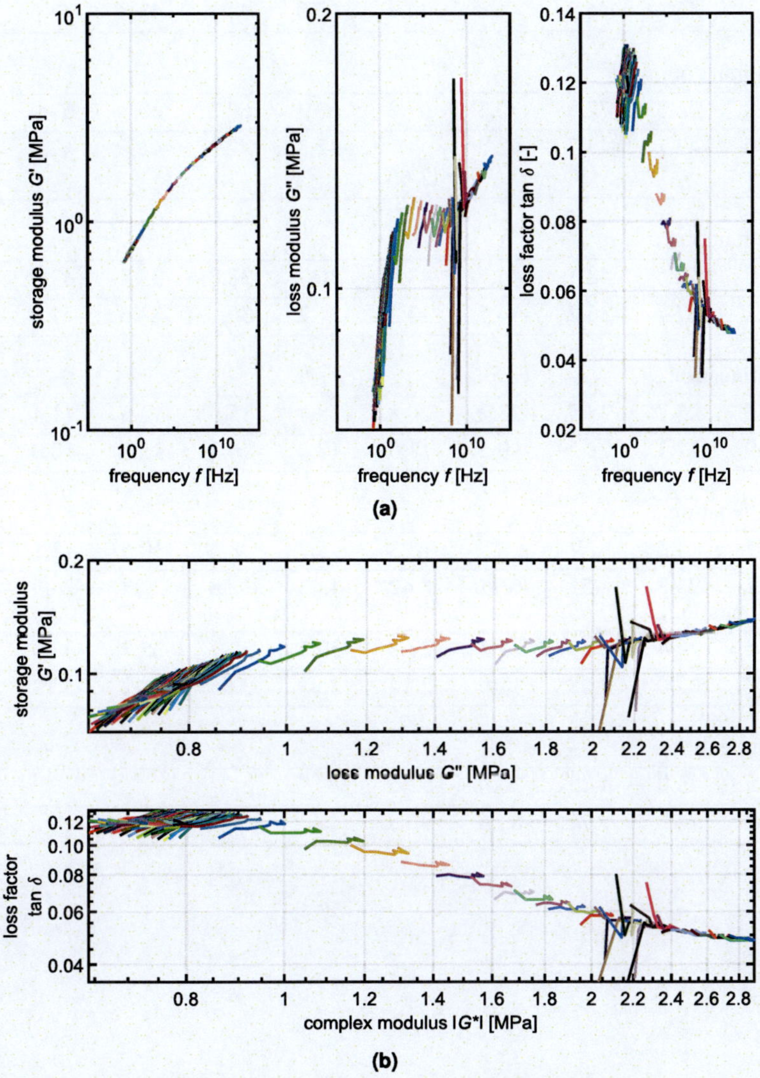

Figure A.10 cEVA FTS cooling (a) mastercurves, (b) control plots: top: Cole-Cole bottom: Van-Gurp Palmen.

Table A.19 Incrementally determined shift factors for cEVA: TFS...

T [°C]	100.0	97.5	95.1	92.6	90.1	87.7	85.2	82.7	80.2
log *a*$_T$	-45.2	-44.9	-44.6	-44.4	-44.0	-44.1	-43.8	-43.6	-43.3

... continuation...

77.8	75.3	72.8	70.3	67.8	65.3	62.8	60.3	57.9	55.5
-43.0	-43.1	-42.7	-42.8	-42.68	-42.3	-42.1	-41.6	-42.0	-40.6

... continuation...

53.0	50.6	48.1	45.7	43.2	40.7	38.3	35.8	33.3	30.2
-39.2	-37.7	-36.2	-34.7	-33.3	-31.7	-30.2	-28.6	-27.0	-25.5

... continuation...

27.6	25.7	25.5	23.2	20.8	18.4	15.9	13.5	11.4	8.9
-23.9	-22.3	-21.2	-19.5	-18.5	-17.4	-16.3	-15.2	-13.9	-12.5

... continuation...

6.5	4.2	1.7	-0.8	-3.3	-5.7	-8.2	-10.6	-13.1	-15.5
-11.2	-9.88	-8.55	-6.71	-5.42	-4.43	-3.58	-2.83	-2.09	-1.38

... continuation...

-18.0	-20.4	-22.9	-25.3	-27.8	-30.2	-32.7	-35.2	-37.6	-40.1
-0.70	0.00	0.68	1.40	2.14	2.89	3.65	4.42	5.19	5.93

... continuation.

-42.5	-45.0	-47.5	-50.0	-52.5
6.62	7.29	7.94	8.56	9.15

Table A.20 Incrementally determined shift factors for cEVA: FTS heating...

T [°C]	100.4	99.4	98.4	97.4	96.4	95.4	94.4	93.4	92.4
$\log a_T$	0.00	0.03	0.06	0.09	0.12	0.15	0.19	0.22	0.25

... continuation...

91.4	90.4	89.4	88.4	87.3	86.3	85.3	84.3	83.3	82.3
0.28	0.31	0.35	0.38	0.41	0.45	0.48	0.52	0.56	0.61

... continuation...

81.3	80.3	79.3	78.3	77.3	76.3	75.3	74.3	73.3	72.3
0.67	0.74	0.83	0.95	1.11	1.30	1.54	1.83	2.16	2.55

... continuation...

71.3	70.3	69.3	68.3	67.3	66.2	65.2	64.1	63	61.8
2.99	3.45	3.93	4.43	4.94	5.46	5.98	6.50	7.05	7.57

... continuation...

60.6	59.4	58.2	57.1	55.9	54.8	53.7	52.5	51.4	50.3
8.11	8.66	9.14	9.62	10.10	10.60	11.13	11.61	12.08	12.61

... continuation...

49.2	48.1	47	45.8	44.7	43.7	42.6	41.5	40.5	39.5
13.06	13.51	14.00	14.45	14.89	15.31	15.71	16.05	16.35	16.60

... continuation.

38.6	37.9	37.6
16.78	16.79	16.68

Table A.21 Incrementally determined shift factors for cEVA: FTS cooling...

T **[°C]**	100.3	99.5	98.5	97.3	96.1	95.0	93.9	92.9	91.9
log a_T	0.00	0.02	0.04	0.07	0.10	0.12	0.14	0.17	0.19

... continuation...

90.8	89.8	88.7	87.6	86.5	85.4	84.3	83.2	82.1	81.1
0.22	0.24	0.27	0.30	0.32	0.35	0.38	0.41	0.44	0.46

... continuation...

78.9	77.8	76.7	75.6	74.5	73.4	72.3	71.3	70.2	80.0
0.49	0.52	0.55	0.58	0.61	0.64	0.66	0.69	0.72	0.75

... continuation...

69.3	68.3	67.3	66.4	65.5	64.6	63.7	62.8	62.0	61.1
0.77	0.80	0.82	0.84	0.87	0.89	0.92	0.94	0.96	0.98

... continuation...

60.3	59.5	58.6	57.9	57.1	56.3	55.6	54.9	54.2	53.5
1.01	1.03	1.01	0.99	1.01	1.02	1.05	1.07	1.08	1.10

... continuation...

52.8	52.1	51.5	50.9	50.3	49.7	49.1	48.5	48.0	47.4
1.12	1.14	1.18	1.23	1.34	1.54	1.88	2.38	3.03	3.74

... continuation...

46.9	46.4	45.9	45.4	45.0	44.5	44.0	43.6	43.1	42.7
4.56	5.23	5.84	6.34	6.78	7.23	7.61	8.02	8.40	8.76

... continuation...

42.5	42.3	42.1	41.9	41.7	41.5	41.3	41.1	40.9	40.7
8.94	9.05	9.17	9.33	9.48	9.70	9.87	1 .00	10.18	10.34

... continuation.

40.4	40.0	39.6	39.3	38.9	38.6	38.3	38.0	37.7
10.66	10.94	11.22	11.50	11.79	12.08	12.37	12.64	12.86

Table A.22 Modified incrementally determined shift factors for cEVA: FTS cooling...

T **[°C]**	100.3	99.5	98.5	97.3	96.1	95.0	93.9	92.9	91.9
log a_T	0.02	0.04	0.06	0.09	0.12	0.14	0.17	0.20	0.23

... continuation...

90.8	89.8	88.7	87.6	86.5	85.4	84.3	83.2	82.1	81.1
0.25	0.29	0.32	0.35	0.38	0.42	0.45	0.48	0.52	0.55

... continuation...

78.9	77.8	76.7	75.6	74.5	73.4	72.3	71.3	70.2	80.0
0.58	0.62	0.66	0.69	0.72	0.76	0.79	0.83	0.87	0.90

... continuation...

69.3	68.3	67.3	66.4	65.5	64.6	63.7	62.8	62.0	61.1
0.93	0.96	1.00	1.03	1.05	1.08	1.11	1.14	1.17	1.20

... continuation...

60.3	59.5	58.6	57.9	57.1	56.3	55.6	54.9	54.2	53.5
1.23	1.26	1.28	1.31	1.33	1.24	1.28	1.32	1.34	1.38

... continuation...

52.8	52.1	51.5	50.9	50.3	49.7	49.1	48.5	48.0	47.4
1.40	1.43	1.48	1.55	1.69	1.97	2.46	3.21	4.17	5.21

... continuation...

46.9	46.4	45.9	45.4	45.0	44.5	44.0	43.6	43.1	42.7
6.22	7.11	7.83	8.42	8.93	9.40	9.85	10.28	10.70	11.12

... continuation...

42.5	42.3	42.1	41.9	41.7	41.5	41.3	41.1	40.9	40.7
11.33	11.53	11.63	11.85	12.05	12.25	12.46	12.66	12.87	13.07

... continuation.

40.4	40.0	39.6	39.3	38.9	38.6	38.3	38.0	37.7	
13.47	13.86	14.09	14.50	14.89	15.29	15.68	16.08	16.45	

A.3.4 Prony series

Table A.23 SG: Reduced Prony series for engineering application, TTSP and TCSP.

Time-Temperature-Superposition for -40°C $\leq T \leq$ 95°C

$$\begin{aligned}\log a_{\mathrm{T}} = & -4.2836\cdot 10^{-11}\cdot T^6 + 6.3631\cdot 10^{-9}\cdot T^5 \\ & -3.5316\cdot 10^{-8}\cdot T^4 - 1.7668\cdot 10^{-5}\cdot T^3 \\ & +6.2796\cdot 10^{-4}\cdot T^2 - 0.17929\cdot T \\ & +11.6502\end{aligned}$$

Time-Crystallinity-Superposition for 0% $\leq \chi \leq$ 8%

$$\log a_{\chi} = 0.3932\cdot \chi^{0.9599}$$

Prony series for T_{ref} = 95°C, χ_{ref} = 0%

G_0	90.52	MPa
i	τ_i [s]	g_i
1	1.00E+00	2.1803E-03
2	1.00E-01	4.1350E-03
3	1.00E-02	5.3377E-03
4	1.00E-03	7.4975E-03
5	1.00E-04	1.2556E-02
6	1.00E-05	2.9382E-02
7	1.00E-06	6.9151E-02
8	1.00E-07	2.6915E-01
9	1.00E-08	2.9989E-01
10	1.00E-09	2.9990E-01

A.3.5 Verification of Prony parameter fits

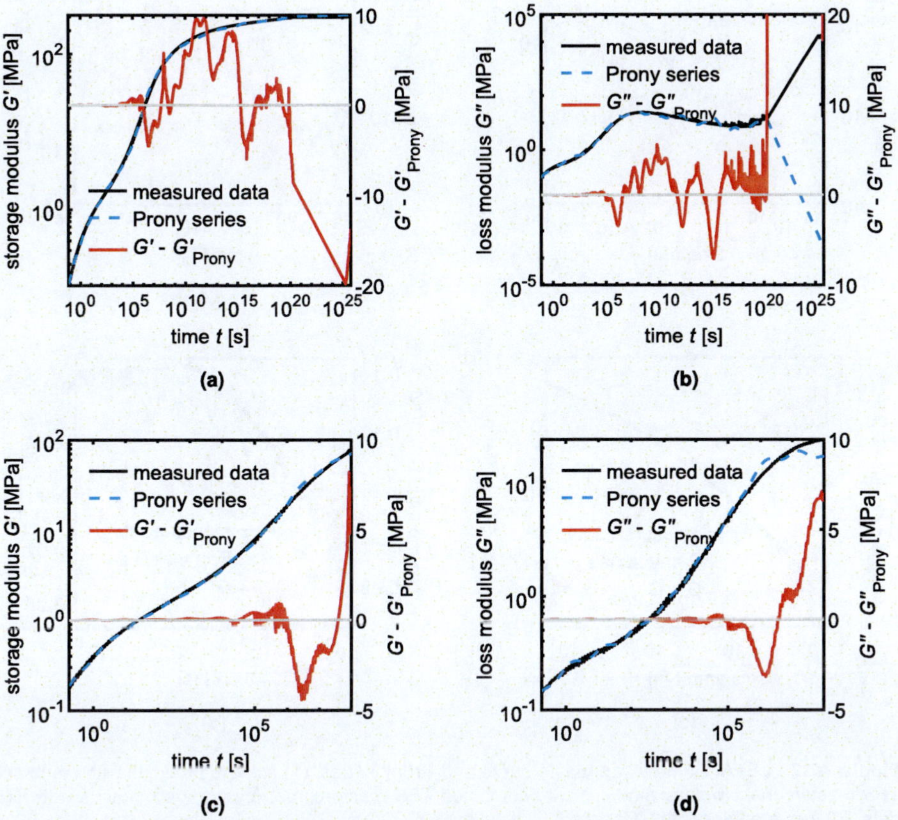

Figure A.11 SG: Comparison mastercurve and Prony series, (a) storage modulus, (b) loss modulus, (c) storage modulus with short Prony series, (d) loss modulus with short Prony series.

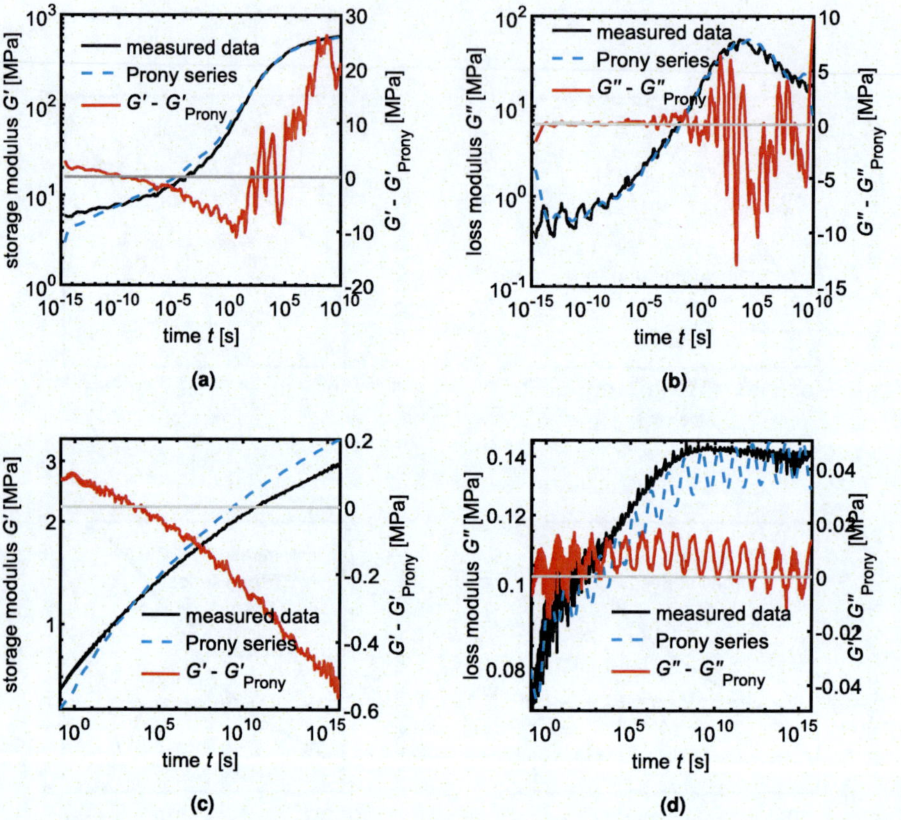

Figure A.12 cEVA: Comparison mastercurve and Prony series, (a) storage modulus with combined TFS mastercurve, (b) loss modulus with combined TFS mastercurve, (c) storage modulus with FTS data, (d) loss modulus with FTS data.

A.4 Linearity limits of PVB

A.4.1 DMTA-TFS test settings

Table A.24 Test matrix: DMTA-TFS.

	Standard	**Structural**	**Acoustic**
test equipment	Anton Paar rheometer MCR 302, tPP configuration		
sample geometry [mm]	⊘ =8, nominal thickness = 0.76		
sample preparation	punched with hole punch iron		
temperature program [°C]	[100:-40] cooling		
	5°C steps		
	convection oven and nitrogen		
frequency [Hz]	[0.1:10]		
excitation	0.1% in T=[100:40]°C		
	0.05% in T=[35:-40]°C		
contact normal force	0.1 N pressure		
tested samples	1	1	1

A.4.2 Mastercurves, control plots and shift factors for standard and structural PVB

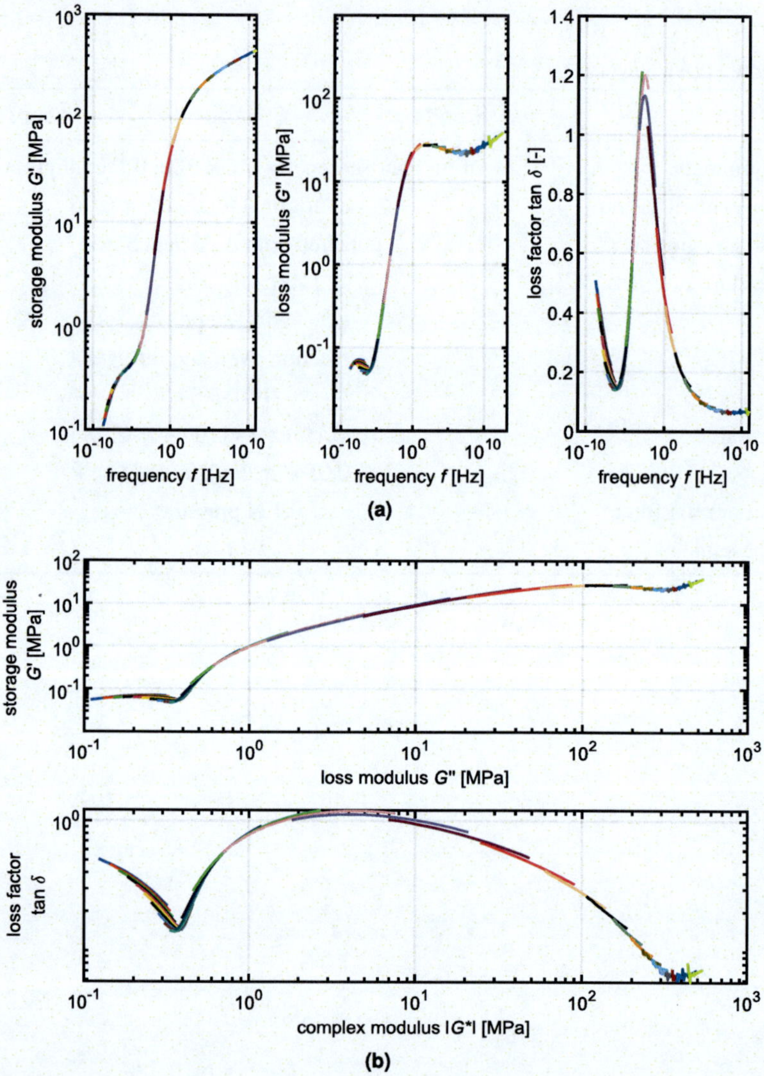

Figure A.13 Standard PVB (a) mastercurves at T_{ref} = 20 °C, (b) control plots: top: Cole-Cole bottom: Van-Gurp Palmen.

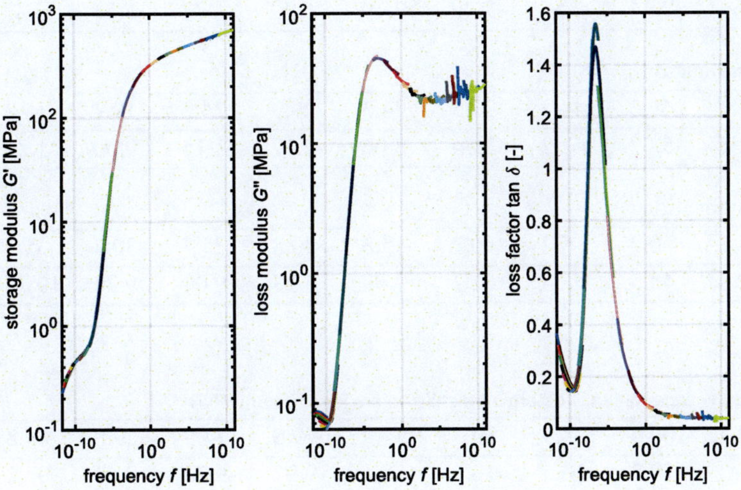

(a)

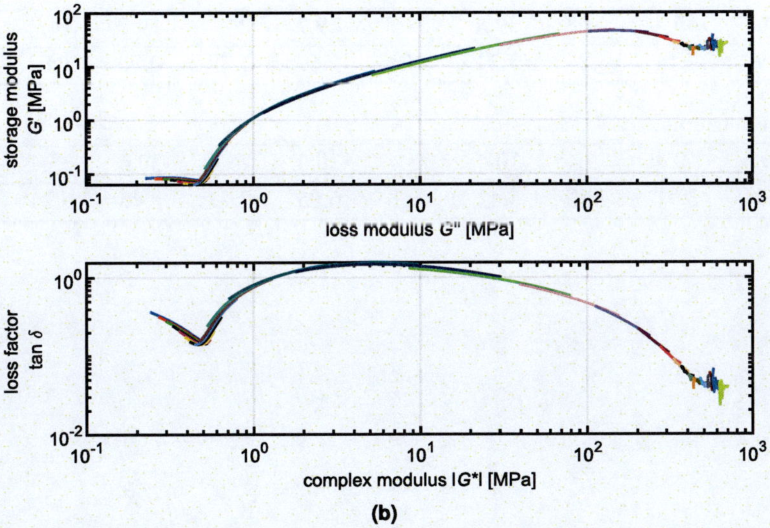

(b)

Figure A.14 Structural PVB (a) mastercurves at $T_{\mathrm{ref}} = 20\,^{\circ}\mathrm{C}$, (b) control plots: top: Cole-Cole bottom: Van-Gurp Palmen.

Table A.25 Incrementally determined shift factors for standard PVB...

T [°C]	99.9	94.9	89.9	84.7	79.7	74.5	69.6	64.5	59.5
log a_T	-7.74	-7.56	-7.40	-7.19	-6.99	-6.77	-6.54	-6.27	-5.95

... continuation...

54.3	49.2	44.3	39.3	34.3	29.3	24.2	19.2	14.2	9.3
-5.55	-5.07	-4.52	-3.84	-3.06	-2.17	-1.14	0.00	1.21	2.42

... continuation.

4.3	-0.7	-5.7	-10.7	-15.8	-20.8	-25.7	-30.7	-35.7	-40.7
3.58	4.65	5.64	6.57	7.43	8.28	9.09	9.88	10.73	11.59

Table A.26 Incrementally determined shift factors for structural PVB...

T [°C]	100.0	94.9	89.9	84.9	79.8	74.9	69.8	64.8	59.8
log a_T	-10.7	-10.5	-10.3	-10.2	-9.94	-9.70	-9.40	-9.02	-8.52

... continuation...

54.8	49.7	44.8	39.8	34.7	29.7	24.7	19.7	14.7	9.7
-7.90	-7.14	-6.26	-5.24	-4.04	-2.69	-1.29	0.00	1.14	2.14

... continuation.

4.6	-0.4	-5.4	-10.4	-15.4	-20.4	-25.4	-30.4	-35.4	-41.0
3.04	3.91	4.70	5.42	6.18	6.95	7.72	8.45	9.24	10.07

A.4.3 Prony series for standard and structural PVB

Table A.27 Reduced Prony series for engineering application for standard PVB at $T_{\mathrm{ref}} = 20\,°\mathrm{C}$.

Time-Temperature-Superposition: WLF		
C_1	33.20	
C_2	212.42	°C
Prony series at $T_{\mathrm{ref}} = 20\,°\mathrm{C}$		
G_0	97.37	MPa
i	τ_i [s]	g_i
1	1.00E+07	2.7681E-03
2	1.00E+06	8.2590E-04
3	1.00E+05	6.0728E-04
4	1.00E+04	1.2505E-03
5	1.00E+03	4.3711E-03
6	1.00E+02	2.0689E-02
7	1.00E+01	9.1238E-02
8	1.00E+00	2.7895E-01
9	1.00E-01	2.9973E-01
10	1.00E-02	2.9949E-01

Table A.28 Reduced Prony series for engineering application for structural PVB at T_{ref} = 20 °C.

Time-Temperature-Superposition: WLF		
C_1	196.91	
C_2	1139.40	°C

Prony series at T_{ref} = 20°C		
G_0	322.68	MPa
i	τ_i [s]	g_i
1	1.00E+09	1.5659E-03
2	1.00E+08	2.5577E-04
3	1.00E+07	1.1927E-03
4	1.00E+06	6.3279E-03
5	1.00E+05	4.1877E-02
6	1.00E+04	1.3957E-01
7	1.00E+03	2.0500E-01
8	1.00E+02	2.1960E-01
9	1.00E+01	1.6199E-01
10	1.00E+00	2.2261E-01

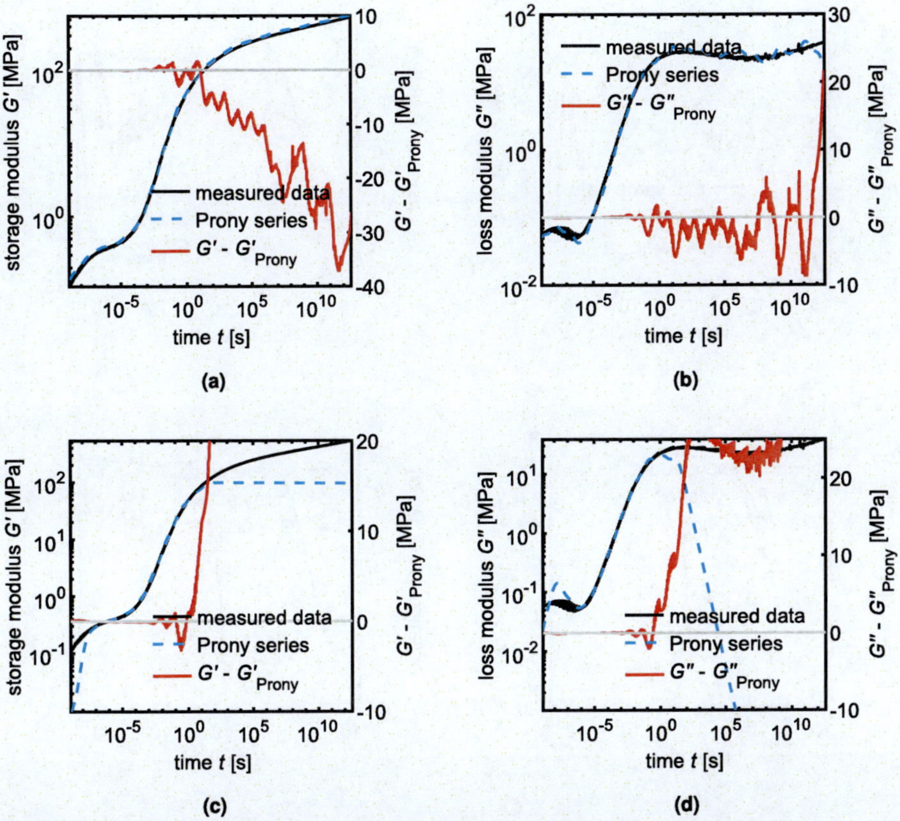

Figure A.15 Standard PVB: Comparison mastercurve and Prony series, (a) storage modulus, (b) loss modulus, (c) storage modulus with short Prony series, (d) loss modulus with short Prony series.

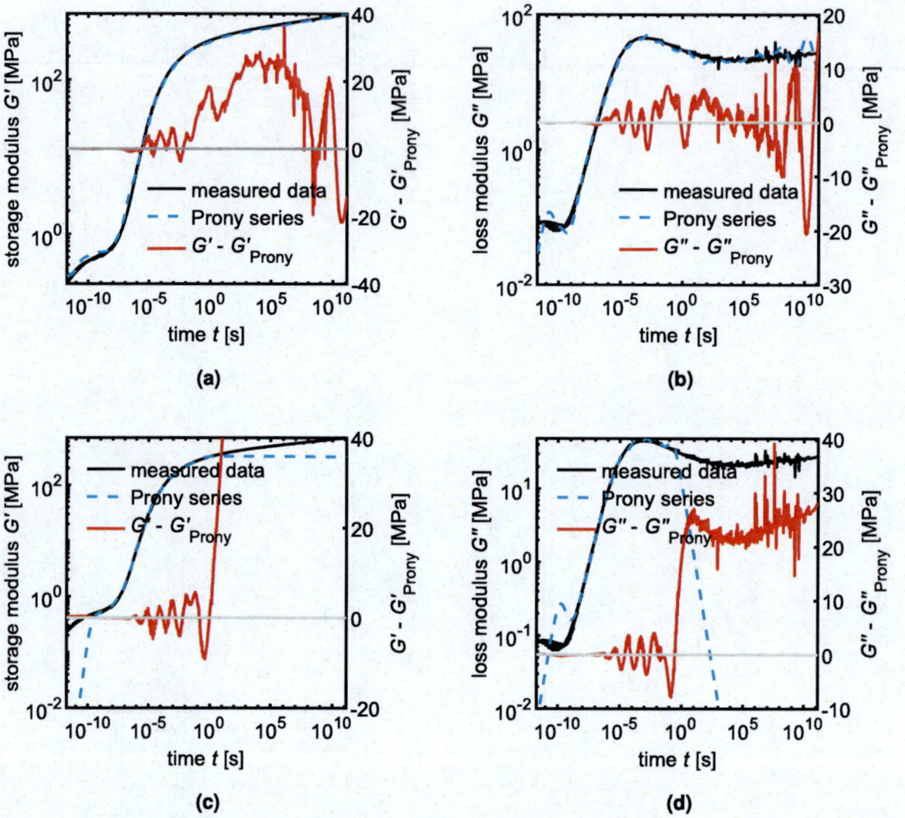

Figure A.16 Structural PVB: Comparison mastercurve and Prony series, (a) storage modulus, (b) loss modulus, (c) storage modulus with short Prony series, (d) loss modulus with short Prony series.

A.4.4 DMTA-AS test settings

Table A.29 Test matrix: DMTA-AS.

	Standard			**Structural**				**Acoustic**	
equipment	Anton Paar rheometer MCR 302, tPP configuration								
sample geometry [mm]	⊘ =8, nominal thickness = 0.76								
contact force [N]	0.1 pressure								
excitation [%]	[0.1:35] dependent on *T*								
tested samples		***f* [Hz]**			***f* [Hz]**				***f*[Hz]**
at temperature	***T* [°C]**	**0.1**	**1**	***T* [°C]**	**0.1**	**1**	**10**	***T* [°C]**	**1**
T [°C]	**0**	-	1	**30**	3	3	3	**20**	1
and frequency	**8**	-	1	**35**	-	3	-	**30**	1
f [Hz]	**10**	-	1	**40**	2	3	3	**40**	1
	15	-	1	**42.5**	-	3	-	**50**	1
	17.5	-	1	**45**	-	3	-		
	20	1	2	**47.5**	-	3	-		
	22.5	-	1	**50**	2	3	3		
	25	-	1	**55**	-	3	-		
	27.5	-	1	**60**	2	2	3		
	30	1	2	**70**	-	2	-		
	32.5	-	2						
	35	-	1						
	37.5	-	1						
	40	-	1						
	45	-	1						

A.4.5 Linearity limits

Table A.30 Linearity limits of standard PVB (mean values) and linear viscoelastic storage modulus in dependency of temperature and frequency.

	f	= 0.1 Hz	f	= 1 Hz
T [°C]	γ_{lin} [%]	G'_{LVE} [MPa]	γ_{lin} [%]	G'_{LVE} [MPa]
8	-	-	0.93	143
10	-	-	0.83	126
15	-	-	0.632	75.8
17.5	-	-	0.833	49.3
20	1.69	12.3	0.905	32.1
22.5	-	-	1.17	18.1
25	-	-	1.47	9.58
27.5	-	-	2.31	4.86
30	15.6	0.965	3.85	2.58
32.5	-	-	8.76	1.51
35	-	-	15	0.989
37.5	-	-	22.8	0.74
40	-	-	30.8	0.608
45	-	-	33	0.486

Table A.31 Linearity limits of structural PVB (mean values) and linear viscoelastic storage modulus in dependency of temperature and frequency.

	f	= 0.1 Hz	f	= 1 Hz	f	= 10 Hz
T [°C]	γ_{lin} [%]	G'_{LVE} [MPa]	γ_{lin} [%]	G'_{LVE} [MPa]	γ_{lin} [%]	G'_{LVE} [MPa]
30	0.609	90.871	0.546	156.33	0.464	225.99
35	-	-	0.601	59.860	-	-
40	3.880	26.852	1.312	12.946	0.667	56.887
42.5	-	-	2.599	55.170	-	-
45	-	-	4.566	28.040	-	-
47.5	-	-	8.528	14.770	-	-
50	24.35	0.6009	12.7	10.238	2.699	38.863
55	-	-	19.98	0.6925	-	-
60	25.05	0.4819	22.48	0.5737	14.6	0.8824
65	-	-	24.08	0.5252	-	-
70	-	-	22.15	0.5062	-	-

A.4.6 Results for acoustic PVB

Glass transition and mastercurves

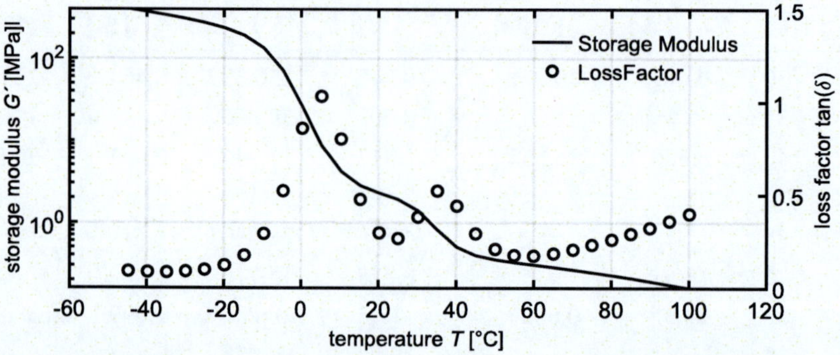

Figure A.17 Temperature dependent material behaviour of acoustic PVB at f = 1 Hz. Two glass transition temperatures appear. One at ≈5.37°C and one at ≈ 35.19°C (values determined with peak values of loss factor).

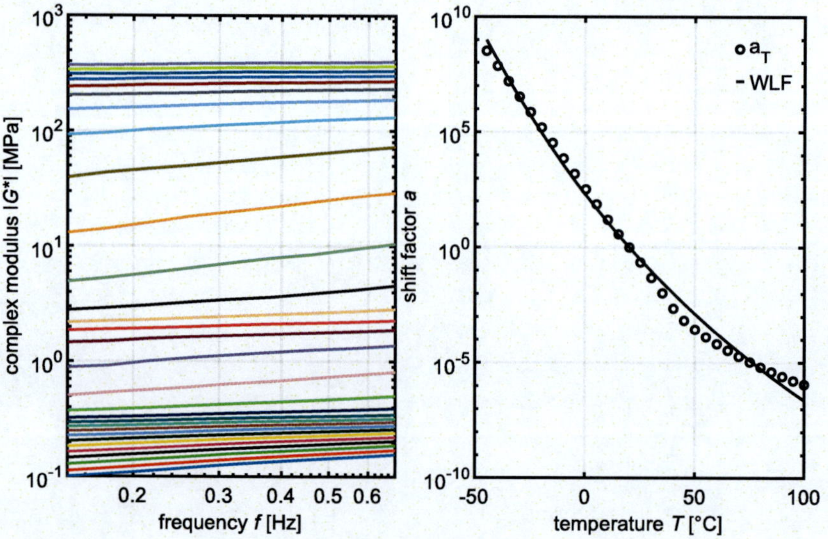

Figure A.18 Raw data and shift factors for mastercurving process.

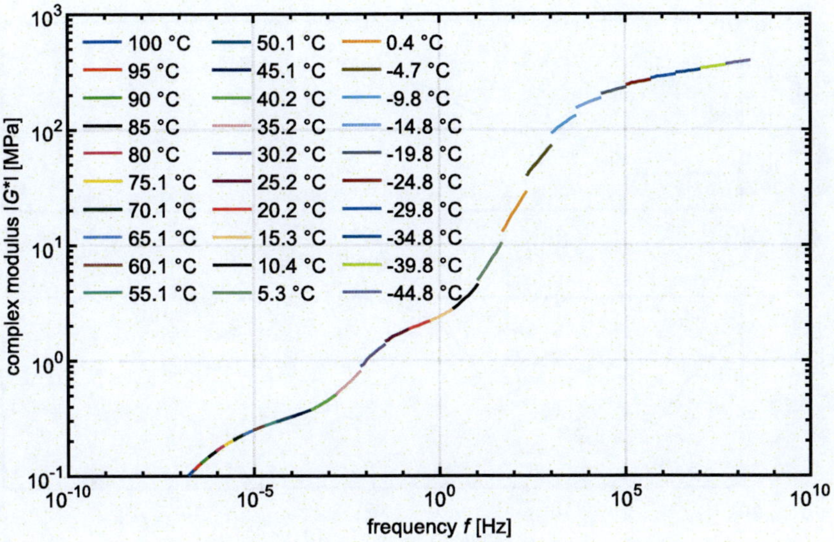

Figure A.19 Mastercurve of the amount of the complex modulus at T_{ref} = 20 °C.

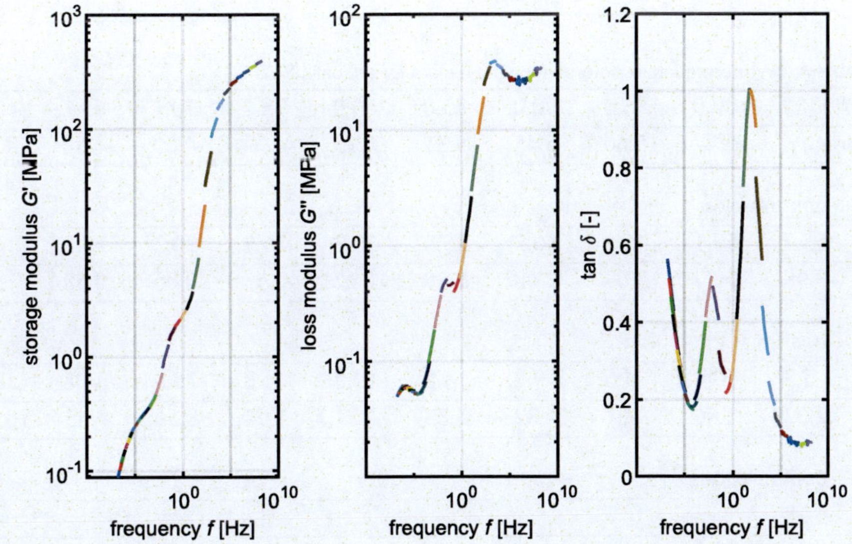

Figure A.20 Mastercurve of the storage modulus, loss modulus and loss factor at T_{ref} = 20 °C.

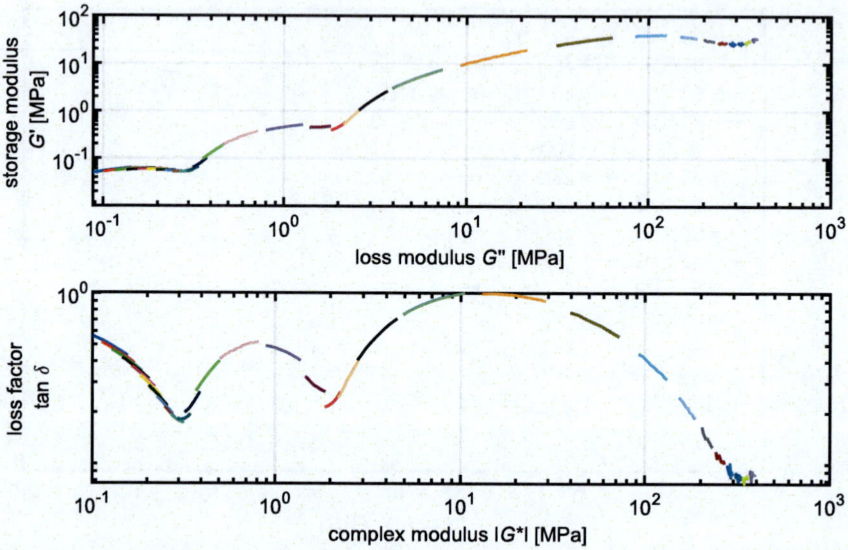

Figure A.21 Control plots: top: Cole-Cole bottom: Van-Gurp Palmen.

Table A.32 Incrementally determined shift factors for acoustic PVB...

T **[°C]**	100.0	95.0	90.0	85.0	80.0	75.1	70.1	65.1	60.1
log a_T	-5.77	-5.60	-5.45	-5.31	-5.15	-4.99	-4.80	-4.56	-4.30

... continuation...

55.1	50.1	45.1	40.2	35.2	30.2	25.2	20.3	15.3	10.3
-3.98	-3.57	-3.09	-2.48	-1.80	-1.12	-0.53	0.00	0.58	1.25

... continuation...

5.3	0.4	-4.6	-9.7	-14.8	-19.8	-24.8	-29.8	-34.8	-39.8
2.02	2.86	3.81	4.82	5.83	6.81	7.74	8.59	9.45	10.31

... continuation.

-44.8
11.11

Prony series

Table A.33 Reduced Prony series for engineering application for acoustic PVB at T_{ref} = 20 °C.

Time-Temperature-Superposition: WLF		
C_1	21.04	
C_2	177.57	°C
Prony series at T_{ref} = 20°C		
G_0	18.18	MPa
i	τ_i [s]	g_i
1	1.00E+05	1.3602E-02
2	1.00E+04	3.1365E-03
3	1.00E+03	4.3629E-03
4	1.00E+02	7.4149E-03
5	1.00E+01	3.9559E-02
6	1.00E+00	4.5144E-02
7	1.00E-01	2.4517E-02
8	1.00E-02	2.8179E-01
9	1.00E-03	2.9300E-01
10	1.00E-04	2.8730E-01

Table A.34 Prony series for acoustic PVB at T_{ref}=20 °C.

Time-Temperature-Superposition: WLF		
C_1	21.04	
C_2	177.57	°C
Prony series at T_{ref} = 20°C		
G_0	437.86	MPa
i	τ_i [s]	g_i
1	1.00E+05	4.9040E+00
2	1.00E+04	1.5856E-01
3	1.00E+03	3.3528E+00
4	1.00E+02	1.3042E+00
5	1.00E+01	1.8315E+01
6	1.00E+00	2.1007E+01
7	1.00E-01	1.6956E+01
8	1.00E-02	8.8642E+01
9	1.00E-03	3.2173E+02
10	1.00E-04	7.1321E+02
11	1.00E-05	1.6076E+03
12	1.00E-06	8.1004E+02
13	1.00E-07	1.2587E+03
14	1.00E-08	1.7323E+02
15	1.00E-09	1.8353E+03
16	1.00E-10	1.5916E+02
17	1.00E-11	5.4258E+02
18	1.00E-12	2.4105E+03
19	1.00E-13	6.0986E+00
20	1.00E-14	5.2953E+00

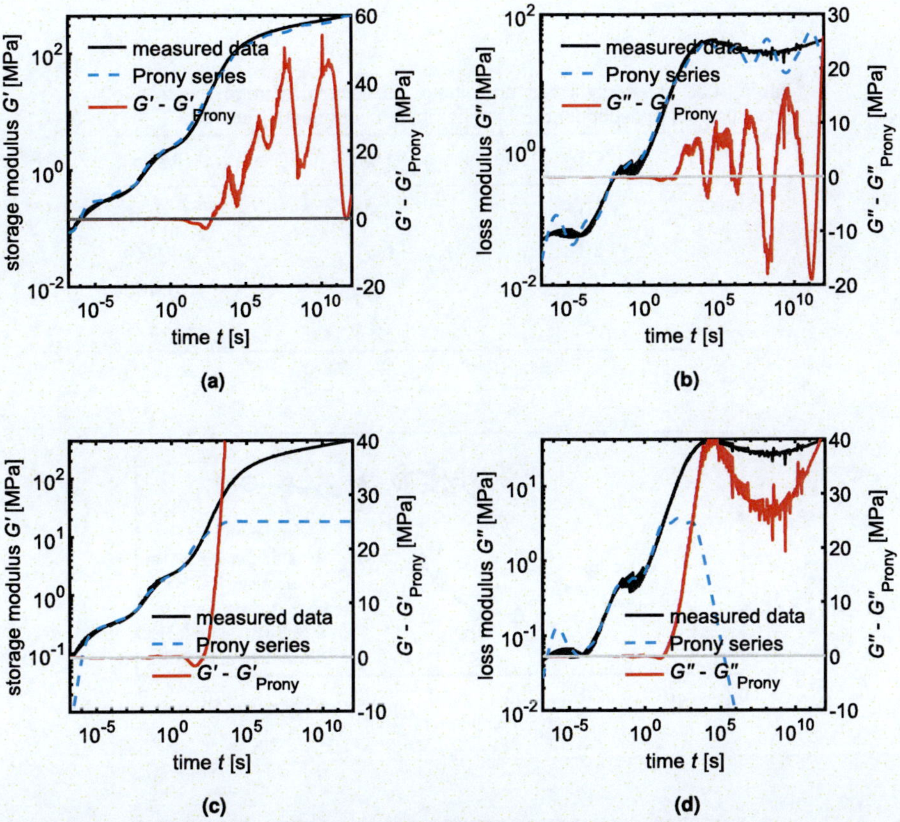

Figure A.22 Acoustic PVB: Comparison mastercurve and Prony series, (a) storage modulus, (b) loss modulus, (c) storage modulus with short Prony series, (d) loss modulus with short Prony series.

Linearity limits

Table A.35 Linearity limits and linear viscoelastic storage modulus of acoustic PVB in dependency of temperature at a frequency of 1Hz.

T [°C]	γ_{lin} [%]	G'_{LVE} [MPa]
20	4.46	2.36
30	6.96	1.24
40	21.9	0.488
45	24.6	0.391

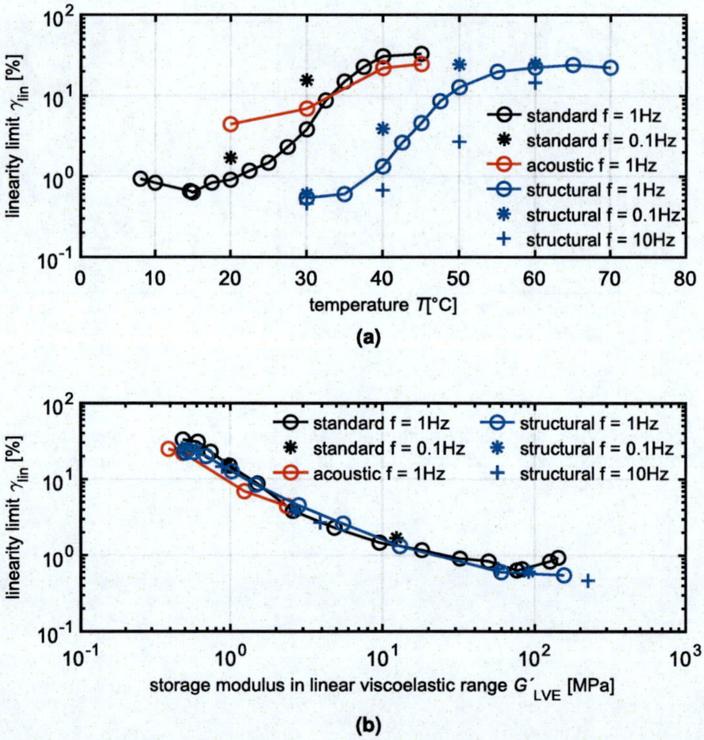

Figure A.23 Comparison of linearity limits from acoustic with standard and structural PVB (a) in dependency of temperature and frequency, (b) in dependency of linear viscoelastic storage modulus. The linearity limit can be expressed independently of the exact interlayer type, solely as a function of the storage modulus.

A.5 Nonlinear viscoelastic behaviour of PVB

A.5.1 Uniaxial tensile relaxation test settings

Table A.36 Test matrix: Uniaxial tensile relaxation tests at large strains.

		Standard			**Structural**			**Acoustic**		
sample		Becker tension rod								
geometry [mm]		0.76 nominal thickness								
		punched								
test equipment		universal testing machine: Zwick Roell Z050								
		climate chamber: Zwick Roell GmbH & Co. KG								
		digital image correlation: Vic-Gauge-2D, isi-sys GmbH								
test mode		tensile relaxation tests, hydraulic clamping jaws								
humidity [% rH]		50			50			50		
temperatures T [°C]		***T* [°C]**			***T* [°C]**			***T* [°C]**		
and displacement		**0**	**20**	**40**	**20**	**40**	**60**	**0**	**20**	**40**
levels u [mm]...	u [mm]	25	50	50	25	50	50	25	50	50
... for a test duration		50	100	100	50	100	100	50	100	100
of 1-2h		75	125	150	75	125	150	75	125	150
...for a test duration of 12h	u [mm]	75	125	150	-	-	-	-	-	-
displacement rate to reach initial disp. [mm/min]		400								

Table A.37 Number of tested samples in large strain relaxation tests.

	u	Standard 50% rH			Structural 50% rH			Acoustic 50% rH		
	[mm]	0°C	20°C	40°C	20°C	40°C	60°C	0°C	20°C	40°C
1-2 h	25	2	-	-	2	-	-	1	-	-
	50	4	3	3	2	2	3	1	1	1
	75	2	-	-	2	-	-	1	-	-
	100	-	2	3	-	2	2	-	1	1
	125	-	2	-	-	2	-	-	1	-
	150	-	-	2	-	-	3	-	-	1
12 h	75	1	-	-	-	-	-	-	-	-
	125	-	1	-	-	-	-	-	-	-
	150	-	-	1	-	-	-	-	-	-

Note: If more than 2 tests were made, the two closest to each other were further evaluated.

A.5.2 Stress, strain and Poisson's ratio for standard and structural PVB

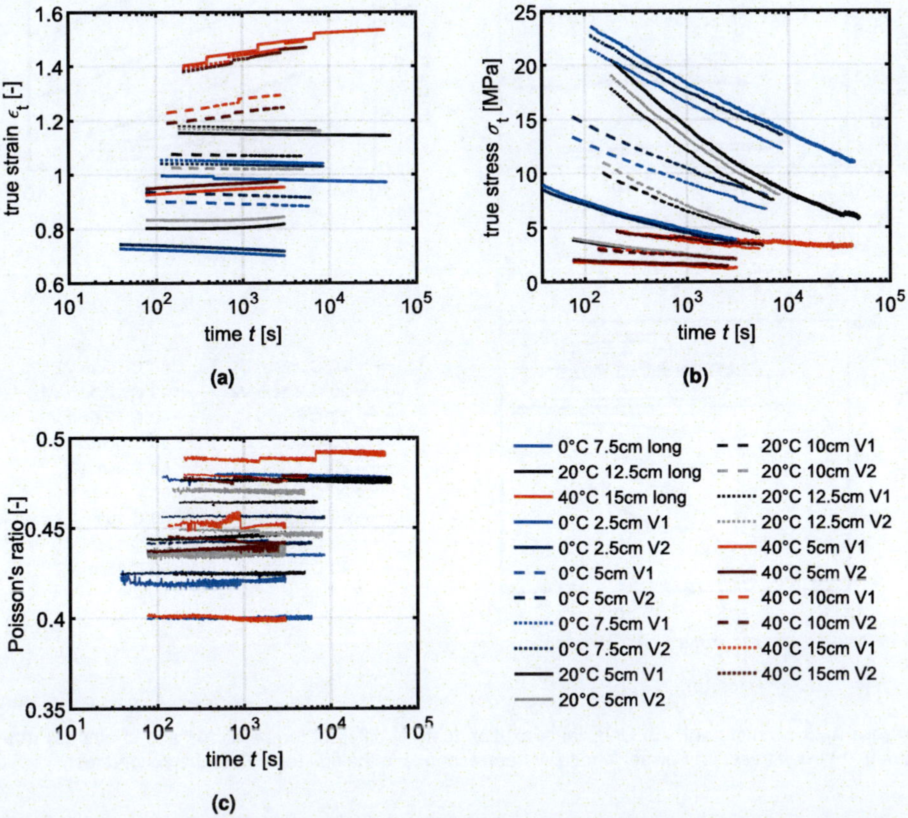

Figure A.24 Standard PVB: Uniaxial relaxation tests at different temperatures and strains (a) true strain, (b) true stress, (c) Poisson's ratio. V1 corresponds to the first test, V2 to the second test, *long* corresponds to a test duration of $\approx$ 12 h.

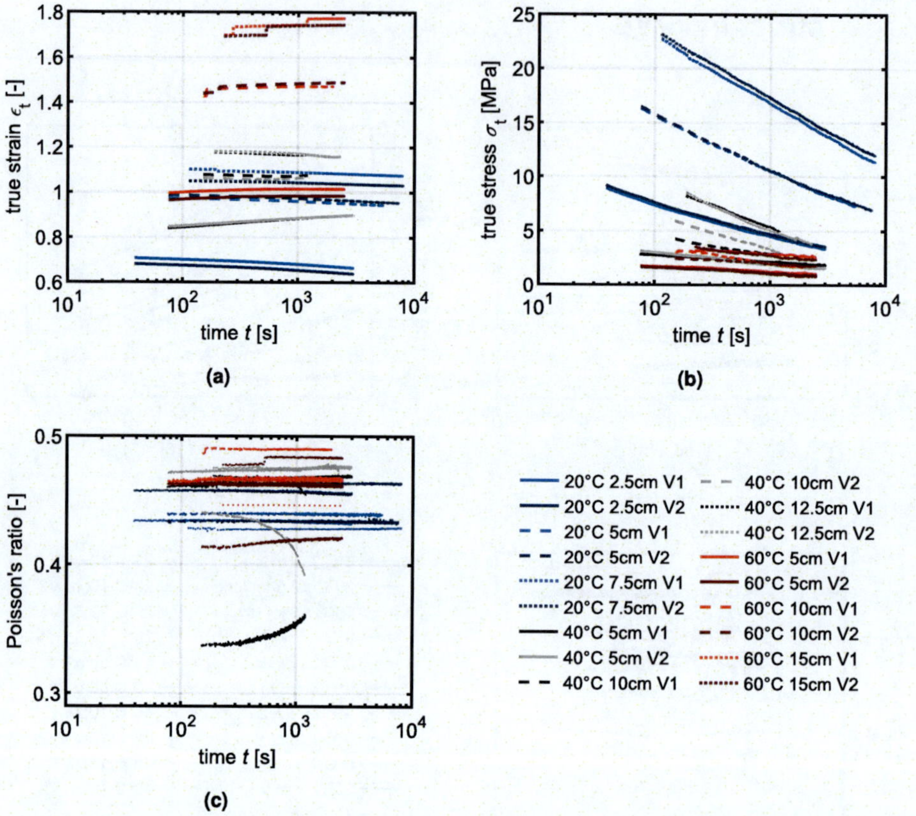

Figure A.25 Structural PVB: Uniaxial relaxation tests at different temperatures and strains (a) true strain, (b) true stress, (c) Poisson's ratio. V1 corresponds to the first test, V2 to the second test.

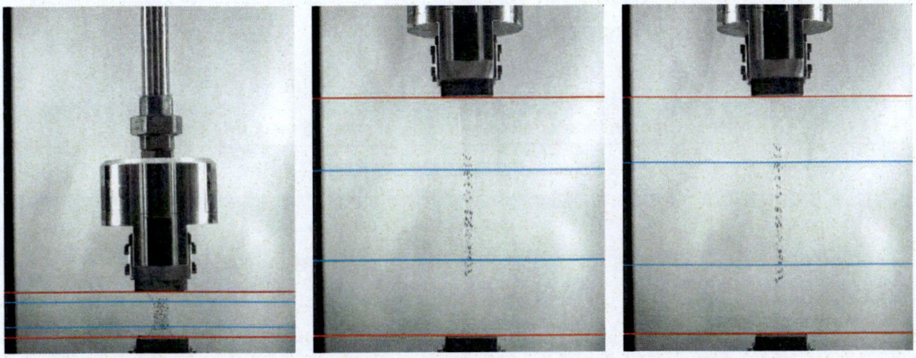

Figure A.26 Standard PVB T = 40 °C, elongation 15 cm, (a) before start of the test, (b) at the begin of the relaxation test, (c) at the end of the relaxation test. The clamp position is marked with red lines. The blue lines indicate the upper resp. lower limit of the vertical strain gauges.

A.5.3 Degree of nonlinearity at T = 0 °C

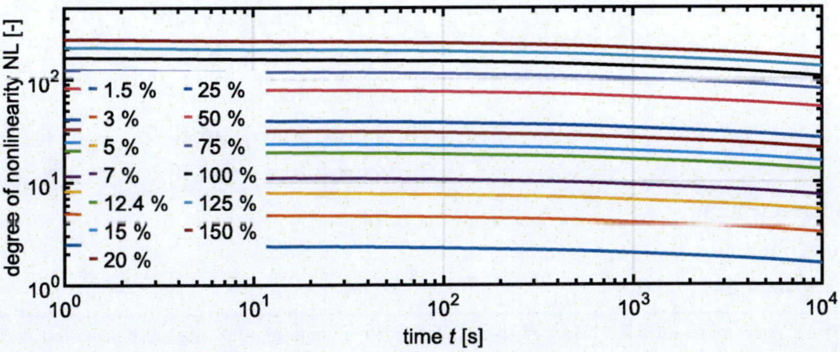

Figure A.27 Degree on nonlinearity NL for standard PVB at 0 °C.

A.5.4 Preliminary uniaxial relaxation tests at small to medium strain levels and 0°C

Table A.38 Test matrix: Uniaxial tensile relaxation tests at small to medium strains

	Standard
sample	Rectangular strip, 12 cm x 1.5 cm
geometry	0.76 mm nominal thickness
test equipment	universal testing machine: Zwick Roell Z050
	climate chamber: Zwick Roell GmbH & Co. KG
	digital image correlation: Vic-Gauge-2D, isi-sys GmbH
test mode	tensile relaxation tests, Mechanical clamping
humidity [% rH]	50
temperature T [°C]	0
test duration t [h]	1-2
displacement u [mm]	1, approx. true strain 2.5%
	2[1], approx. true strain 5%
	4, approx. true strain 10%
	6, approx. true strain 15%
	8, approx. true strain 15%
	10, approx. true strain 20%
	20, approx. true strain 50%
displacement rate to reach initial disp. [mm/min]	400

[1] Two samples were testes for this strain level. For all other strain levels, only one sample was tested.

(a) (b) (c)

Figure A.28 Preliminary uniaxial relaxation tests at small to medium strain levels and at T = 0 ℃, (a) sample geometry with gauges. The upper and lower 4 cm represent the clamping areas. The speckled area represents the tested length. The holes are necessary to fix the clamping device, (b) test set-up with open climate chamber, (c) build-in sample before test start. The white cardboard roll is used to create a uniform background for the DIC.

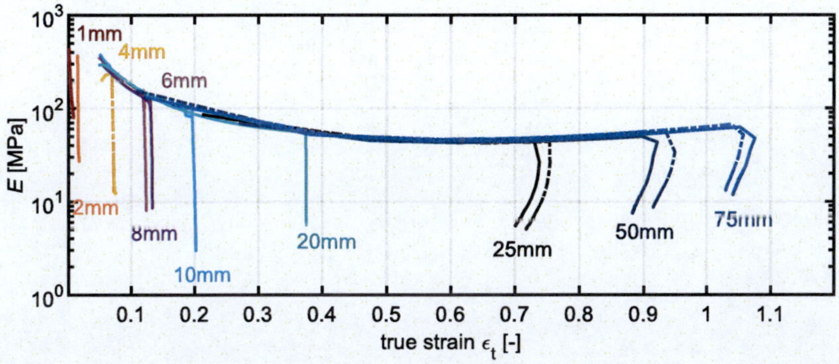

Figure A.29 Comparison of strain application. E is calculated with the true stress and strain data.

Table A.39 Horizontal shift factors - purely horizontal shift.

u **[mm]**	1	2	4	6	8	10	20	25	50	75
log a_ε	0	1.18	2.10	2.76	2.82	3.68	3.82	3.55	2.53	1.23
-	-	-	-	-	-	-	-	3.74	2.22	0.91
-	-	-	-	-	-	-	-	-	-	0.61

Table A.40 Manually determined horizontal and vertical shift factors - horizontal and vertical shift.

u **[mm]**	1	2	4	6	8	10	20	25	50	75
log a_ε	0	1.13	-0.37	-0.74	-1.54	0.20	-2.31	-3.07	-4.08	-5.39
	-	-	-	-	-	-	-	-2.87	-4.39	-5.71
$1/h$	1	1.00	9.96	21.74	36.50	19.90	121.0	149.8	149.8	149.8

A.5.5 Results for acoustic PVB

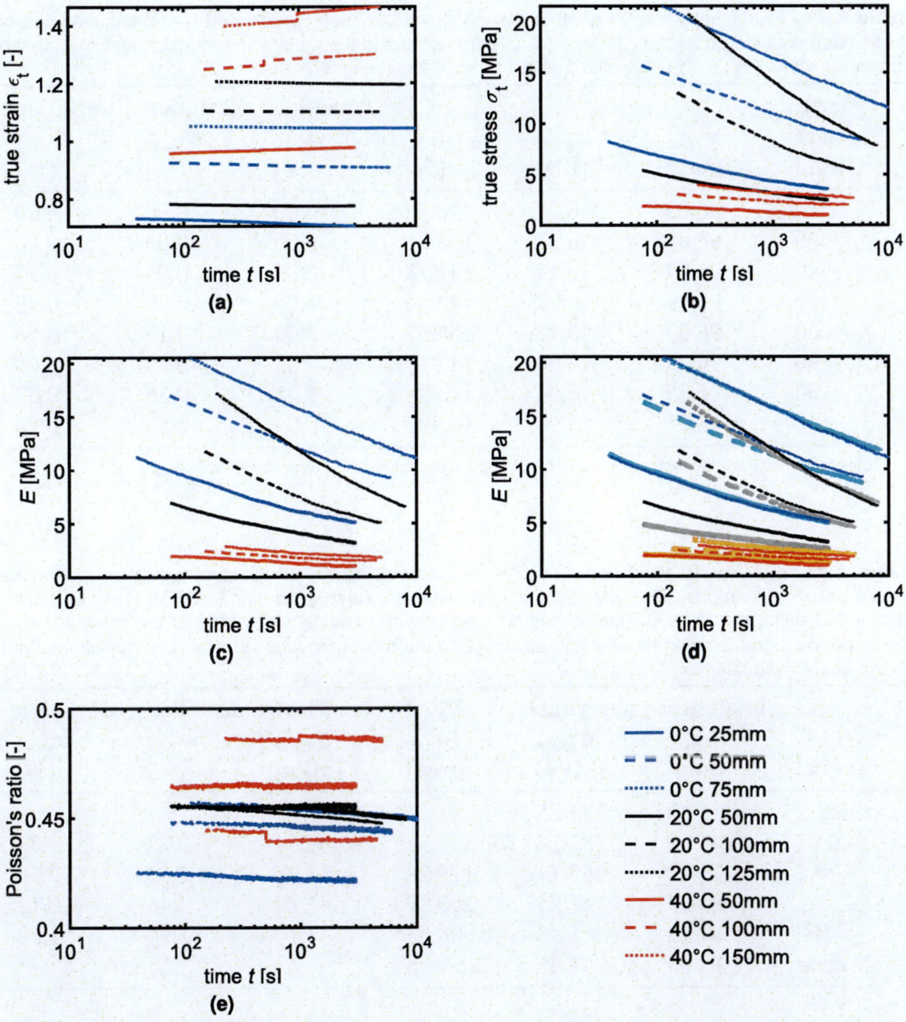

Figure A.30 Acoustic PVB: uniaxial relaxation tests at different temperatures and strains, (a) true strain, (b) true stress, (c) relaxation functions $E(t)$, (d) comparison of E with standard PVB (turquoise: 0°C 2.5 cm V2, 5 cm V2, 7.5 cm V1; grey: 20°C 5 cm V1, 10 cm V2, 12.5 cm V2; and orange: 40°C 5 cm V1, 10 cm V1, 15 cm V1), (e) Poisson's ratio.

A.6 FEM model validation and mesh study

Table A.41 FEM: mesh study. geometrical nonlinearities were considered. A shear modulus of 1 MPa was considered for the interlayer. w_{max} corresponds to the maximum deflection and σ_{glass} to the maximum glass stress. The deviations refer to the smallest element size.

element size [mm]	w_{max} [mm]	σ_{glass} [MPa]	γ[1] [%]	deviation w_{max} [%]	deviation σ_{glass} [%]	deviation γ [%]
10	96.09	46.21	14.878	0.00	0.00	0.00
30	96.09	46.19	14.875	0.00	-0.03	-0.02
40	96.09	46.18	14.918	0.00	-0.05	0.27
50	96.09	46.17	14.946	0.00	-0.08	0.45
60	96.09	46.16	14.957	0.00	-0.11	0.53
90	96.05	46.08	14.979	-0.05	-0.28	0.68
100	96.05	46.04	14.971	-0.05	-0.35	0.62
150	96.05	45.94	14.991	-0.05	-0.58	0.75

[1] elastic shear strain at $x \approx 0$ cm, $y \approx 40$ cm, coordinate system according to Figure 8.4.

Table A.42 FEM model validation. The model was calculated without geometrical nonlinearities (*small deformation mode*). The analytical solution was calculated according to Wölfel (1987). In addition, the analytical results of the limit cases *no shear transfer* and *full shear transfer* are included. w_{max} corresponds to the maximum deflection and σ_{glass} to the maximum glass stress. The deviations refer to the analytical solutions.

G [MPa]	analytical w_{max} [mm]	analytical σ_{glass} [MPa]	FEM w_{max} [mm]	FEM σ_{glass} [MPa]	deviation w_{max} [%]	deviation σ_{glass} [%]
no shear	418.53	93.75				
0.01	345.52	82.24	353.19	83.34	2.22	1.34
0.1	173.72	55.14	178.28	54.97	2.63	-0.30
1	98.38	43.25	100.80	43.20	2.46	-0.12
100	87.16	41.48	88.85	42.14	1.94	1.58
full shear	87.03	41.46				

Appendix B

List of publications, presentations and workshops

The following list contains publications, presentations and workshops in which the author of this thesis was involved and which were published or took place between 2015 and 2022. The list is sorted by date, starting with the most recent.

M. Schuster, K. Thiele, and J. Schneider. Consideration of the nonlinear viscoelasticity of pvb through a time-strain superposition, submitted to *Glass performance days 2021*, not yet published.

M. Schuster. Determination of the linear viscoelastic material behaviour of interlayers with semi-crystalline structures shown by the example of a semi-crystalline ionomer, submitted to *Glass Structures & Engineering, Special issue Challenging Glass Conference 8*, not yet published, presented at *Challenging Glass Conference 8*, Gent Belgium, 2022.

M. Schuster, and J. Schneider. Linear viskoelastisches Materialverhalten von teilkristallinen Zwischenschichten, *ce/papers*, 5: 280-293, 2022. https://doi.org/10.1002/cepa.1675

X. Chen, M. Schuster, J. Schneider. On the Pummel Test and Pummel Value Evaluation of Polyvinyl Butyral 1 Laminated Safety Glass. *Composite Structures*, 280:114878, 2022. https://doi.org/10.1016/j.compstruct.2021.114878

M. Schuster, K. Thiele, and J. Schneider. Investigations on the viscoelastic material behaviour and linearity limits of pvb. *ce/papers*, 4(6), 207-223, 2021. https://doi.org/10.1002/cepa.1633

J. Schneider, M. Schuster, G. Siebert, M. A. Kraus, M. Botz, A. Pauli, J. Weis. REST-VSG: Rechnerische und Experimentelle Studien zur Ermittlung der Resttragfähigkeit von VSG, In.: Bauforschung, Band T 3378, Fraunhofer IRB Verlag Best.-Nr. T 3378 (Kopie des Manuskripts) ISBN 978-3-7388-0591-8, 2020.

M. Schuster, J. Schneider, and T. A. Nguyen. Investigations on the execution and evaluation of the Pummel test for polyvinyl butyral based interlayers. *Glass Structures & Engineering* 5(3), 371–396, 2020. https://doi.org/10.1007/s40940-020-00120-y

M. Schuster, M. A. Kraus, M. Botz, J. Schneider, and G. Siebert. Structural design of laminated glass including the shear coupling effect through the polymeric interlayer: a comparison of different calculation approaches. Workshop at *Glass Performance Days*, Tampere Finland, 2019.

M. Schuster, *Characterization of Laminated Safety Glass Interlayers*, Mechanik, Werkstoffe und Konstruktion im Bauwesen 66,
https://doi.org/10.1007/978-3-658-39821-7

M. Schuster, M. A. Kraus, J. Schneider, and G. Siebert. Investigations on the visco- and hyperelastic behaviour of polyvinylbutyral interlayers at different temperatures. Presented at *Glass Performance Days*, Tampere Finland, 2019. Video available at: Miriam Schuster | Investigations on the Visco - and Hyperelastic Behaviour of PVB Interlayers - YouTube

M. A. Kraus, M. Schuster, M. Botz, J. Schneider and G. Siebert. Zur Normung thermomechanischer Untersuchungen an polymeren Verbundglaszwischenschichten. *ce/papers*, 3(1): 94-106, 2019. https://doi.org/10.1002/cepa.1003

J. K. Kuntsche, M. Schuster, and J. Schneider. Engineering design of laminated safety glass considering the shear coupling: a review. *Glass Structures & Engineering*, 4(2):209–228, 2019. https://doi.org/10.1007/s40940-019-00097-3

M. Schuster, M. A. Kraus, J. Schneider and G. Siebert. Untersuchungen zum Zeit-Temperatur-Superpositions-Prinzip von Steifigkeiten polymerer Verbundglaszwischenschichten. Presented at *DGM FA Polymerwerkstoffe*, Kassel Germany, 2018.

J. Schneider, S. Müller-Braun, T. Peters, N. Pourmoghaddam, and M. Schuster. Neue Dimensionen beim Bauen mit Glas, *22. Dresdner Baustatik-Seminar*, 2018.

M. Schuster, M. A. Kraus, J. Schneider, and G. Siebert. Investigations on the thermorheologically complex material behaviour of the laminated safety glass interlayer ethylene-vinyl-acetate. *Glass Structures & Engineering*, 3(2):373–388, presented at *Challenging Glass Conference 6*, Delft Netherlands, 2018.

M. A. Kraus, M. Schuster, M. Botz, J. Schneider, and G. Siebert. Thermorheologische Untersuchungen der Verbundglaszwischenschichten PVB und EVA. *ce/papers*, 2(1), 159-172, presented at Glasbau, Dresden Germany, 2018. https://doi.org/10.1002/ cepa.639

J. K. Kuntsche, M. Schuster, and J. Schneider. Bemessung von Verbundsicherheitsglas unter Berücksichtigung des Schubverbunds durch polymere Zwischenschichten. *Der Bauingenieur*, 93(1), 28-36, 2018. https://doi.org/10.37544/0005-6650-2018-01-52

S. Lück, J. Max, F. Langner, G. Zimmermann, M. Schuster, L. Riedel, and A. Gilka-Bötzow. Gewächshäuser neu gedacht – vakuumgedämmte, modulare Fertigelementbauweise mit U-Wert 0,2 W / (m² x K). *BHGL-Tagungsband* 33/2018, 2018.

M. Schuster, and J. Schneider. Verbundvorhaben TempergyGlass: Energy efficient tempering of thin glass for solar energy next generation products. Teilvorhaben: Thermomechanical basis for thin glass tempering. Schlussbericht. Laufzeit des Vorhabens: 01.10.2013 - 31.03.2017, Berichtszeitraum: 01.10.2013 - 31.03.2017. Technische Universität Darmstadt, 2017.

J. Schneider, J. Neugebauer, M. Schuster, and T. Peters. Cold-bent thin glass laminates for architectural applications: Computational design and material modeling. *IASS Annual Symposia*, Vol. 2017, No. 8, pp. 1-10. International Association for Shell and Spatial Structures (IASS), Hamburg Germany, 2017.

M. Drass, M. Schuster, and J. Schneider. Comparison of unconventional testing methods for mechanical characterization of polymeric materials in modern glass structures. *IABSE Symposium Report*, Vol. 109, No. 27, pp. 2370-2377. International Association for Bridge and Structural Engineering, presented at IABSE Symposium, Vancouver Canada, 2017. https://doi.org/10.2749/vancouver.2017.2378

M. Schuster, and M. A. Kraus. Testing, rheological modelling and design of interlayer materials in laminated safety glass. Workshop at *Glass Performance Days*, Tampere Finland, 2017.

M. A. Kraus, M. Schuster, J. K. Kuntsche, G. Siebert, and J. Schneider. Parameter identification methods for visco-and hyperelastic material models. *Glass Structures & Engineering*, 2(2):147– 167, presented at *Glass Performance Days*, Tampere Finland, 2017. https://doi.org/10.1007/s40940-017-0042-9

M. Schuster, J. K. Kuntsche, and J. Schneider. Investigation of the viscoelastic material behaviour of laminated safety glass interlayers. *19. Glastechnische Tagung*, Weimar Germany, 2017.

J. Schneider, J. K. Kuntsche, and M. Schuster. Mechanisches Verhalten von polymeren Zwischenschichten in Verbundglas. *Glas im konstruktiven Ingenieurbau 14*, Chapter 16, Munich Germany, 2016.

J. K. Kuntsche, M. Schuster, J. Schneider, and S. Langer. Viscoelastic properties of laminated glass interlayers – theory and experiments. *Glass Performance Days*, pages 143–147, 2015.